Projective Geometry:
From Foundations to Applications

Projective geometry is not only a jewel of mathematics, it also has many applications in modern information and communication science. This book presents the foundations of classical projective and affine geometry as well as its important applications in coding theory and cryptography. It could also serve as a first introduction to diagram geometry.

Written in clear and contemporary language with an entertaining style and around 200 exercises, examples and hints, this book is ideally suited as a textbook for use either in the classroom or on its own.

Albrecht Beutelspacher is Professor of Mathematics and Computer Science at the University of Giessen. He is the author of many entertaining (and successful!) books on mathematics, and has also made contributions in industry; amongst other things, he was responsible for introducing the code for the numbers on the latest German bank notes.

Ute Rosenbaum works in the field of information security for Siemens AG in Munich, where she is responsible for projects concerning the applications of cryptography.

Projective Geometry: From Foundations to Applications

Albrecht Beutelspacher

and

Ute Rosenbaum

CAMBRIDGE
UNIVERSITY PRESS

PUBLISHED BY THE PRESS SYNDICATE OF THE UNIVERSITY OF CAMBRIDGE
The Pitt Building, Trumpington Street, Cambridge, United Kingdom

CAMBRIDGE UNIVERSITY PRESS
The Edinburgh Building, Cambridge CB2 2RU, UK http://www.cup.cam.ac.uk
40 West 20th Street, New York, NY 10011–4211, USA http://www.cup.org
10 Stamford Road, Oakleigh, Melbourne 3166, Australia
Ruiz de Alarcón 13, 28014 Madrid, Spain

First published 1998
Reprinted 2000

Printed in the United Kingdom at the University Press, Cambridge

A catalogue record for this book is available from the British Library

Library of Congress Cataloging in Publication Data
Beutelspacher, A. (Albrecht), 1950–
[Projektive Geometrie. English]
Projective Geometry: from foundations to applications /
A. Beutelspacher, U. Rosenbaum.
p. cm.
Includes bibliographical references (p. –) and indexes.
ISBN 0 521 48277 1. – ISBN 0 521 48364 6 (pbk.)
1. Geometry, Projective. I. Rosenbaum, Ute. II. Title.
QA471.B5613 1998
516′.5–dc21 97-18012 CIP

ISBN 0 521 48277 1 hardback
ISBN 0 521 48364 6 paperback

Content

Preface

It is amusing to browse through the prefaces of randomly selected geometry texts that have been published over the past century. The authors document the ups and downs of their discipline as it moves into and out of fashion. It appears today that geometry's status has reached a new low, as educators, who themselves have had meagre training in the subject, recommend cutting back on a student's exposure to geometry in order to make room in the curriculum for today's more fashionable topics. Of course, to those of us who have studied geometry it is clear that these educators are moving in the wrong direction.

So why should a person study projective geometry?

First of all, projective geometry is a jewel of mathematics, one of the outstanding achievements of the nineteenth century, a century of remarkable mathematical achievements such as non-Euclidean geometry, abstract algebra, and the foundations of calculus. Projective geometry is as much a part of a general education in mathematics as differential equations and Galois theory. Moreover, projective geometry is a prerequisite for algebraic geometry, one of today's most vigorous and exciting branches of mathematics.

Secondly, for more than fifty years projective geometry has been propelled in a new direction by its combinatorial connections. The challenge of describing a classical geometric structure by its parameters – properties that at first glance might seem superficial – provided much of the impetus for finite geometry, another of today's flourishing branches of mathematics.

Finally, in recent years new and important applications have been discovered. Surprisingly, the structures of classical projective geometry are ideally suited for modern communications. We mention, in particular, applications of projective geometry to coding theory and to cryptography.

But what is projective geometry? Our answer might startle the classically trained mathematician who would be steeped in the subject's roots in Renaissance art and would point out that the discipline was first systematised by the seventeenth century architect Girard Desargues. Here we follow the insight provided by the German mathematician David Hilbert in his influential *Foundations of Geometry* (1899): a geometry is the collection of the theorems that follow from its axiom system. Although this approach frees the geometer from a dependence on physical space, it exposes him to the real danger of straying too far from nature and ending up with meaningless abstraction. In this book we avoid that danger by dealing with many applications – among which are some that even the fertile mind of Hilbert could not have imagined. And so we side-step the question of what projective geometry *is*, simply pointing out that it is an extremely good language for describing a multitude of phenomena inside and outside of mathematics. It is our goal in this book to exploit this point of view.

The first four chapters are mainly devoted to pure geometry. In the first chapter we study geometry from a synthetic point of view; here, notions such as basis, dimension, subspace, quotient space, and affine space are introduced. The second chapter presents what will be for us the most important class of projective spaces, namely those that can be constructed using a vector space; in other words, those projective spaces that can be coordinatized using a field. In analytic geometry one usually gets the impression that those are *all* the projective and affine spaces and no other structures are conceivable. In Chapter 3 we deal with precisely that question: A masterpiece of classical geometry is the representation theorem for projective and affine spaces. It says that any projective or affine space that satisfies the theorem of Desargues is coordinatizable. In particular we shall show that any projective or affine space of dimension ≥ 3 can be coordinatized over a vector space. Then we shall be able to describe all collineations (that is automorphisms) of Desarguesian projective spaces. In Chapter 4 we investigate the quadrics, which are probably the most studied objects in classical geometry. We shall look at them from a modern synthetic point of view and try to proceed as far as possible using only the properties of a 'quadratic set'. This has the advantage of a much better insight into the geometric properties of these structures.

We shall consider not only geometries over the reals, but also their *finite* analogues; in particular we shall determine their parameters. Moreover, at the end of each chapter we shall present an application. This makes it clear that some applications are based on remarkably simple geometric structures.

In the two final chapters we shall concentrate on important fields of applications, namely coding theory and cryptography. The aim of *coding theory* is to develop methods that enable the recipient of a message to detect or even correct errors that randomly occur while transmitting or storing data. Many problems of coding theory can be directly translated into geometric problems. As to *cryptography*, one of its tasks is to keep information secret by enciphering it. The other task is to protect data against alteration. Surprisingly, cryptosystems based on geometry have excellent properties: in contrast to most systems used in practice they offer provable security of arbitrarily high level. In Chapter 6 we shall study some of these systems.

From a didactical point of view, this book is based on three axioms.
1. We do not assume that the reader has had any prior exposure to projective or affine geometry. Therefore we present ever the elementary part in detail. On the other hand, we suppose that the reader has some experience in manipulating mathematical objects as found in a typical first or second year at university. Notions such as 'equivalence relation', 'basis', or 'bijective' should not strike terror in your heart.
2. We present those parts of projective geometry that are important for applications.
3. Finally, this book contains material that can readily be taught in a one year course.

These axioms force us to take shortcuts around many themes of projective geometry that became canonized in the nineteenth and twentieth centuries: there are no cross ratios or harmonic sets, non-Desarguesian planes are barely touched upon, projectivities are missing, and collineation groups do not play a central role. One may regret these losses, but, on the other hand, we note the following gains:
– This is a book that can be read independently by students.
– Most of the many exercises are very easy, in order to reinforce the reader's understanding.
– We are proud to present some topics for the first time in a textbook: for instance, the classification of quadrics (Theorem 4.4.4) in finite spaces, which we get by purely combinatorial considerations. Another example is the geometric–combinatorial description of the Reed–Muller codes. Finally we mention the theorem of Gilbert, MacWilliams, and Sloane (see 6.3.1), whose proof is – in our opinion – very illuminating.
– Last but not least, we describe the geometrical structures in Chapters 1 and 4 by diagrams, an approach that has led over the past twenty years to a fundamental restructuring of geometrical research.

To collaborate on a book is a real adventure, much to our surprise. In our case we had throughout an enjoyable collaboration, which was always intense and exciting – even when our opinions were far apart. Many arguments were resolved when one of us asked a so-called 'silly question', and we were forced to thoroughly reexamine seemingly clear concepts. We hope that all this will be an advantage for the reader.

Thanks to our students and assistants who contributed in an essential way to this book, in particular by reading early versions. Particular thanks are due to Uwe Blöcher, Jörg Eisfeld, and Klaus Metsch, who critically read the manuscript and found lots of errors of all kinds.

We could not have done the translation without the generous advice of Chris Fisher and John Lochhas.

Finally we thank our spouses for their extremely generous, yet not always voluntary, help.

Großen-Buseck and Kempten, summer 1996

Albrecht Beutelspacher
Ute Rosenbaum

1 Synthetic geometry

From the ancient beginnings of geometry until well into the nineteenth century it was almost universally accepted that the geometry of the space we live in is the only geometry conceivable by man. This point of view was most eloquently formulated by the German philosopher Immanuel Kant (1724–1804). Ironically, shortly after Kant's death the discovery of non-Euclidean geometry by Gauß, Lobachevski, and Bolyai made his position untenable. Today, we study in mathematics not just one geometry, or two geometries, but an infinity of geometries.

This means: When you start learning geometry (the subject), you are immediately offered geometries (structures) in plural. There is not a unique geometry, but many geometries, and all have equal rights (even though some might be more interesting than others).

1.1 Foundations

It is typical of geometry that we study not only one type of object (such as points), but different types of objects (such as points and lines, points, lines, and planes, etc.) and their relationships. We first define a very general notion of geometry which can be used to describe all possible geometries.

Definition. A **geometry** is a pair $\mathbf{G} = (\Omega, I)$, where Ω is a set and I a relation on Ω that is **symmetric** and **reflexive**; this means the following.

- If $(x, y) \in I$ then also $(y, x) \in I$.
- $(x, x) \in I$ for all $x \in \Omega$.

What is the idea behind this definition? The idea is that the set Ω contains all geometrically relevant objects and I describes their being 'incident'. Let us consider some examples.

Before doing this we note the following. In many situations the 'natural' incidence relation is set-theoretical inclusion. This relation is not symmetric, but can easily be made symmetric by defining that two elements are incident if one is contained in the other. Usually, we shall not mention this explicitly.

Examples. (a) In classical 3-dimensional geometry, Ω is the set of points, lines, and planes of Euclidean space. If one is interested only in the Euclidean plane then Ω is only the set of points and lines. But our imagination should have no limits: Ω could be the set of all subsets of a set, or the set of all lines and circles of the Euclidean plane, etc.
(b) We get another geometry by looking at the cube. In this case the set Ω could consist of the eight vertices, the twelve edges, and the six faces of the cube (see Figure 1.1). The relation I is set-theoretical inclusion.

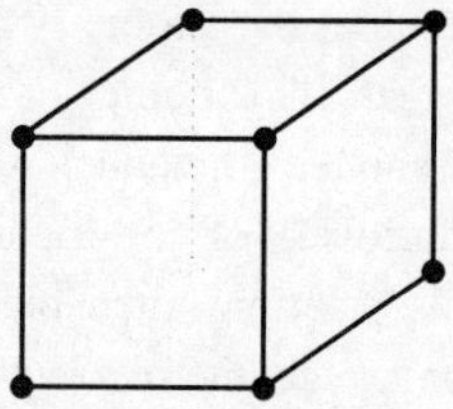

Figure 1.1 The cube

(c) A quite different example is the following. The set Ω is a set of humans; two humans are incident if one is a descendant of the other. For a particular group of persons (family B) this looks as shown in Figure 1.2.

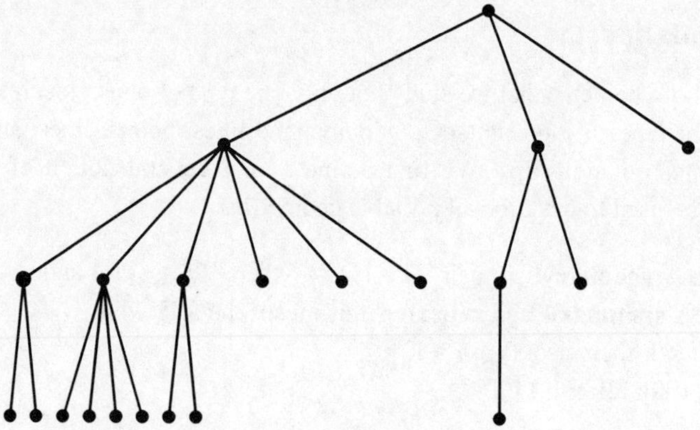

Figure 1.2 Family B

The strokes indicate direct descent; hence two persons are incident if there is an upwards series of strokes leading from one person to the other.

In geometry the relation I describes 'containment' among the elements of Ω – 'planes contain lines and points', 'lines contain points', etc. One also describes this as **incidence** and therefore calls I the **incidence relation** of the geometry **G**.

Two elements of the geometry **G** that are related by I are called **incident**. If x and y are two elements of Ω with $(x, y) \in I$, then one simply writes $x\,I\,y$ or $y\,I\,x$.

If Ω' is a subset of Ω then one can consider the subgeometry **G'** of **G** belonging to Ω': It consists of the elements of Ω' and the relation I' that is the restriction of I to Ω'; one calls I' the incidence relation **induced** by I. Thus, two elements of Ω' are incident in **G'** if they are already incident in **G**.

In our example of 3-dimensional Euclidean space, a point P and a plane π are incident $(P\,I\,\pi)$ if P is a point in the plane π. Similarly, for a line g and a plane π we have $g\,I\,\pi$ if and only if the line g is completely contained in π. (We usually denote a line by g; this is due to the German word 'Gerade' for 'line'.)

If Ω' is the set of points in the interior of the unit circle together with the set of lines that intersect the unit circle in two points, then one can describe the induced incidence in such a way that a point and a line of Ω' are incident if and only if they are incident as elements of Ω.

Definition. Let $\mathbf{G} = (\Omega, I)$ be a geometry. A **flag** of **G** is a set of elements of Ω that are mutually incident. A flag $\mathcal{F}$ is called **maximal** if there is no element $x \in \Omega \setminus \mathcal{F}$ such that $\mathcal{F} \cup \{x\}$ is also a flag.

Examples. (a) Let Q be a point, g a line, and π a plane of 3-dimensional Euclidean space with $Q \in g$ and $g \subseteq \pi$. Then the following sets are flags:

$\{Q\}, \{g\}, \{\pi\}$;

$\{Q, g\}, \{Q, \pi\}, \{g, \pi\}$;

$\{Q, g, \pi\}$.

Only the last flag is maximal (see Figure 1.3).

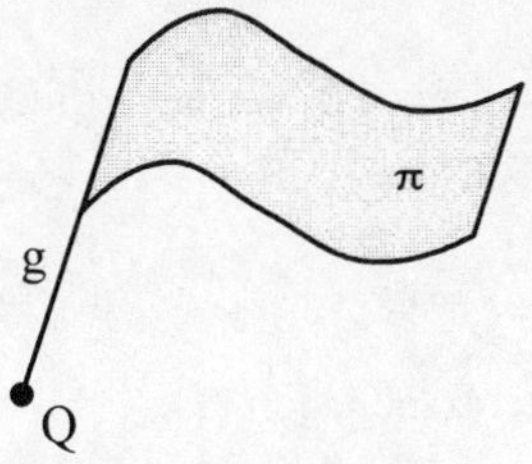

Figure 1.3 A flag

(b) Considering the geometry of the cube one notices that a maximal flag consists of precisely one vertex, one edge, and one face. Thus any maximal flag has precisely three elements.

(c) In the descendant geometry of family B there are maximal flags of 2, 3, or 4 elements.

Definition. We say that a geometry $\mathbf{G} = (\Omega, \mathrm{I})$ has **rank** r if one can partition Ω into sets $\Omega_1, \ldots, \Omega_r$ such that each maximal flag of $\mathbf{G}$ intersects each set Ω_i in exactly one element. In this case the elements of Ω_i are called elements of **type** i.

In particular, in a geometry of rank r each maximal flag has exactly r elements.

Examples. (a) In 3-dimensional Euclidean space we choose as Ω_1 the set of all points, as Ω_2 the set of all lines, and as Ω_3 the set of all planes. This gives a geometry of rank 3.
(b) If we take as Ω_1 the set of vertices, as Ω_2 the set of edges, and as Ω_3 the set of faces of the cube, we get a geometry of rank 3.
(c) The descendant geometry of family B cannot be considered a geometry of rank r for any r, since there are maximal flags of different lengths.
(d) Let $\mathcal{M}$ be a set, and let r be a positive integer with $r < |\mathcal{M}|$. Then we can define a rank r geometry by defining as elements of type i the subsets of $\mathcal{M}$ having exactly i elements $(1 \le i \le r)$; as incidence we choose set-theoretic inclusion. In other words,

$$\Omega_i := \{\mathcal{X} \subseteq \mathcal{M} \mid |\mathcal{X}| = i\} \quad (1 \le i \le r),$$

and

$$\Omega := \Omega_1 \cup \Omega_2 \cup \ldots \cup \Omega_r.$$

(e) Similarly, we can obtain from a vector space V of dimension $\ge r$ a geometry of rank r:

$$\Omega_i := \{\mathrm{U} \mid \mathrm{U} \text{ is a subspace of V with } \dim(\mathrm{U}) = i\} \quad (1 \le i \le r),$$

and

$$\Omega := \Omega_1 \cup \Omega_2 \cup \ldots \cup \Omega_r,$$

where incidence is set-theoretic inclusion.
(f) You will find a further example in exercise 1.

Let $\mathbf{G}$ be a geometry of rank r, and let Ω_i be the set of elements of type i ($i = 1, 2, \ldots, r$). Let $\mathbf{G}' = (\Omega\backslash\Omega_i, \mathrm{I}_{\mathrm{ind}})$ be the geometry on the set $\Omega\backslash\Omega_i$ with the incidence being induced from $\mathbf{G}$. Then $\mathbf{G}'$ is a geometry of rank $r - 1$. In particular one can consider any geometry of rank $r \ge 2$ as a rank 2 geometry. This we will do often. For instance, we shall describe many geometries only by their points and lines or by their points and 'hyperplanes'.

1.1.1 Lemma. *Let* **G** *be a geometry of rank* r*. Then no two distinct elements of the same type are incident.*

Remark. The lemma expresses a natural idea: a line can be incident with a point or a plane, but one usually does not consider two lines to be incident.

Proof. Assume that there exist two distinct elements of the same type that are incident. These elements form a flag. We consider a maximal flag $\mathcal{F}$ that contains these two elements. But $\mathcal{F}$ contains two elements of the same type, contradicting the definition of a geometry of rank r. □

A geometry $\mathbf{G} = (\Omega, \mathrm{I})$ of rank 2 is often called an **incidence structure**. In this case one calls the elements of type 1 **points** and the elements of type 2 **blocks**. If **G** is an incidence structure with point set $\mathcal{P}$ and block set $\mathcal{B}$ then one also writes $\mathbf{G} = (\mathcal{P}, \mathcal{B}, \mathrm{I})$.

In this book we shall mostly deal with incidence structures for which it makes sense to call the blocks **lines**; our fundamental axiom will be that any two distinct points uniquely determine a block.

Now we can describe the aim of the first chapter. We start from a geometry of rank 2, which consists of the points and lines of a 'projective space', a concept defined by very simple axioms. From this we shall develop all of projective geometry – the structure consisting of all subspaces of projective and affine spaces.

1.2 The axioms of projective geometry

From now on, let $\mathbf{G} = (\mathcal{P}, \mathcal{L}, \mathrm{I})$ be a geometry of rank 2; the elements of the block set $\mathcal{L}$ will be called **lines**. Following Euclid, we usually denote points by upper case letters; we denote lines by lower case letters. If $\mathrm{P\,I\,g}$ is true, we shall also say that 'P is incident with g', 'P lies on g', 'g passes through P', and so on.

Later on we shall convince ourselves that in all interesting cases instead of '$\mathrm{P\,I\,g}$' we may also write (and think) '$\mathrm{P} \in \mathrm{g}$'.

Now we introduce the axioms that are fundamental for projective geometry (hence *a fortiori* for this book).

Axiom 1 (line axiom). *For any two distinct points* P *and* Q *there is exactly one line that is incident with* P *and* Q.
This line is denoted by PQ.

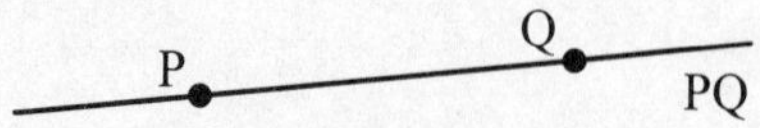

This axiom immediately implies the following assertion.

1.2.1 Lemma. *Let* g *and* h *be two distinct lines. Then* g *and* h *are incident with at most one common point.*
If such a point exists, it will be denoted by $g \cap h$.

Proof. Assume that there exist two distinct points P and Q incident with g and h. By axiom 1 there is just one line through P and Q; hence $g = h$, a contradiction. □

Axiom 2 (Veblen–Young). *Let* A, B, C, *and* D *be four points such that* AB *intersects the line* CD. *Then* AC *also intersects the line* BD.

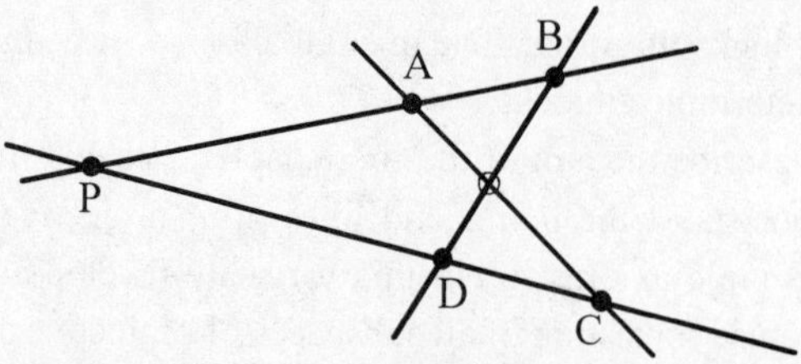

The following formulation of the Veblen–Young axiom is more concise and therefore easier to learn: *if a line* (in our case BD) *intersects two 'sides'* (namely AP and CP) *of a triangle* (in our case APC) *then the line also intersects the third side* (namely AC).

Remarks. 1. The Veblen–Young axiom is a truly ingenious way of saying that any two lines of a plane meet – before one knows what a plane is.
2. Some people call the Veblen–Young axiom the **axiom of Pasch**, because the German geometer Moritz Pasch (1843–1930) used a similar picture.

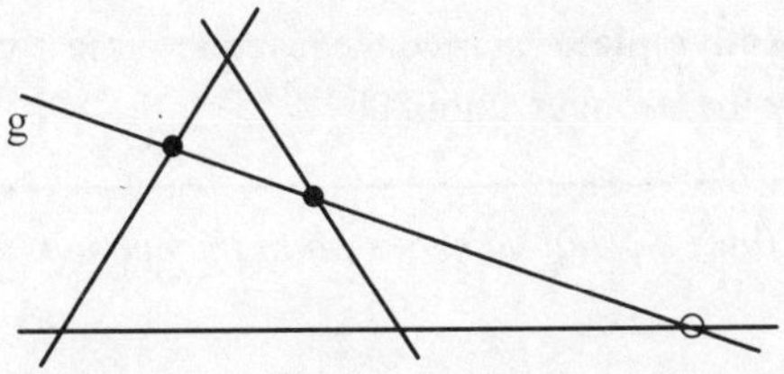

Pasch's aim, however, was different. He used his axiom to introduce *order* in geometry. More precisely, his axiom reads: if a line g intersects one side of a triangle *internally*, then it intersects precisely one other side internally and the third side *externally*.

In introducing the Veblen–Young axiom we took a turn that you probably did not expect: we exclude the existence of parallel lines. You might at this point be so taken aback that you feel like throwing the book aside in disgust – we seem to be studying geometries 'that do not exist'. But this would be too hasty on your part, since we shall soon meet 'affine geometries', for which intuitive geometry serves as a prototype. Furthermore, it will turn out that projective and affine geometry are basically the same thing, even though projective geometry is much easier to work with.

Axiom 3. *Any line is incident with at least three points.*

Definition. A **projective space** is a geometry $\mathbf{P} = (\mathcal{P}, \mathcal{L}, \mathrm{I})$ of rank 2 that satisfies the axioms 1, 2, and 3. A projective space $\mathbf{P}$ is called **nondegenerate** if it also satisfies the following axiom 4.

Axiom 4. *There are at least two lines.*

From now on we suppose that $\mathbf{P} = (\mathcal{P}, \mathcal{L}, \mathrm{I})$
is a nondegenerate projective space.
We usually say, more briefly, that $\mathbf{P}$ is a projective space.

Axiom 2 says that under certain conditions, two lines of a projective space intersect. If each pair of lines intersects (without any further hypothesis) then one has a projective plane.

Definition. A **projective plane** is a nondegenerate projective space in which axiom 2 is replaced by the stronger axiom 2':

Axiom 2'. *Any two lines have at least one point in common.*

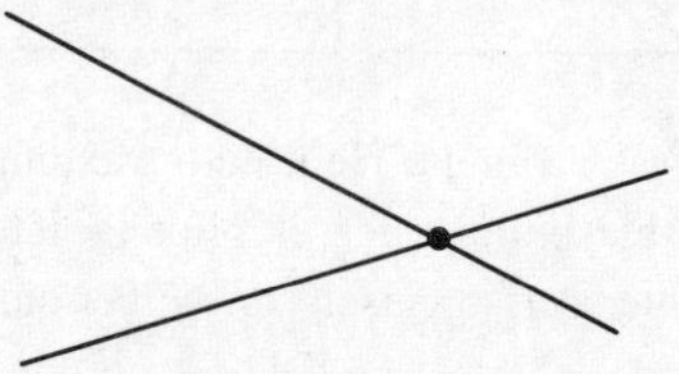

An important property is the **principle of duality** for projective planes. We first formulate this principle in general and shall then prove that the principle of duality holds for projective planes. Later on (in Section **2.3**) we shall also investigate the principle of duality in projective spaces.

Let A be a proposition concerning geometries of rank 2, whose elements we shall call *points* and *blocks*. We get the proposition A^Δ **dual** to A by interchanging the words 'point' and 'block'.

Example. If A is the proposition 'there exist four points no three of which are incident with a common block', then A^Δ is the proposition 'there are four blocks no three of which pass through a common point'.

Let **G** be a geometry of rank 2 with point set Ω_1, block set Ω_2, and incidence set I. Then the geometry $\mathbf{G}^\Delta$ **dual** to **G** has point set Ω_2, block set Ω_1, and two elements of $\mathbf{G}^\Delta$ are incident if and only if they are incident as elements of **G**.

Note that the definition implies that $(\mathbf{G}^\Delta)^\Delta = \mathbf{G}$.

1.2.2 Theorem (principle of duality). *Let* $\mathcal{K}$ *be a class of geometries of rank* 2. *We suppose that* $\mathcal{K}$ *has the property: if* $\mathcal{K}$ *contains the geometry* **G**, *then it also contains the dual geometry* $\mathbf{G}^\Delta$. *Then the following assertion is true:*

if A *is a proposition that is true for all* **G** *in* $\mathcal{K}$, *then* A^Δ *is also true for all* **G** *in* $\mathcal{K}$.

Proof. Let **G** be an arbitrary geometry in $\mathcal{K}$. For $\mathbf{G}' := \mathbf{G}^\Delta$ we have that $\mathbf{G}'^\Delta = \mathbf{G}$. Since $\mathbf{G}'$ is a geometry in $\mathcal{K}$ the proposition A holds for $\mathbf{G}'$. Hence $\mathbf{G}'^\Delta = \mathbf{G}$ satisfies the assertion A^Δ. □

In order to prove the principle of duality for projective planes we need the following lemma.

1.2.3 Lemma. *Any projective plane* **P** *also satisfies the propositions that are dual to the axioms 1, 2', 3, and 4.*

Proof. Axiom 1. The proposition that is dual to axiom 1 is 'any two distinct lines have exactly one point in common'. This is a true statement, since by **1.2.1** any two distinct lines have at most one point in common and by axiom 2' they have at least one point in common.

Axiom 2'. The proposition dual to axiom 2' is 'any two distinct points are incident with at least one common line'. This follows directly from axiom 1.

Axiom 3. The proposition dual to axiom 3 is 'any point is on at least three lines'.

In order to see this we consider an arbitrary point P of **P**. If we already knew that there is a line g not incident with P, then we could proceed as follows. By axiom 3 there are at least three points P_1, P_2, P_3 on g. Then PP_1, PP_2, PP_3 are three distinct lines through P.

Why is there a line g not incident with P? If to the contrary all lines pass through P then, by axiom 4, there are two lines g_1 and g_2 through P. By axiom 3 there is a point $Q_i \neq P$ on g_i $(i = 1, 2)$. Then $g = Q_1Q_2$ is a line not incident with P, a contradiction.

Axiom 4. The proposition dual to axiom 4 is 'there exist two distinct points'. This follows easily from axioms 4 and 3. □

If we consider the lines of a projective plane **P** as new points, and its points as new lines, we again get a projective plane. This plane $\mathbf{P}^\Delta$ is called the **dual plane** of **P**.

The preceding lemma has a remarkable consequence: roughly speaking, for projective planes we have to prove only 'one half' of all assertions.

1.2.4 Theorem (principle of duality for projective planes). *If a proposition* A *is true for all projective planes then the dual proposition* A^Δ *also holds for all projective planes.*

Proof. If A is true for all projective planes **P** then A^Δ holds for all geometries of the form $\mathbf{P}^\Delta$, hence for all dual projective planes. Furthermore we can represent any projective plane $\mathbf{P}_0$ as a dual projective plane: if $\mathbf{P}_1 := \mathbf{P}_0^\Delta$ then $\mathbf{P}_0 = (\mathbf{P}_0^\Delta)^\Delta = \mathbf{P}_1^\Delta$. □

Warning: The principle of duality does not say that **P** and $\mathbf{P}^{\Delta}$ are isomorphic. Of course, sometimes a projective plane might be isomorphic to its dual (see **2.3.5**).

So far we have studied the *axioms* for projective geometry. Now we have to introduce the notions which will help us to investigate projective spaces.

1.3 Structure of projective geometry

Definition. A subset $\mathcal{U}$ of the point set $\mathcal{P}$ is called **linear** if for any two points P and Q that are contained in $\mathcal{U}$ each point of the line PQ is contained in $\mathcal{U}$ as well.

If we denote the set of points incident with a line g by (g), then we can express the above definition as follows: the point set $\mathcal{U}$ is linear if and only if for any two distinct points $P, Q \in \mathcal{U}$ we have $(PQ) \subseteq \mathcal{U}$.

Obviously, for any linear set $\mathcal{U}$ of **P** the geometry $\mathbf{P}(\mathcal{U})$ with

$$\mathbf{P}(\mathcal{U}) = (\mathcal{U}, \mathcal{L}', \mathrm{I}'),$$

where $\mathcal{L}'$ is the set of lines of **P** that are totally contained in $\mathcal{U}$ and I' is the induced incidence, is a – possibly degenerate – projective space. We call $\mathbf{U} = \mathbf{P}(\mathcal{U})$ a (linear) **subspace** of **P**. If no confusion is possible, we shall not distinguish between a linear set and the corresponding subspace.

Examples. The following sets of points are linear sets (or subspaces): the empty set, any singleton, the set of points on a line and the whole point set $\mathcal{P}$. In particular, any subset of $\mathcal{P}$ is contained in at least one linear set.

Since the intersection of arbitrarily many linear sets is again linear, we can define the **span** $\langle \mathcal{X} \rangle$ of a subset $\mathcal{X}$ of $\mathcal{P}$ as follows:

$$\langle \mathcal{X} \rangle := \bigcap \{\mathcal{U} \mid \mathcal{X} \subseteq \mathcal{U}, \mathcal{U} \text{ is a linear set}\}.$$

In other words, $\langle \mathcal{X} \rangle$ is the smallest linear set containing $\mathcal{X}$. We shall also say that $\mathcal{X}$ **spans** or **generates** $\langle \mathcal{X} \rangle$ and shall call $\langle \mathcal{X} \rangle$ the subspace **spanned** by $\mathcal{X}$.

Instead of $\langle \{P_1, P_2, \ldots\} \rangle$ we shall also write $\langle P_1, P_2, \ldots \rangle$; furthermore we shall use 'mixed' expressions: instead of $\langle \mathcal{X} \cup \{P\} \rangle$ we simply write $\langle \mathcal{X}, P \rangle$, etc. We also write $\langle \mathbf{U}, \mathbf{W} \rangle$ for two subspaces **U** and **W**.

Examples. The span of the empty set is the empty set, the span of a singleton $\{P\}$ is again $\{P\}$, the span of a 2-element subset $\{P, Q\}$ is the set (PQ) of the points on PQ.

Definition. A set $\mathcal{M}$ of points is called **collinear** if all points of $\mathcal{M}$ are incident with a common line. The set $\mathcal{M}$ is called **noncollinear** if there is no line that is incident with all points of $\mathcal{M}$.

The span of a set of three noncollinear points is called a **plane**.

We defined subspaces 'top down' as the smallest linear sets containing the generating set. Our first aim is to describe how the subspaces can be built 'bottom up' – how they can be constructed starting with the generating set. It turns out that this can be done in the easiest conceivable way. This is a characteristic property of projective spaces.

The following important theorem describes this recursive construction. Most results in this section depend heavily on this theorem.

1.3.1 Theorem. *Let* $\mathcal{U}$ *be a nonempty linear set of* **P**, *and let* P *be a point of* **P**. *Then*

$$\langle \mathcal{U}, P\rangle = \bigcup \{(PQ) \mid Q \in \mathcal{U}\}.$$

In other words, the span of $\mathcal{U}$ *and* P *can be described easily: it consists just of the points on the lines joining* P *and the points* Q *of* $\mathcal{U}$.

Furthermore, each line of $\langle \mathcal{U}, P\rangle$ *intersects* $\mathcal{U}$.

Before proving the theorem we shall explain the first nontrivial case. If $\mathcal{U}$ is the set of points on a line g then the plane $\langle \mathcal{U}, P\rangle = \langle g, P\rangle$ consists precisely of the points on the lines PQ such that $Q \mathrel{I} g$ (see Figure 1.4).

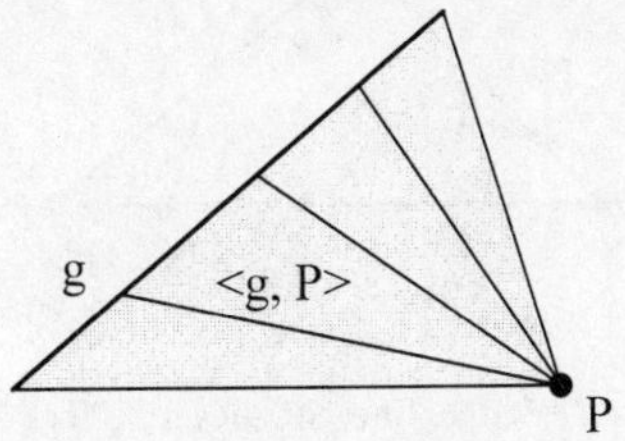

Figure 1.4 Span of a line and a point

Now we shall *prove* **1.3.1**. One inclusion is easy: since $\langle \mathcal{U}, P\rangle$ is a linear set it contains with any two points P and Q also any point on PQ, therefore

$$\bigcup \{(PQ) \mid Q \in \mathcal{U}\} \subseteq \langle \mathcal{U}, P\rangle.$$

We claim: in order to prove the other inclusion it is sufficient to prove that the set

$$\mathcal{N} := \bigcup \{(PQ) \mid Q \in \mathcal{U}\}$$

is linear. For it then follows that

$$\begin{aligned}\langle \mathcal{U}, P\rangle &= \bigcap \{\mathcal{V} \mid P \in \mathcal{V}, \mathcal{U} \subseteq \mathcal{V}, \mathcal{V} \text{ is linear}\}\\ &= \bigcap \{\mathcal{V} \mid P \in \mathcal{V}, \mathcal{U} \subseteq \mathcal{V}, \mathcal{V} \text{ is linear}, \mathcal{V} \neq \mathcal{N}\} \cap \mathcal{N}\\ &\subseteq \mathcal{N},\end{aligned}$$

since $\mathcal{N}$ contains P and $\mathcal{U}$, and is, by assumption, a linear set.

In order to prove that $\mathcal{N}$ is linear we consider a line g that contains two distinct points R, S of $\mathcal{N}$, and show that any point X of g lies in $\mathcal{N}$. We distinguish different cases.

– If R and S are both contained in the linear set $\mathcal{U}$ then we trivially have

$$(g) = (RS) \subseteq \mathcal{U} \subseteq \mathcal{N}.$$

– If the line g contains the point P then it is of the form $g = PQ$ with $Q \in \mathcal{U}$. For g contains two distinct points of $\mathcal{N}$ and passes through P, hence it must be one of the lines through P defining $\mathcal{N}$.

Thus we may assume that g does not contain P. We examine two more difficult cases.

– First, let R be a point of $\mathcal{U}$, but $S \notin \mathcal{U}$ (see Figure 1.5).

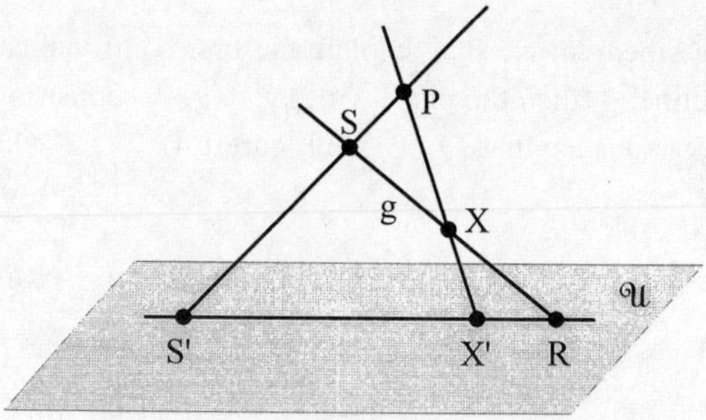

Figure 1.5 The line SR lies in $\langle \mathcal{U}, P\rangle$

Since S is in $\mathcal{N}$, by definition of $\mathcal{N}$ the line PS intersects the set $\mathcal{U}$ in a point S'.

Let X be an arbitrary point on g. In order to show that X is in $\mathcal{N}$ we may assume that $X \neq R, S$. The line PX intersects the sides of the triangle R, S, S' in the two distinct points P and X. By the Veblen–Young axiom, PX also intersects the line RS' in some point X'. Since $X' \in (RS') \subseteq \mathcal{U}$ it follows that $X \in (PX') \subseteq \mathcal{N}$.

– Finally we have to consider the case that neither R nor S lies in $\mathcal{U}$.

Since R and S are points of $\mathcal{N}$ there exist points $R', S' \in \mathcal{U}$ with $R \in (PR')$ and $S \in (PS')$. Since g does not pass through P the line g intersects the triangle P, R', S' in the two distinct points R and S. Again by Veblen–Young the lines g and R'S' meet in some point T. Therefore g = ST with $T \in \mathcal{U}$, and by the preceding case we have $(g) \subseteq \mathcal{N}$.

The proof also shows that each line of $\langle \mathcal{U}, P \rangle$ intersects $\mathcal{U}$. □

Remark. The above theorem says in particular that all subspaces of a projective space are uniquely determined by their sets of points and lines. This means that from the properties of the rank 2 geometry of points and lines the whole geometry can be described.

Using the tools developed so far we shall be able to introduce the notions of 'basis' and 'dimension' of a projective space. The following exchange property is fundamental.

1.3.2 Theorem (exchange property). *Let $\mathcal{U}$ be a linear set of* **P**, *and let* P *be a point of* **P** *that does not lie in $\mathcal{U}$. Then the following implication is true:*

$$\text{if } Q \in \langle \mathcal{U}, P \rangle \setminus \mathcal{U} \text{ then } P \in \langle \mathcal{U}, Q \rangle, \text{ hence also } \langle \mathcal{U}, P \rangle = \langle \mathcal{U}, Q \rangle.$$

One can also express this as follows: any two distinct subspaces through $\mathcal{U}$ that are spanned by $\mathcal{U}$ and a point outside $\mathcal{U}$ intersect only in points of $\mathcal{U}$.

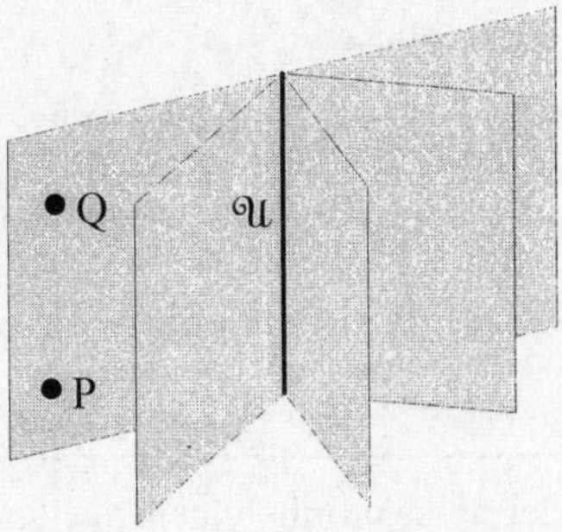

Figure 1.6 Intersection of subspaces

Proof. Since $Q \in \langle \mathcal{U}, P \rangle \setminus \mathcal{U}$, by **1.3.1**, there is a point Q' in $\mathcal{U}$ such that $P \in (QQ') \subseteq \langle \mathcal{U}, Q \rangle$.

This implies immediately that the two subspaces in question are equal: in view of $\mathcal{U} \subseteq \langle \mathcal{U}, Q\rangle$ and $P \in \langle \mathcal{U}, Q\rangle$ it follows that $\langle \mathcal{U}, P\rangle \subseteq \langle \mathcal{U}, Q\rangle$. Similarly, one shows the other inclusion. □

Definition. A set $\mathcal{B}$ of points in the projective space $\mathbf{P}$ is called **independent** if for any subset $\mathcal{B}' \subseteq \mathcal{B}$ and any point $P \in \mathcal{B} \setminus \mathcal{B}'$ we have that

$$P \notin \langle \mathcal{B}' \rangle.$$

In order to show that a set $\mathcal{B}$ of points is independent one has to prove that there is no point $P \in \mathcal{B}$ for which $P \in \langle \mathcal{B} \setminus \{P\}\rangle$ is satisfied. (When $P \in \langle \mathcal{B} \setminus \{P\}\rangle$ we call $\mathcal{B}$ **dependent**.)

An independent set $\mathcal{B}$ of points that spans $\mathbf{P}$ (that is $\langle \mathcal{B}\rangle = \mathbf{P}$) is called a **basis** of $\mathbf{P}$.

Examples of independent point sets are easy to find: any singleton is independent; any two distinct points form an independent set; three points are independent if and only if they are not on a common line. Moreover, it is clear that every subset of an independent set is independent.

1.3.3 Theorem. *A set $\mathcal{B}$ of points of $\mathbf{P}$ is a basis of $\mathbf{P}$ if and only if $\mathcal{B}$ is a* ***minimal spanning set****, that is if $\mathcal{B}$ spans $\mathbf{P}$, but no proper subset of $\mathcal{B}$ spans $\mathbf{P}$.*

Proof. First, let $\mathcal{B}$ be a basis. Then, by definition $\mathcal{B}$ spans $\mathbf{P}$. Assume that there exists a proper subset $\mathcal{B}' \subset \mathcal{B}$ such that $\langle \mathcal{B}'\rangle = \mathbf{P}$. Then there would exist a point $P \in \mathcal{B} \setminus \mathcal{B}'$ which would satisfy $P \in \langle \mathcal{B}'\rangle$, contradicting the independence of $\mathcal{B}$.

Conversely, let $\mathcal{B}$ be a minimal spanning set. We have to show that $\mathcal{B}$ is independent. Assume that $\mathcal{B}$ is not independent. Then there exists a point P in $\mathcal{B}$ such that $P \in \langle \mathcal{B} \setminus \{P\}\rangle$. This implies

$$\mathbf{P} = \langle \mathcal{B}\rangle = \langle \mathcal{B} \setminus \{P\} \cup \{P\}\rangle = \langle \mathcal{B} \setminus \{P\}, P\rangle \subseteq \langle \mathcal{B} \setminus \{P\}\rangle.$$

Hence $\mathcal{B} \setminus \{P\}$ would already span $\mathbf{P}$, contradicting the minimality of $\mathcal{B}$. □

Definition. The projective space $\mathbf{P}$ is called **finitely generated** if there is a finite set of points that spans $\mathbf{P}$.

In the following let $\mathbf{P}$ be a finitely generated projective space.

Warning: This does not imply immediately that any basis of $\mathbf{P}$ is finite. Also, this does not follow from the next theorem. We will later on obtain this property as a corollary of the Steinitz exchange theorem.

1.3.4 Theorem. *Let $\mathcal{E}$ be a finite spanning set of* **P**. *Then there exists a basis $\mathcal{B}$ of* **P** *such that $\mathcal{B} \subseteq \mathcal{E}$. In particular,* **P** *has a finite basis.*

Proof. Define $\mathcal{E}_0 := \mathcal{E}$. If $\mathcal{E}_0$ is a minimal generating system then, by the preceding theorem, $\mathcal{E}_0$ is a basis. Otherwise there is a proper subset $\mathcal{E}_1$ of $\mathcal{E}_0$ that spans **P** as well.

Since $\mathcal{E}$ is finite, after a finite number of steps we must obtain a minimal spanning set $\mathcal{B} = \mathcal{E}_n \subseteq \mathcal{E}$, which is a basis. □

1.3.5 Lemma. *Let $\mathcal{B}$ be an independent set of points of* **P**, *and let $\mathcal{B}_1$, $\mathcal{B}_2$ be subsets of $\mathcal{B}$. If $\mathcal{B}$ is finite then*

$$\langle \mathcal{B}_1 \cap \mathcal{B}_2 \rangle = \langle \mathcal{B}_1 \rangle \cap \langle \mathcal{B}_2 \rangle.$$

Proof. In view of $\mathcal{B}_1 \cap \mathcal{B}_2 \subseteq \mathcal{B}_1, \mathcal{B}_2$ we have that $\langle \mathcal{B}_1 \cap \mathcal{B}_2 \rangle \subseteq \langle \mathcal{B}_1 \rangle \cap \langle \mathcal{B}_2 \rangle$.

The other inclusion is more difficult and will be proved by induction on $|\mathcal{B}_1|$.

If $\mathcal{B}_1 = \varnothing$ then the assertion follows trivially.

Suppose that $|\mathcal{B}_1| \geq 1$ and assume that the assertion is true for sets with $|\mathcal{B}_1| - 1$ elements. We may assume that $\mathcal{B}_1$ is not a subset of $\mathcal{B}_2$. Otherwise $\mathcal{B}_1 \subseteq \mathcal{B}_2$ and so

$$\langle \mathcal{B}_1 \rangle \cap \langle \mathcal{B}_2 \rangle = \langle \mathcal{B}_1 \rangle = \langle \mathcal{B}_1 \cap \mathcal{B}_2 \rangle.$$

We suppose that $\mathcal{B}_1 \not\subset \mathcal{B}_2$. Consider a point P of $\mathcal{B}_1 \setminus \mathcal{B}_2$. By induction, the assertion is true for the set $\mathcal{B}_1' := \mathcal{B}_1 \setminus \{P\}$.

Assume that there is a point X such that

$$X \in (\langle \mathcal{B}_1 \rangle \cap \langle \mathcal{B}_2 \rangle) \setminus \langle \mathcal{B}_1 \cap \mathcal{B}_2 \rangle.$$

If X were a point of $\langle \mathcal{B}_1' \rangle$ then it would be contained in $\langle \mathcal{B}_1' \rangle \cap \langle \mathcal{B}_2 \rangle$, so, by induction, $X \in \langle \mathcal{B}_1' \cap \mathcal{B}_2 \rangle$ and therefore $X \in \langle \mathcal{B}_1 \cap \mathcal{B}_2 \rangle$.

This contradiction shows that the hypothetical point X must be a point of

$$\langle \mathcal{B}_1 \rangle \setminus \langle \mathcal{B}_1' \rangle = \langle \mathcal{B}_1', P \rangle \setminus \langle \mathcal{B}_1' \rangle$$

Using **1.3.2** (exchange property) it then follows that $P \in \langle \mathcal{B}_1', X \rangle$. Since X also lies in $\langle \mathcal{B}_2 \rangle$ we obtain

$$P \in \langle \mathcal{B}_1', X \rangle \subseteq \langle \mathcal{B}_1', \langle \mathcal{B}_2 \rangle \rangle = \langle \mathcal{B}_1', \mathcal{B}_2 \rangle = \langle \mathcal{B}_1' \cup \mathcal{B}_2 \rangle.$$

By the choice of P we have $P \notin \mathcal{B}_2$; moreover, $P \notin \mathcal{B}_1'$. Therefore we get a contradiction to the independence of $\mathcal{B}$: $\mathcal{B}_0 := \mathcal{B}_1' \cup \mathcal{B}_2 \subseteq \mathcal{B}$ satisfies $P \in \langle \mathcal{B}_0 \rangle$ but $P \notin \mathcal{B}_0$. □

The next theorem is the Steinitz exchange theorem for finitely generated projective spaces. It is convenient to handle the most important special case first.

1.3.6 Exchange lemma. *Let* $\mathcal{B}$ *be a finite basis of* $\mathbf{P}$, *and let* P *be a point of* $\mathbf{P}$. *Then there is a point* Q *in* $\mathcal{B}$ *with the property that the set*

$$(\mathcal{B}\setminus\{Q\}) \cup \{P\}$$

is also a basis of $\mathbf{P}$.

Such a point Q *can be found in the following way. There is a subset* $\mathcal{B}'$ *of* $\mathcal{B}$ *such that* $P \in \langle\mathcal{B}'\rangle$, *but* $P \notin \langle\mathcal{B}''\rangle$ *for any proper subset* $\mathcal{B}''$ *of* $\mathcal{B}'$. *Then for any point* $Q \in \mathcal{B}'$ *the set* $(\mathcal{B}\setminus\{Q\}) \cup \{P\}$ *is a basis of* $\mathbf{P}$.

Proof. Since $\mathcal{B}$ is a finite set there exists a subset $\mathcal{B}'$ of $\mathcal{B}$ with the following property:

$$P \in \langle\mathcal{B}'\rangle, \text{ but } P \notin \langle\mathcal{B}''\rangle \text{ for any proper subset } \mathcal{B}'' \text{ of } \mathcal{B}'.$$

Let Q be an arbitrary point of $\mathcal{B}'$.
Claim: The set $\mathcal{B}_1 := (\mathcal{B}\setminus\{Q\}) \cup \{P\}$ *is a basis of* $\mathbf{P}$.

For this we first show that $P \notin \langle\mathcal{B}\setminus\{Q\}\rangle$. Assume that $P \in \langle\mathcal{B}\setminus\{Q\}\rangle$. Then, by **1.3.5** we get

$$P \in \langle\mathcal{B}\setminus\{Q\}\rangle \cap \langle\mathcal{B}'\rangle = \langle(\mathcal{B}\setminus\{Q\}) \cap \mathcal{B}'\rangle = \langle\mathcal{B}'\setminus\{Q\}\rangle,$$

contradicting the minimality of $\mathcal{B}'$.

In view of the exchange property **1.3.2** we obtain

$$\langle\mathcal{B}_1\rangle = \langle\mathcal{B}\setminus\{Q\}, P\rangle = \langle\mathcal{B}\setminus\{Q\}, Q\rangle = \langle\mathcal{B}\rangle = \mathbf{P}.$$

Thus $\mathcal{B}_1$ spans $\mathbf{P}$.

It remains to show that $\mathcal{B}_1$ is independent. Assume that $\mathcal{B}_1$ is dependent. Then there is a point $X \in \mathcal{B}_1$ such that $X \in \langle\mathcal{B}_1\setminus\{X\}\rangle$.

If $X = P$ then $P \in \langle\mathcal{B}_1\setminus\{P\}\rangle$. Since $\mathcal{B}_1\setminus\{P\} = \mathcal{B}\setminus\{Q\}$ and $P \in \langle\mathcal{B}'\rangle$ we get in view of $\mathcal{B}\setminus\{Q\}, \mathcal{B}' \subseteq \mathcal{B}$, and **1.3.5**

$$P \in \langle\mathcal{B}\setminus\{Q\}\rangle \cap \langle\mathcal{B}'\rangle = \langle(\mathcal{B}\setminus\{Q\}) \cap \mathcal{B}'\rangle = \langle\mathcal{B}'\setminus\{Q\}\rangle,$$

contradicting the minimality of $\mathcal{B}'$.

Therefore we have $X \in \mathcal{B}\setminus\{Q\}$. By our assumption, $X \in \langle\mathcal{B}_1\setminus\{X\}\rangle = \langle\mathcal{B}\setminus\{Q, X\}, P\rangle$. Since $\mathcal{B}\setminus\{Q\}$ is independent it follows that $X \notin \langle\mathcal{B}\setminus\{Q, X\}\rangle$. In view of **1.3.2** we now get

$$P \in \langle\langle\mathcal{B}\setminus\{Q, X\}\rangle, X\rangle = \langle\mathcal{B}\setminus\{Q, X\}, X\rangle = \langle\mathcal{B}\setminus\{Q\}\rangle.$$

As in the case $X = P$ we can deduce a contradiction to the minimality of $\mathcal{B}'$ as follows:

$$\mathrm{P} \in \langle \mathcal{B} \setminus \{\mathrm{Q}\} \rangle \cap \langle \mathcal{B}' \rangle = \langle \mathcal{B}' \setminus \{\mathrm{Q}\} \rangle.$$

This shows that $\mathcal{B}_1$ is independent. □

1.3.7 Steinitz exchange theorem for projective spaces [Stei13]. *Let* $\mathcal{B}$ *be a finite basis of* **P**, *and let* $r := |\mathcal{B}|$. *If* $\mathcal{C}$ *is an independent set having* s *points then we have:*

(a) $s \le r$.

(b) *There is a subset* $\mathcal{B}^*$ *of* $\mathcal{B}$ *with* $|\mathcal{B}^*| = r - s$ *such that* $\mathcal{C} \cup \mathcal{B}^*$ *is a basis of* **P**.

Proof. First we suppose that $\mathcal{C}$ is a finite set and prove the assertion by induction on s.

If $s = 1$ then the assertion follows directly from the exchange lemma.

Suppose now $s > 1$ and assume that the assertion is true for $s - 1$. Let P be an arbitrary point of $\mathcal{C}$. Then $\mathcal{C}' := \mathcal{C} \setminus \{\mathrm{P}\}$ is an independent set with $s - 1$ points. Thus, by induction, the following assertions are true:

– $s - 1 \le r$.

– There is a subset $\mathcal{B}'$ of $\mathcal{B}$ with $|\mathcal{B}'| = r - (s - 1)$ such that $\mathcal{C}' \cup \mathcal{B}'$ is a basis of **P**.

Claim: We also have $s \le r$. Otherwise, $s - 1 = r$, so $\mathcal{B}' = \varnothing$. Thus $\mathcal{C}'$ would be a basis of **P**. In particular, $\mathrm{P} \in \langle \mathcal{C}' \rangle$ contradicting the independence of $\mathcal{C}$. Thus (a) is proved.

(b) Since $\mathcal{B}$ is finite there exists a subset $\mathcal{B}''$ of $\mathcal{C}' \cup \mathcal{B}'$ such that $\mathrm{P} \in \langle \mathcal{B}'' \rangle$, but P is not contained in the span of any proper subset of $\mathcal{B}''$. Then $\mathcal{B}'' \cap \mathcal{B}' \ne \varnothing$. Otherwise, in view of **1.3.5** we would obtain the following contradiction to the independence of $\mathcal{C}$:

$$\mathrm{P} \in \langle \mathcal{B}'' \rangle \cap \langle (\mathcal{C}' \cup \mathcal{B}') \rangle = \langle \mathcal{B}'' \cap (\mathcal{C}' \cup \mathcal{B}') \rangle \subseteq \langle \mathcal{B}'' \cap \mathcal{C}' \rangle \subseteq \langle \mathcal{C}' \rangle = \langle \mathcal{C} \setminus \{\mathrm{P}\} \rangle.$$

By the exchange lemma, for each $\mathrm{Q} \in \mathcal{B}'' \cap \mathcal{B}'$ the set $(\mathcal{C}' \cup (\mathcal{B}' \setminus \{\mathrm{Q}\})) \cup \{\mathrm{P}\}$ is a basis. Since

$$(\mathcal{C}' \cup (\mathcal{B}' \setminus \{\mathrm{Q}\})) \cup \{\mathrm{P}\} = (\mathcal{C}' \cup \{\mathrm{P}\}) \cup (\mathcal{B}' \setminus \{\mathrm{Q}\}) = \mathcal{C} \cup (\mathcal{B}' \setminus \{\mathrm{Q}\}),$$

the assertion (b) follows if we define $\mathcal{B}^* := \mathcal{B}' \setminus \{\mathrm{Q}\}$.

Finally we consider the case that $\mathcal{C}$ is an infinite set. Then $\mathcal{C}$ would contain a finite subset having exactly $s' = r + 1$ elements, and by the above discussion we would get

$$r + 1 = s' \le r,$$

a contradiction. □

An easy, but important corollary is the following assertion.

1.3.8 Basis extension theorem. *Let* $\mathbf{P}$ *be a finitely generated projective space. Then any two bases of* $\mathbf{P}$ *have the same number of elements. Moreover, any independent set (in particular any basis of a subspace) can be extended to a basis of* $\mathbf{P}$.

Proof. By **1.3.4** the projective space $\mathbf{P}$ has a finite basis; let r be the number of elements in this basis. By the Steinitz exchange theorem **1.3.7** any independent set, in particular any basis, has at most r elements. In particular any basis of $\mathbf{P}$ is finite. If $\mathcal{B}_1$ and $\mathcal{B}_2$ are two bases then **1.3.7**(a) implies $|\mathcal{B}_1| \leq |\mathcal{B}_2|$ as well as $|\mathcal{B}_2| \leq |\mathcal{B}_1|$.

The second assertion directly follows from **1.3.7**(b). □

Definition. Let $\mathbf{P}$ be a finitely generated projective space. If $d + 1$ denotes the number of elements in a basis (which, by **1.3.8**, is constant) then we call d the **dimension** of $\mathbf{P}$ and write $d = \dim(\mathbf{P})$.

1.3.9 Lemma. *Let* $\mathbf{U}$ *be a subspace of the finitely generated projective space* $\mathbf{P}$. *Then the following assertions are true:*

(a) $\dim(\mathbf{U}) \leq \dim(\mathbf{P})$.

(b) *We have* $\dim(\mathbf{U}) = \dim(\mathbf{P})$ *if and only if* $\mathbf{U} = \mathbf{P}$.

Proof. (a) Any basis of $\mathbf{U}$ is an independent set of $\mathbf{P}$. By **1.3.8**, it contains at most $\dim(\mathbf{P}) + 1$ elements. Hence $\dim(\mathbf{U})$ is finite, and we have $\dim(\mathbf{U}) \leq \dim(\mathbf{P})$.

(b) One direction is trivial. If, conversely, $\dim(\mathbf{U}) = \dim(\mathbf{P})$, then any basis $\mathcal{B}$ of $\mathbf{U}$ is a basis of $\mathbf{P}$ as well; thus $\mathbf{U} = \langle \mathcal{B} \rangle = \mathbf{P}$. □

Definition. Let $\mathbf{P}$ be a projective space of finite dimension d. The subspaces of dimension 2 are called **planes**, and the subspaces of dimension $d - 1$ are called **hyperplanes** of $\mathbf{P}$.

We denote the set of all subspaces of $\mathbf{P}$ by $\mathcal{U}(\mathbf{P})$. We call $\mathcal{U}(\mathbf{P})$ together with the subset relation $\subseteq$ the **projective geometry** belonging to the projective space $\mathbf{P}$.

The empty set and the whole space are called the **trivial** subspaces. The set of all **nontrivial** subspaces is denoted by $\mathcal{U}^*(\mathbf{P})$.

It will not be necessary to distinguish between a 'projective space' and its corresponding 'projective geometry', since each uniquely determines the other.

Using the notions presented in Section **1.1** one can say: $\mathfrak{U}^*(\mathbf{P})$ is, together with the relation $\subseteq$, a geometry of rank d (see exercise 24).

1.3.10 Lemma. *Let* **P** *be a d-dimensional projective space, and let* **U** *be a t-dimensional subspace of* **P** $(-1 \leq t \leq d)$. *Then there exist* $d-t$ *hyperplanes of* **P** *such that* **U** *is the intersection of these hyperplanes.*

Proof. Let $\{P_0, P_1, \ldots, P_t\}$ be a basis of **U**. Using **1.3.7** we can extend it to a basis $\mathcal{B} = \{P_0, P_1, \ldots, P_t, P_{t+1}, \ldots, P_d\}$ of **P**. If we define

$$\mathbf{H}_i := \langle \mathcal{B} \setminus \{P_{t+i}\} \rangle \quad (i = 1, \ldots, d-t),$$

then we obtain $d-t$ hyperplanes $\mathbf{H}_1, \ldots, \mathbf{H}_{d-t}$. By **1.3.5** (see also exercise 25) their intersection can be computed as follows:

$$\begin{aligned}\mathbf{H}_1 \cap \ldots \cap \mathbf{H}_{d-t} &= \langle \{P_0, P_1, \ldots, P_t, P_{t+2}, \ldots, P_d\} \rangle \cap \ldots \\ &\quad \cap \langle \{P_0, P_1, \ldots, P_t, P_{t+1}, \ldots, P_{d-1}\} \rangle \\ &= \langle \{P_0, P_1, \ldots, P_t, P_{t+2}, \ldots, P_d\} \cap \ldots \\ &\quad \cap \{P_0, P_1, \ldots, P_t, P_{t+1}, \ldots, P_{d-1}\} \rangle \\ &= \langle \{P_0, P_1, \ldots, P_t\} \rangle = \mathbf{U}.\end{aligned}$$

□

1.3.11 Theorem (dimension formula). *Let* **U** *and* **W** *be subspaces of* **P**. *Then*

$$\dim(\langle \mathbf{U}, \mathbf{W} \rangle) = \dim(\mathbf{U}) + \dim(\mathbf{W}) - \dim(\mathbf{U} \cap \mathbf{W}).$$

Before the proof we shall discuss an *example.* Two distinct lines g and h generate a plane if and only if they have a point in common (for then $\dim(g \cap h) = 0$); otherwise they generate a 3-dimensional subspace.

Proof. We choose a basis $\mathcal{A} = \{P_1, \ldots, P_s\}$ of $\mathbf{U} \cap \mathbf{W}$ and extend it (by **1.3.8**) to a basis $\mathcal{B}$ of **U** and to a basis $\mathcal{C}$ of **W**:

$$\mathcal{B} = \{P_1, \ldots, P_s, P_{s+1}, \ldots, P_{s+t}\}, \mathcal{C} = \{P_1, \ldots, P_s, Q_{s+1}, \ldots, Q_{s+t'}\}.$$

It suffices to show that $\mathcal{B} \cup \mathcal{C}$ is a basis of $\langle \mathbf{U}, \mathbf{W} \rangle$. For then

$$\begin{aligned}\dim(\langle \mathbf{U}, \mathbf{W} \rangle) &= |\mathcal{B} \cup \mathcal{C}| - 1 \\ &= s + t + s + t' - s - 1 = s + t - 1 + s + t' - 1 - (s - 1) \\ &= \dim(\mathbf{U}) + \dim(\mathbf{W}) - \dim(\mathbf{U} \cap \mathbf{W}).\end{aligned}$$

In view of $\langle \mathbf{U}, \mathbf{W} \rangle = \langle \mathcal{B}, \mathcal{C} \rangle = \langle \mathcal{B} \cup \mathcal{C} \rangle$, the set $\mathcal{B} \cup \mathcal{C}$ certainly spans $\langle \mathbf{U}, \mathbf{W} \rangle$.

It remains to show that $\mathcal{B} \cup \mathcal{C}$ is an independent set. In order to do this we first convince ourselves of the following *claim:*

$$\mathbf{U} \cap \langle \mathcal{C} \setminus \mathcal{A} \rangle = \varnothing.$$

Assume that there exists a point $X \in \mathbf{U} \cap \langle \mathcal{C} \setminus \mathcal{A} \rangle$. Since $\langle \mathcal{C} \setminus \mathcal{A} \rangle \subseteq \mathbf{W}$, the point X would be contained in $\mathbf{W}$ and hence in $\mathbf{U} \cap \mathbf{W}$. Using **1.3.5** we would get a contradiction as follows:

$$X \in (\mathbf{U} \cap \mathbf{W}) \cap \langle \mathcal{C} \setminus \mathcal{A} \rangle = \langle \mathcal{A} \rangle \cap \langle \mathcal{C} \setminus \mathcal{A} \rangle = \langle \varnothing \rangle = \varnothing.$$

Now assume that $\mathcal{B} \cup \mathcal{C}$ is not independent. Then there is a point $P \in \mathcal{B} \cup \mathcal{C}$ with $P \in \langle (\mathcal{B} \cup \mathcal{C}) \setminus \{P\} \rangle$. We may assume that $P \in \mathcal{B}$. It follows that

$$P \in \langle (\mathcal{B} \cup \mathcal{C}) \setminus \{P\} \rangle = \langle \mathcal{B} \setminus \{P\}, \mathcal{C} \setminus \mathcal{A} \rangle = \langle \langle \mathcal{B} \setminus \{P\} \rangle, \langle \mathcal{C} \setminus \mathcal{A} \rangle \rangle.$$

The 'join theorem' (exercise 16) says that there are points $T \in \langle \mathcal{C} \setminus \mathcal{A} \rangle$ and $S \in \langle \mathcal{B} \setminus \{P\} \rangle \subseteq \mathbf{U}$ such that $P \in (ST)$.

We claim: the line ST *intersects* $\mathbf{U}$ *in just one point, namely* S. If to the contrary ST were to contain two distinct points of $\mathbf{U}$ then any point of ST, in particular T, would be contained in $\mathbf{U}$. In view of $T \in \langle \mathcal{C} \setminus \mathcal{A} \rangle$ this contradicts the above assertion.

Since the point P satisfies $P \in \mathbf{U} \cap ST$ we have $P = S \in \langle \mathcal{B} \setminus \{P\} \rangle$, contradicting the independence of $\mathcal{B}$. □

1.3.12 Corollary. *Let* $\mathbf{P}$ *be a projective space, and let* $\mathbf{H}$ *be a hyperplane of* $\mathbf{P}$. *Then for any subspace* $\mathbf{U}$ *of* $\mathbf{P}$ *we have the following alternatives:*
- $\mathbf{U}$ *is contained in* $\mathbf{H}$, *or*
- $\dim(\mathbf{U} \cap \mathbf{H}) = \dim(\mathbf{U}) - 1$.

In particular we have the following assertion: any line that is not contained in $\mathbf{H}$ *intersects* $\mathbf{H}$ *in precisely one point.*

Proof. Let $d = \dim(\mathbf{P})$. Suppose that $\mathbf{U}$ is not contained in $\mathbf{H}$. Then $\langle \mathbf{U}, \mathbf{H} \rangle$ is a subspace that contains $\mathbf{H}$ properly. It follows that $\langle \mathbf{U}, \mathbf{H} \rangle = \mathbf{P}$. Using the dimension formula we get

$$\begin{aligned}\dim(\mathbf{U} \cap \mathbf{H}) &= \dim(\mathbf{U}) + \dim(\mathbf{H}) - \dim(\langle \mathbf{U}, \mathbf{H} \rangle) \\ &= \dim(\mathbf{U}) + (d-1) - d = \dim(\mathbf{U}) - 1.\end{aligned}$$ □

1.4 Quotient geometries

In many situations one has to study the 'local' structure of a geometry. In this book we feature two different local structures: the subspaces (or more generally, the objects of a geometry together with the induced incidence) and the quotients. In general, these local structures are geometries of smaller rank and therefore, at least in principle, easier to handle. Often, one can deduce global properties by looking at these local structures. An example of a quotient is the set of all subspaces passing through a fixed point. In this section we shall discuss quotients of projective geometries. Historically, this was one of the structures that led to the development of projective geometry.

Definition. Let Q be a point of **P**. The rank 2 geometry $\mathbf{P}/Q$ whose *points* are the lines of **P** through Q, whose *lines* are the planes of **P** through Q, and whose *incidence* is the incidence induced by **P** is called the **quotient geometry** of Q. We also speak of '**P** modulo the point Q'.

We shall show that the quotient geometry is a projective space. This task requires us to 'identify' the quotient geometry with a suitable projective space; such an 'identifying' will be formally described by the notion of an 'isomorphism'.

Definition. Let $\mathbf{G} = (\Omega, I)$ and $\mathbf{G}' = (\Omega', I')$ be rank 2 geometries; let $\mathcal{P}$, $\mathcal{B}$ and $\mathcal{P}'$, $\mathcal{B}'$ the sets of points and blocks of **G** and **G'**. We call **G** and **G'** **isomorphic** if there exists a map $\alpha: \mathcal{P} \cup \mathcal{B} \to \mathcal{P}' \cup \mathcal{B}'$ with the following properties:
– α maps $\mathcal{P}$ onto $\mathcal{P}'$ and $\mathcal{B}$ onto $\mathcal{B}'$ and the restrictions of α to $\mathcal{P}$ and to $\mathcal{B}$ are bijections.
– For all $P \in \mathcal{P}$ and all $B \in \mathcal{B}$ the following equivalence is true:

$$P\,I\,B \Leftrightarrow \alpha(P)\,I'\,\alpha(B).$$

Such a map α is called an **isomorphism** from **G** onto **G'**.

An **automorphism** is an isomorphism from a rank 2 geometry **G** onto itself. If the blocks **G** are called lines then an automorphism is often called a **collineation**. This applies in particular for projective spaces.

1.4.1 Theorem. *Let* **P** *be a d-dimensional projective space, and let* Q *be a point of* **P**. *Then the quotient geometry* $\mathbf{P}/Q$ *of* **P** *modulo* Q *is a projective space of dimension* $d-1$.

Proof. We shall show that $\mathbf{P}/Q$ is isomorphic to a hyperplane not through Q. First we shall show that such a hyperplane exists.

By Lemma **1.3.8** one can extend Q to a basis $\{Q, P_1, P_2, \ldots, P_d\}$ of **P**. The subspace $\mathbf{H} = \langle P_1, P_2, \ldots, P_d\rangle$ is spanned by d independent points, hence **H** has dimension $d-1$ and is therefore a hyperplane. Since the set $\{Q, P_1, P_2, \ldots, P_d\}$ is independent, it follows that $Q \notin \mathbf{H}$.

Now we show that the geometry $\mathbf{P}/Q$ is isomorphic to any hyperplane **H** not through Q. In order to do this we define the map α from the points and lines of

$\mathbf{P}/Q$ onto the points and lines of $\mathbf{H}$ in the simplest possible way (see Figure 1.7):

$$\alpha: g \mapsto g \cap \mathbf{H},$$
$$\alpha: \pi \mapsto \pi \cap \mathbf{H}.$$

Since the points of $\mathbf{P}/Q$ are the lines g of $\mathbf{P}$ through Q and since Q is not incident with $\mathbf{H}$, $g \cap \mathbf{H}$ is a point of $\mathbf{H}$. Similarly one argues for the lines of $\mathbf{P}/Q$.

We have to show that this map is bijective when restricted to the points and lines and preserves the incidence.

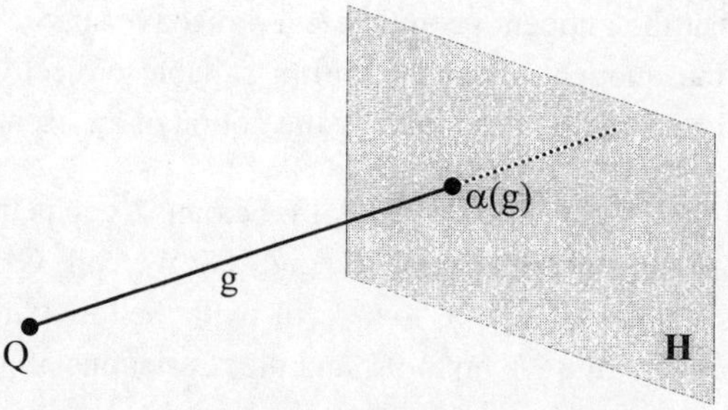

Figure 1.7 $\mathbf{P}/\mathbf{Q}$ is isomorphic to $\mathbf{H}$

α *is injective:* Let g and h be two distinct points of $\mathbf{P}/Q$, that is, two distinct lines through Q. If g and h were to intersect $\mathbf{H}$ in the same point X then g and h would have the points Q and X in common. Since $X \in \mathbf{H}$, but $Q \notin \mathbf{H}$, we have that $X \neq Q$; this is a contradiction.

α *is surjective:* If X denotes a point of $\mathbf{H}$ then QX is a line through Q, hence a point of $\mathbf{P}/Q$.

Similarly one shows that α acts bijectively on the set of lines.

α *preserves incidence:* Let g be a point and π a plane of $\mathbf{P}/Q$. Then

$$g \subseteq \pi \Leftrightarrow g \cap \mathbf{H} \subseteq \pi \cap \mathbf{H} \Leftrightarrow \alpha(g) \subseteq \alpha(\pi).$$

Thus the theorem is proved completely. □

As a corollary we note the following assertion.

1.4.2 Corollary. *Let* $\mathbf{P}$ *be a d-dimensional projective space, and let* Q *be a point of* $\mathbf{P}$. *Then there is a hyperplane of* $\mathbf{P}$ *that does not pass through* Q. □

1.5 Finite projective spaces

In this section we shall determine the numbers of points, lines, and hyperplanes of finite projective spaces. For this, the following lemma is fundamental, although it holds in arbitrary, not necessarily finite, projective spaces.

1.5.1 Lemma. *Let* g_1 *and* g_2 *be two lines of a projective space* $\mathbf{P}$. *Then there exists a bijective map*

$$\varphi: (g_1) \to (g_2)$$

from the set (g_1) *of points on* g_1 *onto the set* (g_2) *of points on* g_2.

Proof. W.l.o.g. we have $g_1 \neq g_2$.
Case 1. The lines g_1 and g_2 intersect in a point S.

Let P_1 be a point on g_1 and P_2 a point on g_2 such that $P_1 \neq S \neq P_2$. By axiom 3 there is a third point P on the line P_1P_2. By definition, P is neither on g_1 nor on g_2. By axiom 2 any line through P that contains a point $X \neq S$ of g_1 intersects the line g_2 in a uniquely determined point $\varphi(X)$ with $\varphi(X) \neq S$ (see Figure 1.8).

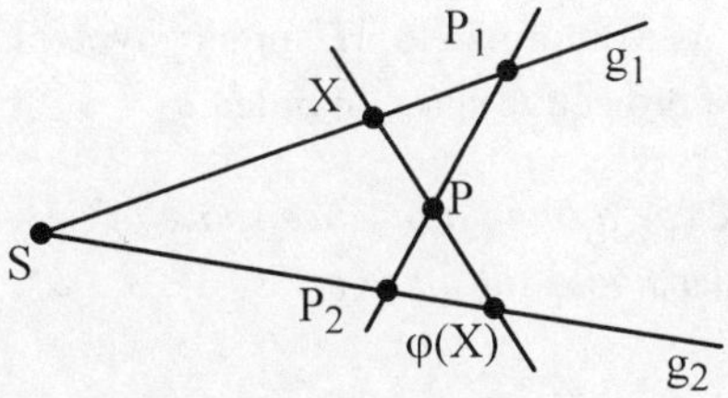

Figure 1.8 Bijective map of the points on g_1 onto the points of g_2

This means that the map φ defined by

$$\varphi: X \mapsto XP \cap g_2$$

is a bijection from $(g_1)\setminus\{S\}$ onto $(g_2)\setminus\{S\}$. Trivially, this can be extended to a bijection φ from (g_1) onto (g_2), by defining $\varphi(S) := S$.
Case 2. The lines g_1 and g_2 have no point in common.

Let h be a line connecting some point of g_1 to some point of g_2. By the first case, there are bijections

$$\varphi_1: (g_1) \to (h) \quad \text{and} \quad \varphi_2: (h) \to (g_2).$$

Then $\varphi := \varphi_2 \circ \varphi_1$ is a bijective map from (g_1) onto (g_2). □

Definition. A projective space $\mathbf{P}$ is called **finite** if its point set is a finite set. (If $\dim(\mathbf{P}) > 1$ then $\mathbf{P}$ is finite if and only if the set of lines of $\mathbf{P}$ is finite; see exercise 31.)

If $\mathbf{P}$ is a finite projective space then, by **1.5.1**, each line of $\mathbf{P}$ is incident with the same number of points. Therefore there exists a positive integer q such that each line of $\mathbf{P}$ is incident with precisely $q + 1$; in view of axiom 3 we have $q \geq 2$. The so defined integer q is called the **order** of the finite projective space $\mathbf{P}$.

For a finite projective space $\mathbf{P}$, its dimension d and its order q are the fundamental parameters. This can, for instance, be seen from the fact that all parameters of $\mathbf{P}$ (e.g. its numbers of points, lines, and hyperplanes) are functions of d and q.

We shall compute some parameters of projective spaces. First we shall determine the order of the quotient geometry.

1.5.2 Lemma. *Let $\mathbf{P}$ be a finite projective space of dimension $d \geq 2$ and order q. Then for each point Q of $\mathbf{P}$ the quotient geometry $\mathbf{P}/Q$ has order q as well.*

Proof. By **1.4.1**, $\mathbf{P}/Q$ is isomorphic to $\mathbf{H}$ for any hyperplane $\mathbf{H}$ not through Q. In particular, $\mathbf{P}/Q$ is a projective space of order q. □

1.5.3 Theorem. *Let $\mathbf{P}$ be a finite projective space of dimension d and order q, and let $\mathbf{U}$ be a t-dimensional subspace of $\mathbf{P}$ $(1 \leq t \leq d)$. Then the following assertions are true:*

(a) *The number of points of $\mathbf{U}$ is*

$$q^t + q^{t-1} + \ldots + q + 1 = \frac{q^{t+1} - 1}{q - 1}.$$

In particular, $\mathbf{P}$ has exactly $q^d + \ldots + q + 1$ points.

(b) *The number of lines of $\mathbf{U}$ through a fixed point of $\mathbf{U}$ equals*

$$q^{t-1} + \ldots + q + 1.$$

(c) *The total number of lines of $\mathbf{U}$ equals*

$$\frac{(q^t + q^{t-1} + \ldots + q + 1) \cdot (q^{t-1} + \ldots + q + 1)}{q + 1}.$$

Proof. We prove (a) and (b) simultaneously by induction on t. If $t = 1$ then (a) is nothing else than **1.5.1** together with the definition of the order of **P**. In this case (b) holds trivially, since **U** is a line.

Suppose now that (a) and (b) are true for projective spaces of dimension $t-1 \geq 1$. We consider an arbitrary point Q of **U**. Since by **1.4.1** and **1.5.2** the quotient geometry $\mathbf{U}/\mathrm{Q}$ is a $(t-1)$-dimensional projective space of order q, by induction, its number of points is $q^{t-1} + \ldots + q + 1$. Since, by definition of $\mathbf{U}/\mathrm{Q}$, this is the number of lines of **U** through Q, we have already shown (b).

Since each of these lines carries exactly q points different from Q and since each point $\mathrm{R} \neq \mathrm{Q}$ of **U** is on exactly one of these lines (namely on QR) **U** contains exactly

$$1 + (q^{t-1} + \ldots + q + 1) \cdot q = q^t + q^{t-1} + \ldots + q + 1$$

points. This shows (a).

(c) Let b be the number of all lines of **U**. Since **U** has exactly $q^t + \ldots + q + 1$ points and any point of **U** is on exactly $q^{t-1} + \ldots + q + 1$ lines of **U**, one could be tempted to think that $(q^t + \ldots + q + 1) \cdot (q^{t-1} + \ldots + q + 1)$ is the number of lines of **U**. But since any line has exactly $q + 1$ points it is counted exactly $q + 1$ times. Thus we get

$$b = \frac{(q^t + \ldots + q + 1) \cdot (q^{t-1} + \ldots + q + 1)}{q + 1}. \qquad \square$$

Definition. The numbers $q^t + \ldots + q + 1$ are often denoted by $\Theta_t = \Theta_t(q)$ ('theta').

1.5.4 Theorem. *Let* **P** *be a finite projective space of dimension* d *and order* q. *Then*

(a) *the number of hyperplanes of* **P** *is exactly*

$$q^d + \ldots + q + 1;$$

(b) *the number of hyperplanes of* **P** *through a fixed point* **P** *equals*

$$q^{d-1} + \ldots + q + 1.$$

Proof. We show (a) by induction on d. For $d = 1$ the theorem says only that any line has $q + 1$ points, for $d = 2$ the assertion follows from **1.5.3**.

Suppose that the assertion is true for projective spaces of dimension $d - 1 \geq 1$. Consider a hyperplane **H** of **P**. By the dimension formula **1.3.11**, every hyperplane different from **H** intersects **H** in a subspace of dimension $d - 2$. Thus, any hyperplane $\neq$ **H** of **P** is spanned by a $(d-2)$-dimensional subspace of **H** and a point outside **H**.

For each $(d-2)$-dimensional subspace **U** of **H** and each point $P \in \mathbf{P}\backslash\mathbf{H}$, the subspace $\langle \mathbf{U}, P \rangle$ is a hyperplane, which contains exactly $(q^{d-1} + \ldots + 1) - (q^{d-2} + \ldots + 1) = q^{d-1}$ points outside **H**. Since there exist q^d points of **P** outside **H** there are q hyperplanes $\neq$ **H** through **U**. By induction, there are exactly $q^{d-1} + \ldots + q + 1$ hyperplanes of **H**, which are subspaces of dimension $d-1$ of **P**. Thus, the total number of hyperplanes of **P** is $q \cdot (q^{d-1} + \ldots + q + 1) + 1$.
(b) Let P be a point of **P** and **H** be a hyperplane not through P. Then any hyperplane of **P** through P intersects **H** in a hyperplane of **H**. Thus, by (a), there are exactly $q^{d-1} + \ldots + 1$ such hyperplanes. □

1.5.5 Corollary. *Let* **P** *be a finite projective plane. Then there exists an integer* $q \geq 2$ *such that any line of* **P** *has exactly* $q+1$ *points, and* **P** *the total number of points is* $q^2 + q + 1$. □

For instance, there is no projective plane having exactly 92, 93, . . ., or 110 points.

The question of which positive integers q can be the order of a finite projective plane is among the most discussed questions in finite geometry. Here are some facts.

order q	2	3	4	5	6	7	8	9	10	11	12
existence	yes	yes	yes	yes	no	yes	yes	yes	no	yes	?

The fact that the 'yes'-entries are true will be proved in the next chapter. Although the question of existence is very interesting, we will not pursue it in great detail. We shall simply summarize what is known.

1. The order of any *known* finite projective plane is a prime power, that is a positive integer of the form p^e, where p is a prime and $e \geq 1$.
2. The theorem of Bruck and Ryser [BrRy49] says the following: if q is a positive integer of the form $q = 4n + 1$ or $q = 4n + 2$ $(n \in \mathbf{N})$ and if there is a projective plane of order q then q is the sum of two square numbers, one of which might be 0. From this one can in particular infer that there does not exist a projective plane whose order is $q = 8n + 6$ $(n \in \mathbf{N})$; so there does not exist a projective plane of order 6, 14, ...
3. The only other integer q that has been excluded as the order of a projective plane is $q = 10$. This result has been obtained by methods which are in marked contrast to the methods used for the Bruck–Ryser theorem. While the Bruck–Ryser theorem was proved using sophisticated ideas of number theory, the non-

existence of a projective plane of order 10 was proved by massive computer power: just to exclude a certain configuration (a '12-arc') the computer program had to run for 183 days! (Cf. [Lam91].)

The projective planes of order 4 and 5 are true gems and contain many beautiful geometric structures. Elementary treatments of these gems can be found in [Beu86], [Beu87], [Cox74].

1.6 Affine geometries

When one learns for the first time of projective spaces one could perhaps think that these structures are of a purely theoretical nature, without any real meaning. One might believe that 'affine' geometry, where one has parallel lines, is much more natural.

In this section we will show, however, that these structures are basically the same – distinguished only by different points of view. Why does one concentrate on projective geometry and not affine geometry? The reason is that projective geometries have 'homogeneous' properties and there is no need to distinguish many special cases, which is unavoidable when studying affine geometries.

Definition. Let $\mathbf{P}$ be a projective space of dimension $d \geq 2$, and let $\mathbf{H}_\infty$ be a hyperplane of $\mathbf{P}$. We define the geometry $\mathbf{A} = \mathbf{P} \setminus \mathbf{H}_\infty$ as follows:

- The *points* of $\mathbf{A}$ are the points of $\mathbf{P}$ that are not in $\mathbf{H}_\infty$.
- The *lines* of $\mathbf{A}$ are those lines of $\mathbf{P}$ that are not contained in $\mathbf{H}_\infty$.
- In general, the *t-dimensional subspaces* of $\mathbf{A}$ are those t-dimensional subspaces of $\mathbf{P}$ that are not contained in $\mathbf{H}_\infty$.
- The *incidence* of $\mathbf{A}$ is induced by the incidence of $\mathbf{P}$.

The rank 2 geometry consisting of the points and lines of $\mathbf{A}$ is called an **affine space of dimension** d; we often also denote this rank 2 geometry by $\mathbf{A} = \mathbf{P} \setminus \mathbf{H}_\infty$. An affine space of dimension 2 is called an **affine plane**.

The set of all subspaces of $\mathbf{A}$ is called an **affine geometry**.

Finally, for a fixed integer $t \in \{1, \ldots, d-1\}$ we denote the rank 2 geometry consisting of the points and the t-dimensional subspaces of $\mathbf{A}$ by $\mathbf{A}_t$. So an affine space is the geometry $\mathbf{A}_1$.

The subspaces of an affine space are often called **flats**.

We call the hyperplane $\mathbf{H}_\infty$ the **hyperplane at infinity** and the points of $\mathbf{H}_\infty$ the **points at infinity** of $\mathbf{A}$. Sometimes these points are also called 'improper' points. But this is only a convenient name and does not imply that there is something wrong with these points.

The projective space **P** is also called the **projective closure** of the affine space **A**.

The essential difference between projective and affine geometries is that affine geometries have a natural parallelism.

Definition. Let $\mathbf{G} = (\mathcal{P}, \mathcal{B}, \mathrm{I})$ be a rank 2 geometry. A **parallelism** of **G** is an equivalence relation $\|$ on the block set $\mathcal{B}$ satisfying the following **parallel axiom**:

If $\mathrm{P} \in \mathcal{P}$ is a point and $\mathrm{B} \in \mathcal{B}$ is a block of **G**, then there is precisely one block $\mathrm{C} \in \mathcal{B}$ through P such that $\mathrm{C} \| \mathrm{B}$.

The blocks B and C are called **parallel** if $g \| h$ holds.

1.6.1 Theorem. *For* $t \in \{1, \ldots, d-1\}$, *the geometry* $\mathbf{A}_t$ *has a parallelism.*

Proof. Let **P** be the projective space belonging to **A**, and let $\mathbf{H}_\infty$ be the hyperplane at infinity. Consider two t-dimensional subspaces **U**, **W** of **A**. By definition, these are t-dimensional subspaces of **P** that are not contained in $\mathbf{H}_\infty$. By **1.3.12**, $\mathbf{U} \cap \mathbf{H}_\infty$ and $\mathbf{W} \cap \mathbf{H}_\infty$ are subspaces of dimension $t-1$ of $\mathbf{H}_\infty$. We define

$$\mathbf{U} \| \mathbf{W} :\Leftrightarrow \mathbf{U} \cap \mathbf{H}_\infty = \mathbf{W} \cap \mathbf{H}_\infty$$

and shall show that $\|$ is a parallelism. It is clear that $\|$ is an equivalence relation, since it is defined via an equality relation.

Now we shall show the parallel axiom. Let **U** be a t-dimensional subspace of **A**, and let P be a point of **A**. We denote the $(t-1)$-dimensional subspace $\mathbf{U} \cap \mathbf{H}_\infty$ of $\mathbf{H}_\infty$ by **V**.

We see that any t-dimensional subspace **W** through P that is parallel to **U** must contain P and **V**. Since $\langle \mathbf{V}, \mathrm{P} \rangle$ is already a t-dimensional subspace of **P** which is contained in **W** it follows by **1.3.9** that $\mathbf{W} = \langle \mathbf{V}, \mathrm{P} \rangle$. Thus we have proved the existence as well as the uniqueness of a parallel to **U** through P. □

Remark. We call the parallelism constructed in **1.6.1** the **natural** parallelism of **A**. (This name indicates of course that there might also exist nonnatural parallelisms.)

For the natural parallelism we have that any two distinct, parallel t-dimensional subspaces of **A** span a subspace of dimension $t+1$.

We call two subspaces of arbitrary dimension **parallel** if one is parallel to a subspace of the other. This means: if $\mathbf{H}_\infty$ denotes the hyperplane at infinity of **P** then the subspaces **U** and **W** are parallel if and only if we have either $\mathbf{U} \cap \mathbf{H}_\infty$

$\subseteq \mathbf{W} \cap \mathbf{H}_\infty$ or $\mathbf{W} \cap \mathbf{H}_\infty \subseteq \mathbf{U} \cap \mathbf{H}_\infty$. In particular we say that a line g of **A** is parallel to a hyperplane **H** if g is parallel to some line h of **H**; this means that g intersects **H** in a point at infinity.

1.6.2 Lemma. *Let* $\mathbf{A} = \mathbf{P} \setminus \mathbf{H}_\infty$, *where* **P** *is a d-dimensional projective space and* $\mathbf{H}_\infty$ *is the hyperplane at infinity of* **A**.
(a) *Each line that is not parallel to a hyperplane* **H** *intersects* **H** *in precisely one point of* **A**.
(b) *If* $d = 2$ *then any two nonparallel lines intersect in a point of* **A**.

Proof. (a) Let g be a line and **H** a hyperplane of **A** that are not parallel. Then g intersects $\mathbf{H}_\infty$ in a point outside $\mathbf{H} \cap \mathbf{H}_\infty$. By **1.3.12**, g and **H** intersect in some point of **P**; since they don't meet in $\mathbf{H}_\infty$ they have a point of **A** in common.
(b) follows trivially from (a). □

1.6.3 Corollary. *Any affine plane* **A** *has the following properties:*
(1) *Through any two distinct points there is exactly one line.*
(2) (**Playfair's parallel axiom**) *If* g *is a line and* P *a point outside* g *then there is precisely one line through* P *that has no point in common with* g.
(3) *There exist three points that are not on a common line.*

Proof. Let $\mathbf{A} = \mathbf{P} \setminus g_\infty$, where **P** is a projective plane and g_∞ is the line at infinity of **A**.
(1) By axiom 1 any two points of **P**, in particular any two points of **A**, lie on precisely one line.
(2) follows by **1.6.1** and **1.6.2**.
(3) By axiom 3 g_∞ has at least three points. Through any point P of **P** outside g_∞ there are at least three lines g_i. These have, apart from P, another point P_i in **A** ($i = 1, 2, 3$). Hence P, P_1, P_2 are three noncollinear points of **A**. □

1.6.4 Theorem. *Let* $\mathbf{S} = (\mathcal{P}, \mathcal{G}, \mathrm{I})$ *be a geometry that satisfies the conditions* (1), (2), *and* (3) *of* **1.6.3**. *Then* **S** *is an affine plane.*

Proof. We have to show that there are a projective plane **P** and a line g_∞ of **P** such that $\mathbf{S} = \mathbf{P} \setminus g_\infty$.

For this we have to extend **S** by additional points ('points at infinity') and by a 'line at infinity'. The essential tool for this is the parallel classes.

We define a relation $\|$ on the set of lines of **S** by

$$g \| h :\Leftrightarrow g = h \text{ or } (g) \cap (h) = \varnothing.$$

Step 1. $\|$ *is a parallelism.*

For this we show first that $\|$ is an equivalence relation: reflexivity and symmetry of $\|$ follow directly from the definition. In order to show that $\|$ is transitive let g, h, and k be lines with $g \| h$ and $h \| k$. If g and k are disjoint then they are parallel. Hence let g and k have a point P in common. Then we see that P is on two distinct lines, namely g and k, which are both parallel to h. By Playfair's parallel axiom (applied to P and h) we have $g = k$, and so in particular $g \| k$.

By Playfair's parallel axiom it follows that $\|$ is a parallelism.

By definition of $\|$ and in view of (1) we alsohave

Step 2. Any two nonparallel lines intersect each other in (precisely) one point.

Since $\|$ is an equivalence relation we may consider the corresponding equivalence classes; we call them **parallel classes**. A parallel class consists of a set of disjoint lines; in view of Playfair's parallel axiom any point of **S** is on (precisely) one line of each parallel class.

Now we regard any parallel class Π as a 'new' point. (Behind this there is the intuitive idea that 'parallel lines meet at infinity'.) Furthermore, we collect all new points into a new line g_∞.

More precisely we define the geometry **P** as follows.

- The *points* of **P** are the points of **S** and the parallel classes of **S**.
- The *lines* of **P** are the lines of **S** and one further line g_∞.
- The *incidence* I* of **P** is defined as follows:

$P\, I^*\, g :\Leftrightarrow P\, I\, g$	for $P \in \mathcal{P}$ and $g \in \mathcal{G}$,
$P\, I^*\, g_\infty$	for no point $P \in \mathcal{P}$,
$\Pi\, I^*\, g :\Leftrightarrow g \in \Pi$	for any parallel class Π and all $g \in \mathcal{G}$,
$\Pi\, I^*\, g_\infty$	for any parallel class Π.

Step 3. **P** *is a projective plane.*

For this we show that axioms 1, 2', and 3' (see exercise 7) hold.

Axiom 1. Any two points of **S** are joined by a line of **S** and by no other line since g_∞ has only new points. Let P be a point of **S** and let Π be a parallel class. If g denotes the line of Π through P then in **P** the line g is incident with P and Π. Since by definition any line of Π is incident with Π, g is the only line incident with Π and P.

Axiom 2'. The line at infinity intersects any line g of **S**, precisely in the parallel class containing g. By Step 2, any two nonparallel lines of **S** intersect each other in a point of **S**, and any two parallel lines of **S** are incident with a common point of **P**, namely with the parallel class containing them.

Axiom 3'. By (3) there are three points P_0, P_1, and P_2 of **S** that are not on a common line. Let Π_1 and Π_2 be the parallel classes of **S** containing P_0P_1 and P_0P_2. Then $\{P_1, P_2, \Pi_1, \Pi_2\}$ is a quadrangle of **P**, that is a set of four points no three of which are on a common line.

Thus we have proved Step 3.

Since, by construction, $\mathbf{S} = \mathbf{P}\backslash g_\infty$, **S** is an affine plane with g_∞ as line at infinity. □

Theorem **1.6.4** says that the Euclidean plane, which we all know and love, is an affine plane. This also means that projective planes are not so unfamiliar to us as it may seem at first glance. The smallest affine planes (having four and nine points) are shown in Figure 1.9.

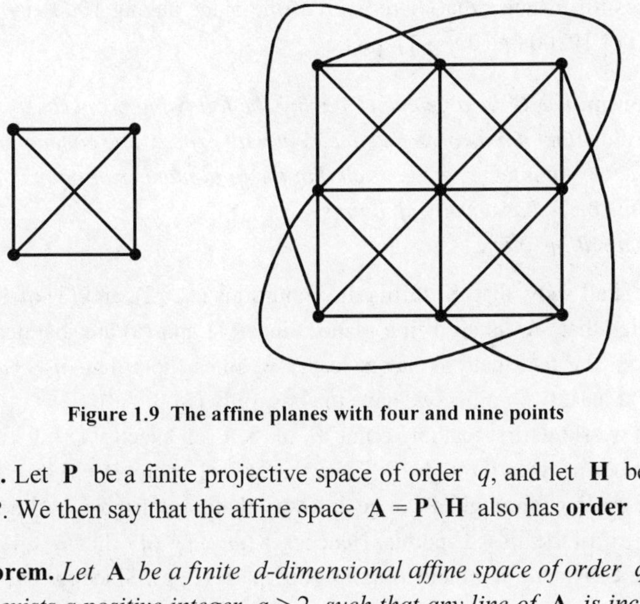

Figure 1.9 The affine planes with four and nine points

Definition. Let **P** be a finite projective space of order q, and let **H** be a hyperplane of **P**. We then say that the affine space $\mathbf{A} = \mathbf{P}\backslash\mathbf{H}$ also has **order** q.

1.6.5 Theorem. *Let* **A** *be a finite* d*-dimensional affine space of order* q.
(a) *There exists a positive integer* $q \geq 2$ *such that any line of* **A** *is incident with exactly* q *points.*
(b) *If* **U** *denotes a* t*-dimensional subspace* $(1 \leq t \leq d)$ *of* **A** *then* **U** *has precisely* q^t *points.*

Proof. Let **P** be the projective space belonging to **A**, and let $\mathbf{H}_\infty$ be the hyperplane at infinity of **A**.
(a) We consider an arbitrary line g of **A**, and let q be its number of points (in **A**). Then g has – considered as a line of **P** – exactly $q + 1$ points. Hence **P** has order q. Therefore any line of **A** has exactly q points of **A**.

(b) Any t-dimensional subspace **U** of **A** intersects – when considered as a subspace of **P** – the hyperplane at infinity in a $(t-1)$-dimensional subspace $\mathbf{U} \cap \mathbf{H}_\infty$. Thus we have

number of points of **A** in **U**
= number of points of **U** in **P** – number of points in $\mathbf{U} \cap \mathbf{H}_\infty$
$= q^t + \ldots + q + 1 - (q^{t-1} + \ldots + q + 1) = q^t.$ □

1.6.6 Corollary. *Let* **A** *be a finite affine plane. Then there exists an integer* $q \geq 2$ *such that any line of* **A** *is incident with exactly* q *points, and the total number of points of* **A** *is* q^2. □

This implies for instance that there is no affine plane having 10001 ($= 100^2 + 1$), 10002, . . ., or 10200 ($= 101^2 - 1$) points.

1.6.7 Theorem. *Let* **S** *be a geometry having the following properties:*
(a) *Any two distinct points of* **S** *are incident with exactly one common line.*
(b) *There exists an integer* $q \geq 2$ *such that the total number of points of* **S** *is* q^2.
(c) *Each line of* **S** *has exactly* q *points.*
Then **S** *is an affine plane.*

Proof. We shall show that **S** fulfils the conditions (1), (2), and (3) of **1.6.3**. Then **1.6.4** implies that **A** is an affine plane. Since (1) and (a) are identical, and (3) follows from the fact that **S** has at least two lines (note that $q^2 > q$), each of which has at least two points, we have to show only (2).

For this we show first that any point P of **S** is on exactly $q + 1$ lines: Let r be the number of lines through P. Then the $q^2 - 1$ points $\neq$ P are distributed among the r lines through P in such a way that any of these r lines contains exactly $q - 1$ of the $q^2 - 1$ points. Hence $r = (q^2 - 1)/(q - 1) = q + 1$.

Now we consider a nonincident point–line pair (P, g). Since g has exactly q points, the point P is joined to the points of g by q lines in total. Hence there remains just one line through P that has no point in common with g. □

1.7 Diagrams

The aim of this section is to unify into a coherent theory the results on projective and affine spaces developed in the previous sections. Our new point of view is provided by the ‘diagram geometries’. We shall present a method to describe a geometry (in the sense of Section **1.1**) very economically and effectively. This will be done by the so-called diagrams. The theory of diagrams has recently undergone an explosive growth. It all began with a seminal paper [Tits74] by Jaques Tits (born 1930). The theory was provided with its first systematic treatment by Francis Buekenhout (born 1937) in [Buek79], [Buek81]; there are already books devoted to diagrams, for instance [Pas94]. For the foundations of diagram geometry see [BuBu88].

This section provides an important insight into a current area of research. We shall describe the diagrams for projective and affine spaces, and shall describe how to read a diagram and how to go from a geometry to its diagram. You should try to understand at least the fundamental ideas.

The main idea is that one introduces a symbolic notation for the most important rank 2 geometries, and tries to describe the whole geometry using these symbols.

Definition. Figure 1.10 presents the projective and the affine planes.

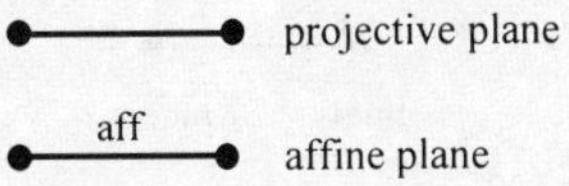

Figure 1.10 Diagrams for projective and affine planes

Here is how to interpret these little pictures. The left-hand node is a symbol for the points, and the right-hand node represents the lines of a projective or an affine plane. At this point it is not yet obvious why diagrams are more than just another symbolism for an already known structure, but this will soon become clear.

The description for an affine plane by this symbol is merely an abbreviation for its definition. For geometries of higher rank it is much more: the corresponding diagrams provide a 'code' for a great many nontrivial geometrical properties. First let us restrict ourselves to the 3-dimensional case. The diagram of a 3-dimensional projective space is as follows:

How should we read this? We shall try to understand this diagram intuitively to prepare us for the formal definition.

Any **diagram** consists of **nodes** and **edges**. The nodes are easy: any geometry that has a diagram is a rank r geometry, hence a geometry with r different types of objects in which any maximal flag contains one object of each type. The nodes of a diagram are chosen in such a way that each type of the geometry corresponds to one node. In particular, there are exactly r nodes. In our case we have a rank 3 geometry.

The edges are obviously more difficult to explain, since the edges must describe the structure of the geometry. Roughly speaking, the edge joining two nodes describes the structure of the corresponding objects inside the whole geometry. In order to see this in our diagram we assign the traditional names to the objects of the different types:

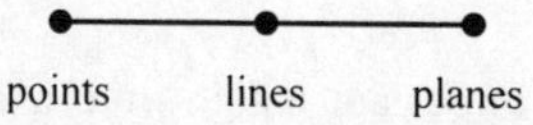

Our diagram is composed of two 'elementary diagrams', each of which represents a projective plane: the symbol

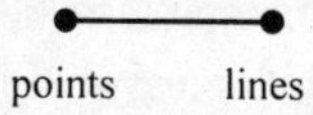

means that the points and lines *of any plane* form a projective plane. (This is in fact true: any plane of a projective space is a projective plane.)

The right-hand side of the diagram

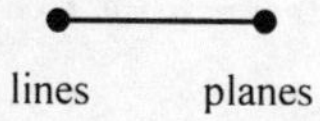

means that the lines and planes *through any point* form a projective plane. (This is also true, since by **1.4.1** the quotient geometry modulo any point is a projective plane.)

Now we can read the diagram

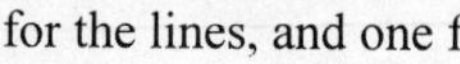

much better: a geometry belonging to this diagram is a rank 3 geometry with the property that the structure of points and lines in any plane is a projective plane and that the structure of lines and planes through any point is also a projective plane.

Before defining a diagram geometry in general we shall consider the diagram of a 3-dimensional affine space. Since this geometry has rank 3 we need three nodes, one for the points, one for the lines, and one for the planes.

In order to determine the edge between the first two nodes we must know the structure of the set of points and lines in a plane. By **1.6.4**, this is an affine plane. In order to describe the edge between the last two nodes we must know the structure of lines and planes through a point. It is easy to show that this is a projective plane (see exercise 36). Thus, the diagram belonging to a 3-dimensional affine space looks as follows:

Now we describe in general how one gets the diagram of a geometry. Here, the crucial notion is the 'residue of a flag'.

Definition. Let $\mathbf{G} = (\Omega, \mathrm{I})$ be a geometry, and let $\mathfrak{F}$ be a flag of $\mathbf{G}$. The **residue** of $\mathfrak{F}$ is the geometry $\mathrm{Res}(\mathfrak{F}) = (\Omega', \mathrm{I}')$ whose elements are those elements of $\mathbf{G}$ that are not contained in the flag $\mathfrak{F}$ but are incident with each element of $\mathfrak{F}$; the incidence I' is induced by I.

We shall consider the residue only if the geometry is a rank r geometry. For those geometries the residues can be described as follows:

Let $\mathfrak{F}$ be a flag of a rank r geometry $\mathbf{G}$. The residue of $\mathfrak{F}$ is the set $\mathrm{Res}(\mathfrak{F})$ of those elements of $\mathbf{G}$ that are incident with any element of $\mathfrak{F}$ but whose type does not show up in $\mathfrak{F}$. In order to determine the residue of a flag one can therefore proceed as follows. First one determines the types of the elements in $\mathfrak{F}$ and looks for all the elements in $\mathbf{G}$ that are incident with all elements in $\mathfrak{F}$ but do not have a type occurring in $\mathfrak{F}$.

Obviously $\mathrm{Res}(\mathfrak{F})$ is a geometry; if $\mathfrak{F}$ has s elements and $\mathbf{G}$ is a rank r geometry then $\mathrm{Res}(\mathfrak{F})$ is a geometry of rank $r - s$.

Examples. Let $\mathbf{G}$ be the geometry consisting of all nontrivial subspaces of a d-dimensional projective space $(d \geq 3)$.

First, let $\mathfrak{F}$ consist of only one element. If $\mathfrak{F}$ consists only of a hyperplane $\mathbf{H}$ then $\mathrm{Res}(\mathfrak{F})$ is the set of all subspaces incident with $\mathbf{H}$. We know that this is a projective space of dimension $d - 1$.

If $\mathfrak{F}$ consists only of one point P then $\mathrm{Res}(\mathfrak{F})$ is the set of all lines, planes, . . ., hyperplanes through P. By **1.4.1** this structure is a projective space of dimension $d - 1$.

If $\mathfrak{F}$ consists of a line g then the elements of $\mathrm{Res}(\mathfrak{F})$ are the points on g and the planes, . . . that pass through g.

If $\mathfrak{F} = \{P, g\}$ with $P \mathrm{I} g$ then $\mathrm{Res}(\mathfrak{F})$ consists of the planes, 3-dimensional subspaces, . . . through P and g. By Theorem **1.4.1** (see also exercise 34), $\mathrm{Res}(\mathfrak{F})$ is a $(d - 2)$-dimensional projective space.

For the definition of a diagram we consider only residues of flags $\mathfrak{F}$ having exactly $r - 2$ elements; in this case there are integers i and j such that $\mathfrak{F}$ contains an element of any type $\neq i, j$. By definition, $\mathrm{Res}(\mathfrak{F})$ consists of elements of just two types, namely i and j. In particular, $\mathrm{Res}(\mathfrak{F})$ is a rank 2 geometry. In this case we shall also say that $\mathrm{Res}(\mathfrak{F})$ is a **rank 2 residue**.

Now we are ready to define the **diagram** of a rank r geometry. To each type of objects we assign one **node**. The **edge** between two nodes i and j describes the rank 2 geometry that is the residue of all flags consisting of $r-2$ elements none of which is of type i or j.

In principle one can invent for each class of rank 2 geometries a special symbol and label the corresponding edge with this symbol. In this way we can describe all geometries of rank r.

Remark. In a geometry that belongs to a certain diagram, the rank 2 residues of a certain type need not be isomorphic. Let us consider, for instance, the following diagram:

This means in particular that the residue of any plane is a 'linear space', that is a geometry in which any two distinct points are on exactly one common line. (Cf. Section **2.8.**) The residue of one plane could be a projective plane, the residue of another plane could be an affine plane, etc.

In geometric research the following question has proved extremely stimulating. Draw r nodes and between the nodes any symbols for rank 2 geometries. Thus one obtains a hypothetical 'molecule'. Question: can one 'synthesize' a corresponding geometry, and can one describe all geometries belonging to that diagram in a satisfactory way? We have so far introduced only three symbols, namely 'projective plane', 'affine plane', and 'linear space'. This is sufficient for our purposes. The reader who wants to know more about these topics should look up the papers of Buekenhout or the book by Pasini.

We repeat: between any two nodes there is an edge. But the above diagrams for 3-dimensional spaces did not show an edge between the node representing the points and the node representing the planes. Which is the corresponding geometry? It is the residue of a line g, which consists of all points on g and all planes through g. This is a rank 2 geometry with the property that any element of one type is incident with any element of the other type. Such rank 2 geometries are called **trivial**. Therefore we also have to introduce a diagram for the trivial rank 2 geometries. The diagram of a 3-dimensional projective space should really look as follows:

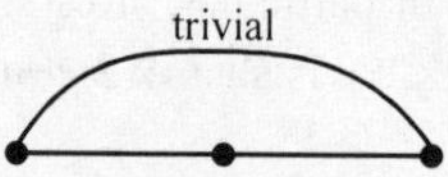

On the other hand, this looks very strange and would lead to complete confusion when dealing with higher-dimensional geometries so we agree that a trivial rank 2 geometry is represented by an **invisible** edge.

One also calls a geometry that can be described by a diagram a **Buekenhout–Tits geometry**.

We now discuss the diagrams for d-dimensional projective and affine geometries. Both are geometries of rank d, so we need d nodes. The diagram for projective spaces looks as follows:

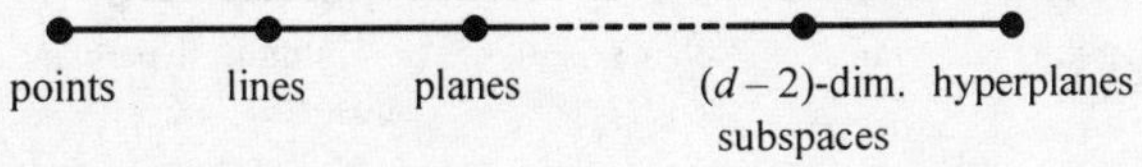

How can we read such a diagram?

The 'strokes' indicate that some rank 2 residues are projective planes. Which are those residues? For the first 'stroke' we have to consider the residue of a flag consisting of a plane π, a 3-dimensional subspace, . . . and a hyperplane. By definition, this residue consists of all points and lines of π, and we know that this is in fact a projective plane.

What is the meaning of the second 'stroke' between the node for lines and the node for planes? Let us consider a corresponding residue, which consists of a point P, a 3-dimensional subspace **U** and higher-dimensional subspaces. Such a residue contains all lines and planes through P that lie in **U**. In other words: this residue is the quotient geometry $\mathbf{U}/\mathrm{P}$, which, by **1.4.1**, is a projective plane.

Here, we can already see the general scheme: the 'stroke' between node i and node $i+1$ indicates the set of i- and $(i+1)$-dimensional subspaces through a subspace $\mathbf{U}_{i-1}$ of dimension $i-1$ and contained in a subspace $\mathbf{U}_{i+2}$ of dimension $i+2$, where $\mathbf{U}_{i-1} \subseteq \mathbf{U}_{i+2}$. This is the quotient geometry $\mathbf{U}_{i+2}/\mathbf{U}_{i-1}$, which is a projective plane.

All other rank 2 residues are trivial. Let us consider an example. Let $\mathcal{F}$ be a flag that contains an element of each type except for a point and a plane. Let g be the line in $\mathcal{F}$. Then $\mathrm{Res}(\mathcal{F})$ consists of all points on g and all planes through g; this residue is in fact trivial.

But there is yet further information coded in our diagrams. What is the residue of a hyperplane? In order to see this, you only have to cover with your finger the node belonging to the hyperplanes, and look at the remaining diagram. We see

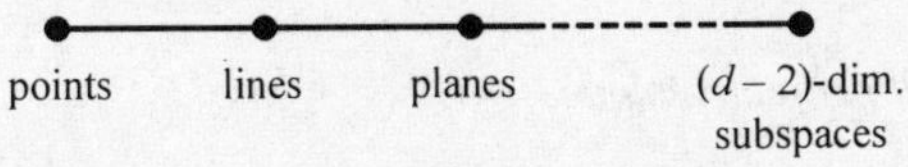

We recognize the diagram of a projective space of dimension $d-1$. We know that this is the geometry induced by a hyperplane.

To sum up: In order to determine the residue of a flag $\mathcal{F}$ one deletes the nodes corresponding to the types showing up in $\mathcal{F}$ and all edges incident with these nodes. Then Res($\mathfrak{F}$) is a Buekenhout–Tits geometry corresponding to the flag considered.

Now we study the diagram of a d-dimensional affine space. It has d nodes and looks as follows:

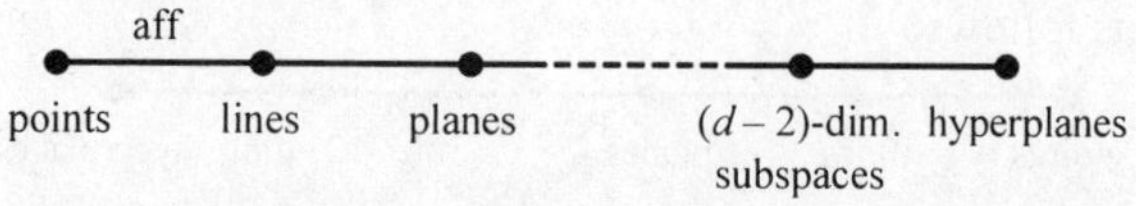

There is only one place in this diagram that differs from the diagram of a projective geometry. The residue of a flag consisting of a plane, a 3-dimensional subspace, . . . is an affine plane, while all other 'visible' rank 2 residues are projective planes. This shows in a spectacular way the structural kinship of affine and projective spaces.

So far we have seen that projective and affine spaces have diagrams. But the converse is also true, and only this justifies the claim that the diagram describes the geometry. In order to formulate the respective theorems we need the notion of a 'connected geometry'.

Definition. A geometry $\mathbf{G} = (\Omega, \mathrm{I})$ is called **connected** if for any two elements $X, Y \in \Omega$ there is a sequence $X = X_0, X_1, X_2, \ldots, X_n = Y$ of elements of Ω such that X_i is incident with X_{i+1} $(i = 0, 1, \ldots, n-1)$.

For instance each projective or affine geometry is connected. If, for example, g and h are two lines then there is a sequence as required in the definition: If g and h intersect each other in a point P then (g, P, h) is such a sequence. If g and h have no point in common then one chooses a point P on g and a point Q on h; then (g, P, PQ, Q, h) is a sequence that shows that g and h are connected.

1.7.1 Theorem. *Each connected geometry with diagram*

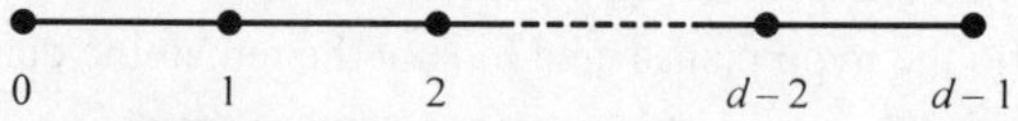

is a d-dimensional projective space.

Proof. Let **G** be a rank d geometry with the above diagram. We shall show by induction on d that the sets of points and lines of **G** fulfil the axioms of a projective space. For $d = 2$ this is precisely the definition of the diagram. We suppose therefore that $d > 2$ and assume that the assertion is true for $d - 1$.

Axiom 1. Let P and Q be two distinct points. We have to show that there exists precisely one line through P and Q.

The most difficult part is the existence of a connecting line: Since **G** is connected there is a sequence $(X_1, X_2, \ldots, X_n)$ of elements of **G** such that P is incident with X_1, Q is incident with X_n, and two consecutive elements have a point in common. Let $(X_1, X_2, \ldots, X_n)$ be such a sequence with a minimum number of elements. We first show that $n = 1$. Assume that $n > 1$. Let R_1 be a point incident with X_1 and X_2, and let R_2 be a point incident with X_2 and X_3; if $n = 2$ then define $R_2 = Q$. The element X_1 has a type i with $1 \le i \le d - 1$. We consider a flag $\mathfrak{F} = \{Y_i = X_1, Y_{i+1}, \ldots, Y_{d-1}\}$, where the index of Y_j denotes its type. We see that Res($\mathfrak{F}$) is a rank i geometry with the following diagram:

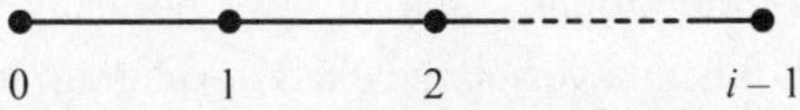

Since $i < d$ we may apply induction. It follows that Res($\mathfrak{F}$) is a projective geometry of rank i containing the points P and R_2. Since the points P and R_2 are joined in the projective space Res($\mathfrak{F}$) by a line, they are also joined in **G** by a line g. Thus we could shorten the sequence $(X_1, X_2, \ldots, X_n)$ to obtain $(g, X_3, \ldots, X_n)$, a contradiction.

Therefore we have P I X I Q, where X is an element of some type i. We have to show that $i = 1$. In order to do this we again consider the residue Res($\mathfrak{F}$) of a flag $\mathfrak{F} = \{X = Y_i, \ldots, Y_{d-1}\}$. This is a geometry of rank $< d$ that contains the points P and Q. By induction, Res($\mathfrak{F}$) is a projective space. Hence P and Q are connected in Res($\mathfrak{F}$) by a line, therefore they are also joined in **G** by a line.

The *uniqueness* of the line through P and Q is easier to prove: Assume that there exist two lines g, h through P and Q. The residue Res(P) of P is a geometry having the diagram

of rank $d - 1$. By induction, it is a projective geometry of dimension $d - 1$. Hence the lines g and h (which are points of Res(P)) are contained in precisely one plane (line of Res(P)). Since the geometry of points and lines incident with a plane is a projective plane, the lines g and h cannot intersect in two distinct points.

Axiom 2 is easy to prove. Let g and h be two lines that intersect in some point P. We have to show that any two lines g' and h' that both meet g and h in points $\neq$ P intersect each other.

Since g and h pass through P and since the residue of P is a projective geometry there is a plane π containing g and h. Now we show that g' and h' also lie in π: Let P' and Q' be the intersections of g' with g and h. If g' were not a line of π then there would exist a second line g'' (of π) through P' and Q', contradicting axiom 1.

Hence g' and h' are contained in the projective plane π and therefore intersect.

The other axioms follow easily.

The projective geometry has dimension d since the maximal flags have d elements. □

Remarks. 1. In order to prove axiom 1 we have only used that the residues of a point and a plane have the property that any two points are joined by exactly one line. Hence axiom 1 holds in a very large class of diagram geometries.
2. In order to prove axiom 2 also we used only a weak property, namely that the residue of any point P has the property that any two lines through P lie in a common plane, and that the residue of any plane is a projective plane.
3. The theorem remains true if one does not claim that **G** is connected; see exercise 46.

Similarly we get the following theorem for affine spaces.

1.7.2 Theorem. *Each geometry of rank d with diagram*

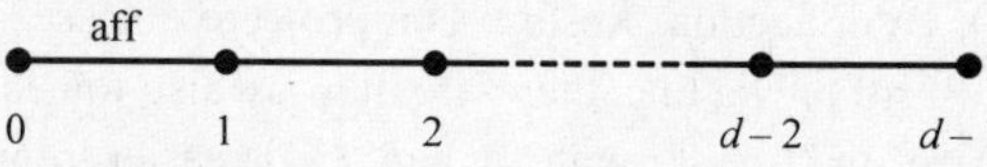

that contains a line with at least four points is a d-dimensional affine space.

Proof. The property that any two points are on a unique line is shown in the above proof.

Since any plane is an affine plane it follows from a rather deep theorem of Buekenhout ([Buek69b], see also [KaPi70]) that the geometry belonging to this diagram is in fact an affine space. □

1.8 Application: efficient communication

Before defining the problem in general we consider an example. We imagine that eight students form a study group in order to do their homework economically: each solves part of the exercises and copies the rest from the others. We imagine moreover that the students do their part of the homework at home and that the solutions are communicated by telephone.

They agree that the solutions must be exchanged at a certain time, let's say one hour before they have to hand in a complete set of exercises to the University. How should they proceed? One possibility is that all phone one another; then each one has to partake in seven phone calls, and clearly the telephone lines will be occupied constantly. Another possibility is that one of the students is the communication centre. Then this student receives seven phone calls, collects the solutions and has to phone back at least six of his colleagues. Also this procedure has the disadvantage that it takes a very long time until every participant has the complete information, and also it has a very unpractical central structure.

A much better solution is the following. We represent the eight students by the eight vertices of a cube (see Figure 1.11).

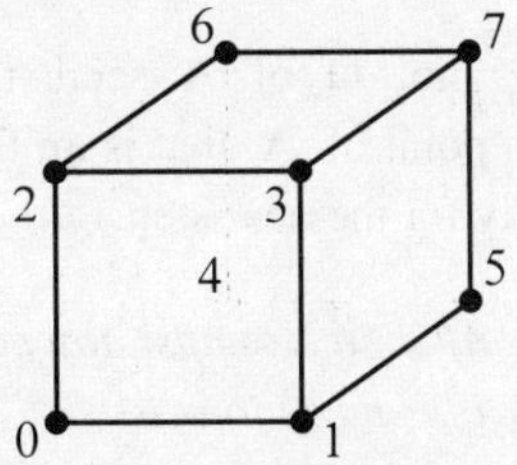

Figure 1.11 The cube as a model for efficient communication

The students agree upon a precise communication strategy: First, those students who are connected by a *vertical* edge exchange their solutions (so, 0 and 2, 1 and 3, 4 and 6, and 5 and 7 communicate). Then those students who are connected by a *horizontal* left–right edge talk to each other (so 0 and 1, 2 and 3, 4 and 5 and 6 and 7 communicate). In this step they exchange both the solutions they know so far. For instance, 2 already has the solution of 0, therefore he also reports that solution to 3. Thus, after the second phone call, 3 knows all solutions of 0, 1, 2 (and 3). Similarly, 7 (as well as 4, 5, and 6) knows all solutions of 4, 5, 6, and 7.

Finally, there is a last round of phone calls; they use the third dimension, that is 0 and 4, 1 and 5, 2 and 6, and 3 and 7 talk to each other. It is clear that now each student knows all solutions.

Note that each student has to make only three phone calls and that the total number $8 \cdot 3 / 2 = 12$ of phone calls is remarkably small.

In general, N nodes would like to exchange information in such a way that all information eventually reaches every node. An algorithm for the nodes to exchange information among themselves is as follows: Let q be a positive integer such that there exist projective spaces of any dimension of order q. (The next chapter will show that for q we can, for instance, choose any prime.) Then one determines the positive integer d in such a way that it satisfies

$$q^{d-1} < N \leq q^d.$$

We will realize the algorithm using the affine space $\mathbf{A} = \mathbf{P} \setminus \mathbf{H}_\infty$, where $\mathbf{P}$ is a d-dimensional projective space of order q and $\mathbf{H}_\infty$ is its hyperplane at infinity. By the choice of d we may identify the communication nodes with points of $\mathbf{A}$. For the sake of simplicity we suppose that each point of $\mathbf{A}$ represents a node.

Let $\mathbf{H}_\infty$ be the hyperplane at infinity of $\mathbf{A}$. The trick consists simply of considering a basis $\{P_1, \ldots, P_d\}$ of $\mathbf{H}_\infty$. The algorithm works in d rounds:
1st round. Each node X sends its information to any point of $\mathbf{A}$ that is on the line SP_1. Then each point evaluates the received information and is ready for the second round.
ith round ($2 \leq i \leq d$). Each point Q of $\mathbf{A}$ sends the information it has received and evaluated so far to each point of $\mathbf{A}$ that is on the line QP_i. Then it evaluates the received data and is ready for the next step.

1.8.1 Theorem ([Beu90b]). *After d rounds, any point of $\mathbf{A}$ knows the total information. The total number of transactions is*

$$d \cdot (q-1) \cdot q^d = (q-1) \cdot d \cdot N \leq c \cdot \log_q(N) \cdot N.$$

Proof. We consider a point X of $\mathbf{A}$ and trace which nodes have, after i rounds, received the information of X.

After the first round any affine point (that is any node) on the line XP_1 knows the information of X. In the second round any affine point Y on XP_1 sends the information of X (perhaps in another form) to the points on the line YP_2. By Theorem **1.3.1** in such a way all points on the plane $\langle X, P_1, P_2 \rangle$ will be reached.

And so it goes. After round i all nodes in the subspace $\langle X, P_1, P_2, \ldots, P_i \rangle$ know the information of X. In particular, after round d, the information of X is known in the subspace $\langle X, P_1, P_2, \ldots, P_d \rangle$; since this is the whole space any node knows the information of X. Since X was arbitrarily chosen, the first assertion is proved.

In any round, each of the q^d nodes sends exactly $q-1$ messages; so in d rounds we have exactly $d \cdot (q-1) \cdot q^d$ transactions. □

Exercises

1 A **tiling** of the Euclidean plane consists of a set of polygons (regions of the plane that are bounded by straight line segments) with the following properties:

– any point of the plane is in the interior or on the boundary of at least one of the polygons,

– no point is in the interior of more than one polygon.

See Figure 1.12.

(a) Let Ω be the set of 'vertices', 'edges', and polygons of a tiling. Then Ω, together with set-theoretical inclusion is a geometry.

(b) Is this a geometry of rank r? If yes, what is r?

(c) We get another geometry if we define two elements of Ω to be incident if their intersection is not empty. Is this in any case a rank r geometry?

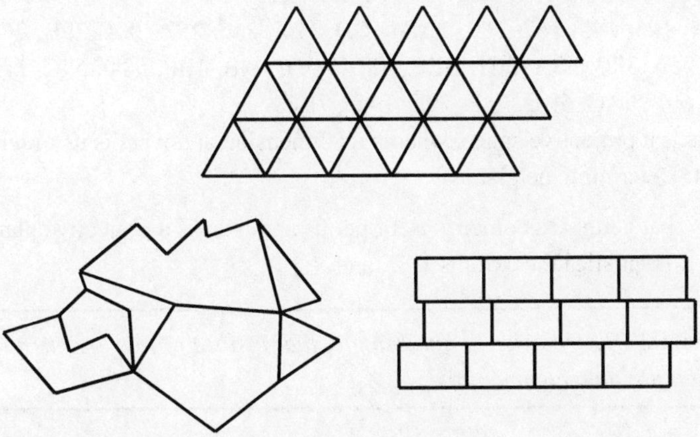

Figure 1.12 Three tilings

2 Let **G** be a finite rank 2 geometry having v points, b blocks, a constant number r of blocks per point, and a constant number k of points per block. Show that

$$v \cdot r = b \cdot k.$$

3 A tiling is called **regular** if there is a positive integer n such that any polygon is a regular n-gon, and if any two distinct edges intersect at most in a vertex. Determine all regular tilings.

4 Interpret the following picture as a projective plane:

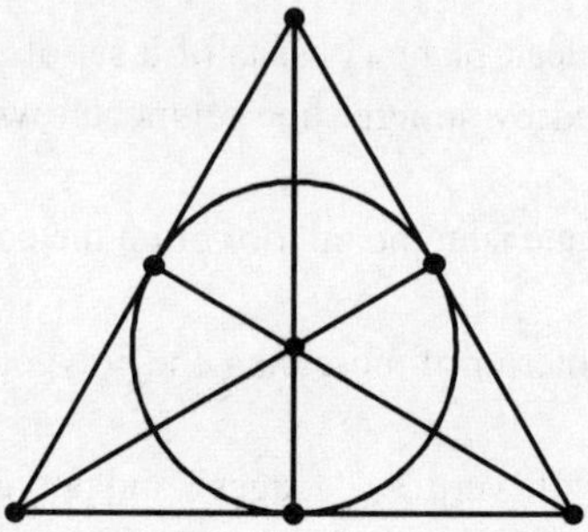

5 Construct a projective plane having exactly 13 points. [It is useful to compute first the number of points on a line.]

6 Check whether the following rank 2 geometry is a projective space:
Points: 1, 2, 3, 4, 5, 6, 7, 8, 9, A, B, C, D, E, F.
Lines: 123, 145, 167, 189, 1AB, 1CD, 1EF, 246, 257, 28A, 29B, 2CE, 2DF, 347, 356, 38B, 39A, 3CF, 3DE, 48C, 49D, 4AE, 4BF, 58D, 59C, 5AF, 5BE, 68E, 69F, 6AC, 6BD, 78F, 79E, 7AD, 7BC.
If this is a projective space, what is its dimension and what is its order?
[Hint: Determine the planes.]

7 Show that a rank 2 geometry **S** of points and lines is a projective plane if and only if it satisfies the axioms 1, 2', and 3':

Axiom 3'. *There exists a quadrangle, that is a set of four points no three of which are on a common line.*

8 Let X and Y be subsets of the point set of a projective space **P**. Prove the following assertions:
(a) $X \subseteq Y \Rightarrow \langle X \rangle \subseteq \langle Y \rangle$,
(b) $\langle\langle X \rangle\rangle = \langle X \rangle$,
(c) $\langle X, Y \rangle = \langle\langle X \rangle, Y \rangle$.

9 Show that each plane of a projective space is a projective plane.

10 Let **P** be a projective plane such that any line has exactly five points. Show that any point of **P** is on exactly five lines and that **P** has a total of 21 points and 21 lines.

11 Try to construct a projective plane described in the preceding exercise in the following way:

– Points are the numbers 1, 2, . . ., 21.

– The lines are constructed one after the other, beginning with the lines through point 1, by taking at each step the point with the smallest number that is not yet forbidden.

12 Similarly, construct a projective space with 15 points, and three points on any line.
[Hint: Cf. exercise 6.]

13 Formulate the algorithm of exercise 11 as a recursive procedure that depends on n and determine the next projective plane that can be constructed using this procedure.

14 Let **U** be a subspace of a projective space **P**, and let P be a point of **P** that is not contained in **U**. Show that for any point Q of **U** the span ⟨**U**, P⟩ of **U** and P consists precisely of the points in the planes ⟨g, P⟩, where g runs over all lines of **U** through Q (see Figure 1.13).

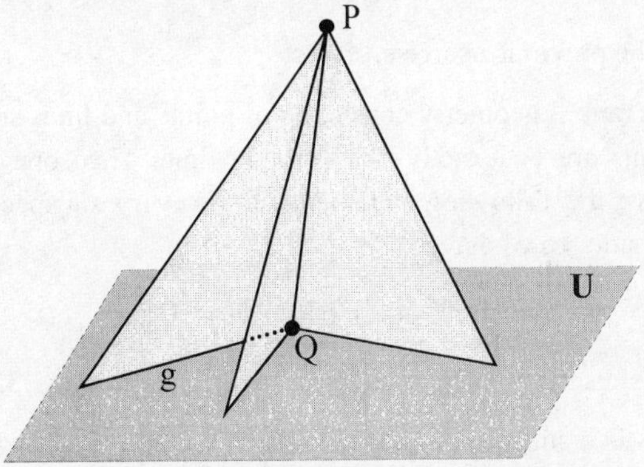

Figure 1.13 The span of U and P

15 Let g_1, g_2 be two skew lines of a projective space **P**. Let X_1, Y_1 be points on g_1, X_2, Y_2 be points on g_2. Furthermore, let X be a point on X_1X_2 and Y be a point on Y_1Y_2 such that XY is skew to g_1 and g_2 (see Figure 1.14).
Using only **1.3.1** show that each point Z on XY is on some line (transversal) that intersects g_1 and g_2.
[Hint: Show that the line g_2 intersects the plane $\pi := \langle g_1, Z\rangle$.]

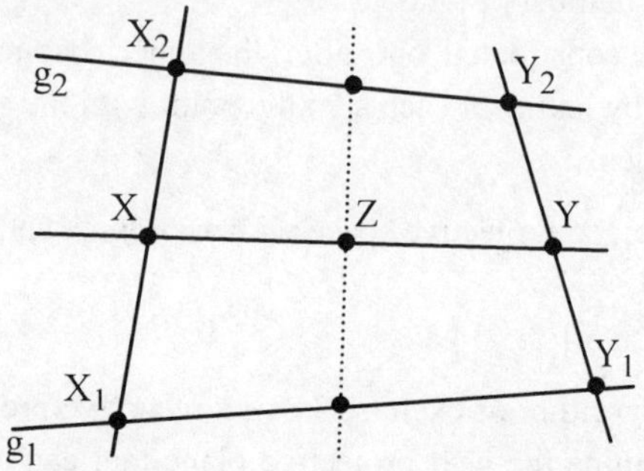

Figure 1.14 The transversal through Z

16 ('Join theorem') Let **U** and **V** be nonempty subspaces of **P**. Then the span $\langle \mathbf{U}, \mathbf{V} \rangle$ can be described as follows:

$$\langle \mathbf{U}, \mathbf{V} \rangle = \bigcup \{(PQ) \mid P \in \mathbf{U}, Q \in \mathbf{V}, P \neq Q\}.$$

[Hint: Use the previous exercise.]

17 Let **D** be a rank 2 geometry consisting of points and lines such that any two distinct points are on exactly one common line. Then one can define subspaces. Prove the following statement: if for every subspace **U** and each point P outside **U** we have

$$\langle \mathbf{U}, P \rangle = \bigcup \{(PQ) \mid Q \in \mathbf{U}\}$$

then **D** satisfies the Veblen–Young axiom. [Cf. Theorem **1.3.1**.]

18 Show: if **U** is a subspace of a projective space and if g is a line that intersects **U** in just one point then there is a hyperplane through **U** that does not contain g.

19 What are the dimensions of the following subspaces: the empty set, a point, a line, a plane, a hyperplane, the whole space?

20 Show that a set $\mathcal{B}$ of points of a projective space is a basis if and only if it is a maximal independent set.

21 Generalize **1.3.5** to arbitrarily many subsets $\mathcal{B}_1, \ldots, \mathcal{B}_s$ of an independent set $\mathcal{B}$.

22 A **generalized** projective space is a rank 2 geometry satisfying the axioms 1, 2, and 3'':

Axiom 3". *Every line has at least two points.*

Clearly, an ordinary projective space is a generalized projective space, but a generalized projective space might have lines with just two points.

Let $\mathbf{P}_1, \mathbf{P}_2, \ldots$ be projective spaces with disjoint point sets. Define the geometry $\mathbf{P}$ as follows:
the *points* of $\mathbf{P}$ are all the points of $\mathbf{P}_1, \mathbf{P}_2, \ldots$;
the *lines* of $\mathbf{P}$ are all the lines of $\mathbf{P}_1, \mathbf{P}_2, \ldots$ and the point sets $\{P, Q\}$, where P and Q are points in different $\mathbf{P}_i$s.
Show that P is a generalized projective space.
$\mathbf{P}$ is called the **direct product** of the projective spaces $\mathbf{P}_1, \mathbf{P}_2, \ldots$.

23 Show that any generalized projective space is a direct product of ordinary projective spaces.

24 Show that the set of all nontrivial subspaces of a d-dimensional projective or affine space is a geometry of rank d.

25 Let $\mathbf{P}$ be a d-dimensional projective space.
(a) Show: if $\mathbf{H}_1, \ldots, \mathbf{H}_s$ are hyperplanes then

$$\dim(\mathbf{H}_1 \cap \ldots \cap \mathbf{H}_s) \geq d - s.$$

(b) Prove that a t-dimensional subspace cannot be represented as the intersection of fewer than $d - t$ hyperplanes.

26 Let $\mathbf{D}$ be an affine or projective space, and let α be a bijective map of the point set of $\mathbf{D}$ onto itself. Show that α can be extended to a collineation of $\mathbf{D}$ if and only if for any three points P, Q, R of $\mathbf{D}$ we have

$$\text{P, Q, R are collinear} \Leftrightarrow \alpha(\text{P}), \alpha(\text{Q}), \alpha(\text{R}) \text{ are collinear.}$$

27 Let $\mathbf{D} = (\mathcal{P}, \mathcal{G}, \text{I})$ be an affine or projective space. We define the rank 2 geometry $(\mathbf{D})$ by $(\mathbf{D}) := (\mathcal{P}, \{(g) \mid g \in \mathcal{G}\}, \in)$. Show that $\mathbf{D}$ and $(\mathbf{D})$ are isomorphic.
This means that everybody who has solved this exercise may w.l.o.g. suppose that the incidence relation of a projective or affine space is $\in$.

28 Consider real 3-dimensional affine space (intuitive geometry) and fix the origin O. Convince yourself that the following geometries are isomorphic:
(a) *Points:* lines through O. *Lines:* planes through O.
(b) *Points:* pairs of antipodal points on the unit sphere. *Lines:* great circles on the unit sphere.

29 Show that the geometries considered in the preceding exercise are projective planes.

30 Convince yourself that the following diagram also represents an affine plane of order 3:

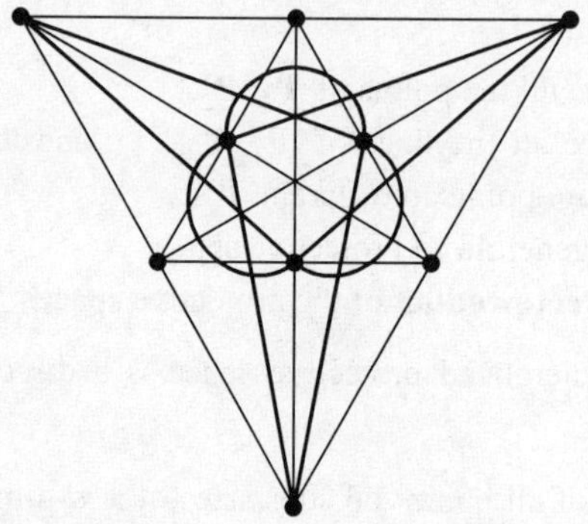

31 Show that a projective space of dimension > 1 is finite if and only if its set of lines is finite.

32 Let **P** be a finite d-dimensional projective space of order q $(d \geq 2)$.
(a) Compute the number of all planes through a point.
(b) Compute the number of all planes of **P**.

33 Compute the number of all t-dimensional subspaces of a finite projective space.

34 Define the quotient geometry $\mathbf{P}/g$ of **P** modulo a line g. Prove a theorem that is similar to **1.4.1**.

35 Generalize the preceding exercise to quotient geometries $\mathbf{P}/\mathbf{U}$, where **U** is an arbitrary subspace of **P**.

36 Let **A** be a d-dimensional affine geometry, and let Q be a point **A**. Show that the set of subspaces of **A** through Q (the **quotient geometry** $\mathbf{A}/Q$) is a projective geometry of dimension $d-1$.

37 Let $\mathbf{U}_{i-1}$ and $\mathbf{U}_{i+2}$ be two subspaces of dimension $i-1$ and $i+2$ of a projective or an affine geometry such that $\mathbf{U}_{i-1} \subseteq \mathbf{U}_{i+2}$. Show that the quotient geometry $\mathbf{U}_{i+2}/\mathbf{U}_{i-1}$ is a projective plane.

38 Let **P** be a projective geometry, and let Q be a point and g a line of **P** with Q I g. Describe precisely the differences between
– the residue of g,

- the residue of $\{Q, g\}$,
- the residue of Q,
- the quotient geometry $\mathbf{P}/Q$,
- the quotient geometry $\mathbf{P}/g$.

39 One says that $d+1$ hyperplanes of a d-dimensional projective space are **independent** if they have no common point. In other words, the hyperplanes $\mathbf{H}_0, \mathbf{H}_1, \ldots, \mathbf{H}_d$ are independent if $\bigcap \mathbf{H}_i = \varnothing$.
Let $\mathbf{H}_0, \mathbf{H}_1, \ldots, \mathbf{H}_d$ be independent hyperplanes of a d-dimensional projective space $\mathbf{P}$. For $s \le d$ compute the dimension

$$\dim(\mathbf{H}_0 \cap \mathbf{H}_1 \cap \ldots \cap \mathbf{H}_s).$$

40 Let $\mathbf{P}$ be a finite d-dimensional projective space of order q. Show that the number of points outside a system of $d+1$ independent hyperplanes is $(q-1)^d$.

41 Let g and h be two skew lines of a 3-dimensional projective space $\mathbf{P}$ (this means that g and h span $\mathbf{P}$), and let P be a point on neither g nor h. Show that there is a unique line through P that intersects g and h.

42 Generalize the preceding exercise as follows: Let $\mathbf{P}$ be a $(2t+1)$-dimensional projective space, and let $\mathbf{U}$ and $\mathbf{W}$ be two skew t-dimensional subspaces of $\mathbf{P}$. If P is a point outside $\mathbf{U}$ and $\mathbf{W}$ then there is a unique line through P that intersects $\mathbf{U}$ and $\mathbf{W}$.

43 Let $\mathbf{S}$ be a geometry of rank 2 with the following properties:
(1) Through any two distinct points there is precisely one line.
(2) There is an integer $q \ge 2$ such that any line has exactly q points.
(3) Every point is on exactly $q+1$ lines.
Show that $\mathbf{S}$ is an affine plane.

44 Let $\mathbf{S}$ be a geometry of rank 2 with the following properties:
(1) Through any two distinct points there is precisely one line.
(2) There is an integer $q \ge 2$ such that any line has exactly $q+1$ points.
(3) $\mathbf{S}$ has a total number of q^2+q+1 points.
Show that $\mathbf{S}$ is a projective plane.

45 Interpret the **Platonic solids** (regular tetrahedron, cube, octahedron, dodecahedron, and icosahedron, see [Cox69]) as geometries of rank 3. (The elements of Ω are the vertices, edges, and faces of a Platonic solid.) In each case, determine the residue of a point.

46 Is there a geometry with diagram

that is not a projective space?

47 Let **G** be a geometry of rank d with diagram

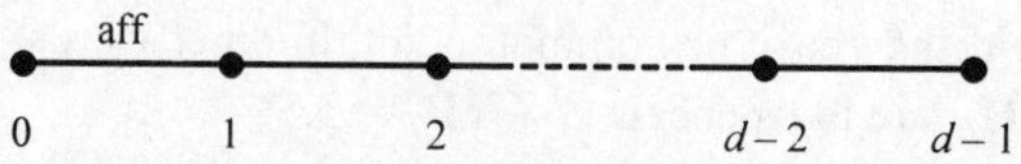

and let X be an element of type $i \geq 2$. Show that the geometry of points and lines that are incident with X is an affine space of dimension i.

48 How does one have to modify the algorithm described in Section **1.8** if the number of nodes is smaller than q^d?

49 Explain how the example with the cube is a special case of Theorem **1.8.1**.

50 Formulate Theorem **1.8.1** explicitly in the case $q = 2$.

51 Another algorithm to solve the problem of distributing information (see Section **1.8)** is the following.
Let **P** be a projective plane, and let φ be a map of the points of **P** onto the lines of **P** such that no point X is incident with its image φ(X). The algorithm works as follows:
In each round each point X sends its information to all points on the line φ(X) and to the points Y, where φ(Y) is a line through X.
Show that already after two rounds every point has all the information; moreover, the number of transactions is $c \cdot N \cdot \sqrt{N}$, where N is the number of nodes.
[This is difficult and needs some time, cf. [AgLa85]).]

True or false?

☐ Any projective plane is a projective space.

☐ Any projective space has at least seven points.

☐ Any projective space of dimension ≥ 3 has infinitely many points.

☐ Four points are independent if and only if no three of them are on a common line.

☐ There is no projective space whose dimension is infinite.

☐ The points $P_1, \ldots, P_t$ are independent if and only if

$$\dim(\langle P_1, \ldots, P_t \rangle) = t - 1.$$

☐ Any finite projective plane has an odd number of points.

☐ Any finite projective space has an even number of points.

☐ The quotient geometry of an affine space is an affine space.

☐ The quotient geometry of an affine space is a projective space.

Project

One can also introduce *affine spaces* axiomatically and then show that they can be completed to projective spaces. A first step in this direction is to imitate the procedure for projective spaces, that is to prove the analogous statements to the theorems in Section **1.3**. A particularly important axiom system for affine spaces is the following, which appeared for the first time in Tamaschke [Tam72]): Let **S** = $(\mathcal{P}, \mathcal{G}, \mathrm{I})$ be a geometry of rank 2 having the following properties:

(1) Any two distinct points are on precisely one common line.

(2) **S** has a parallelism.

(3) (**Triangle axiom**): Let A, B, C be three noncollinear points, and let A', B' be points such that AB ∥ A'B'. If g is the line through A' parallel to AC, and h is the line through B' parallel to BC then g and h intersect in a point C' (see Figure 1.15).

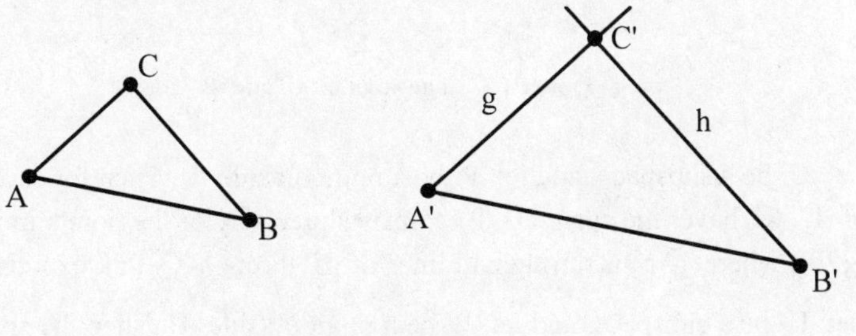

Figure 1.15 The triangle axiom

(4) There is a line with at least two points; there are at least two lines.

One can show that such a geometry is an affine space as we have defined it. Try to proceed as far as possible in this direction. In particular, you should prove the following exercises.

1 **(Trapezoid axiom)** [Lenz54]: Let A, B, C be noncollinear points. Then for any point B' on AB, the line through B' parallel to BC intersects AC (see Figure 1.16).

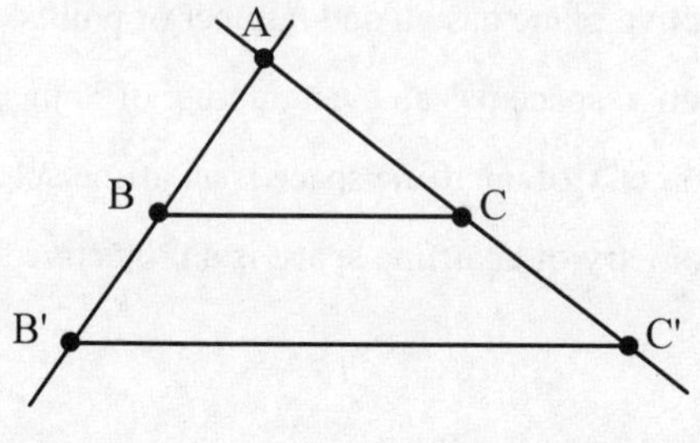

Figure 1.16 The trapezoid axiom

2 Let **U** be a subspace (what is that?), and let P be a point outside **U**. Then one can describe the span ⟨**U**, P⟩ (what is that?) of **U** and P as follows: Let Q be an arbitrary point of **U**. Then ⟨**U**, P⟩ consists of the points on the lines through the points X of **U** that are parallel to PQ (see Figure 1.17).

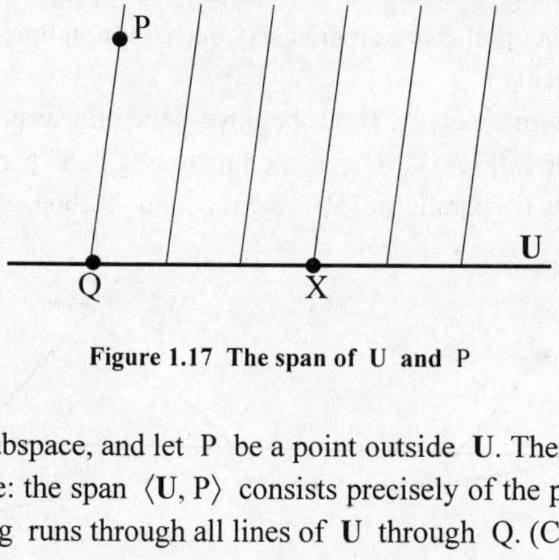

Figure 1.17 The span of U and P

3 Let **U** be a subspace, and let P be a point outside **U**. Then for each point Q of **U** we have: the span ⟨**U**, P⟩ consists precisely of the points in the planes ⟨g, P⟩, where g runs through all lines of **U** through Q. (Cf. exercise 14.)

4 Let **U** be a subspace, and let P be a point outside **U**. Then the span ⟨**U**, P⟩ consists precisely of the points on the lines PX with X ∈ **U** and the points on the lines through P that are parallel to some line of **U** (see Figure 1.18).

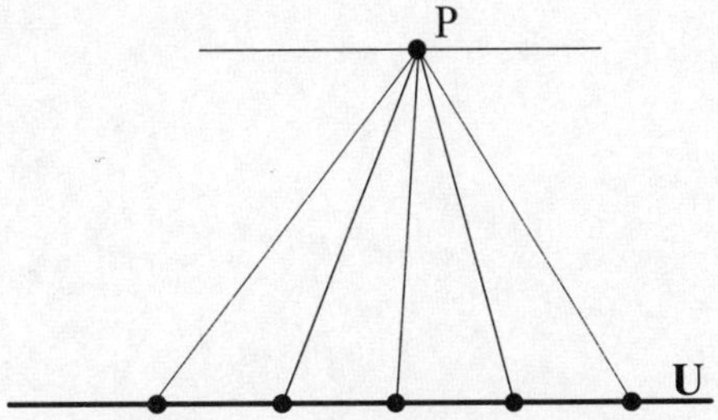

Figure 1.18 The span of U and P

5 Define independence of points and the notions 'basis' and 'dimension'.

6 We say that two subspaces **U** and **U'** are parallel if each line of **U'** is parallel to some line of **U**. Show that the parallelism defined in this way is an equivalence relation satisfying Playfair's parallel axiom.

7 Let **U**, **U'** be distinct t-dimensional subspaces. If **U**, **U'** are parallel, then $\langle \mathbf{U}, \mathbf{U}' \rangle$ has dimension $t+1$.

8 Let g, g' be two skew lines. (What are those?) Prove that $\langle g, g' \rangle$ has dimension 3.
Generalize this to skew t-dimensional subspaces.

9 Formulate a dimension formula. Prove that your formula is correct.

You should know the following notions

Geometry, incidence structure, rank of a geometry, projective space, projective plane, duality, linear set, subspace, span, independent set, basis, dimension, hyperplane, quotient geometry, order, affine space, affine plane, point at infinity, hyperplane at infinity, parallelism, Playfair's parallel axiom, parallel class, flag, residue, diagram, Buekenhout–Tits geometry.

2 Analytic geometry

The introduction of coordinates (numbers) in geometry goes back to the French philosopher ('cogito, ergo sum'), mathematician, and scientist René Descartes (1596–1650). The aim of this 'analytic geometry' is to develop a method by which one gets geometrical results by mechanically manipulating numbers. While in synthetic geometry one would like to get *insight*, in analytic geometry one wants to get *results*. An advantage of analytic geometry is that one can efficiently apply methods of algebra in general and linear algebra in particular.

The importance of analytic geometry should not be underestimated. The description of space by numbers is on everybody's list of the important milestones in the history of mathematics. It is often possible to grind out results for which a synthetic approach is unavailable or unsuitable.

2.1 The projective space P(V)

All considerations in this chapter are based on vector spaces. But in contrast to linear algebra, in geometry we use vector spaces not only over fields, but also over noncommutative structures, the so-called skewfields.

In order to clarify matters we shall repeat some definitions. By a division ring we mean a set F together with two binary operations $+$ and $\cdot$, called addition and multiplication, that fulfils all axioms of a field except possibly the commutativity of multiplication. More precisely, a **division ring** is a triple $(F, +, \cdot)$, where F is a set and $+, \cdot$ are binary operations on F such that $(F, +)$ is an abelian group with identity element 0, $(F\setminus\{0\}, \cdot)$ is a (not necessarily commutative) group, and both right and left distributive laws hold.

Although the terminology we shall use here is not universally agreed upon, the following convention seems to be the most widely used. A **field** is a division ring with commutative multiplication. A **skewfield** is a division ring such that the multiplication is not commutative; so in a skewfield there is at least one pair of elements x, y such that $x\cdot y \neq y\cdot x$.

One can define vector spaces over division rings in the same way as vector spaces over rings. If we say 'V is a left vector space over the division ring F' we mean that if v is a vector and a an element of F, then $a v$ is again a vector in

V. We shall deal only with left vector spaces, unless otherwise specified, and hence call them simply **vector spaces**.

The basic theory of vector spaces over division rings can be developed in just the same way as for vector spaces over fields. So all elementary properties of independence, basis, dimension, subspaces, and linear maps also hold in the more general context.

Definition. Let V be a vector space of dimension $d+1 \geq 3$ over a division ring F. We define the rank 2 geometry **P**(V) as follows:
- the *points* of **P**(V) are the 1-dimensional subspaces of V,
- the *lines* of **P**(V) are the 2-dimensional subspaces of V,
- the *incidence* of **P**(V) is set-theoretical containment.

Warning. You will no doubt have noticed that there is an ambiguity that arises from calling 1-dimensional subspaces of a vector space 'lines of the vector space', while the 2-dimensional subspaces are called 'lines of P(V)'. Accept the definition as it is; before long you will get used to it.

Remark. We shall have to distinguish between the dimension of a projective space and the dimension of a vector space. Therefore we shall use the symbol $\dim_V$ to denote the vector space dimension.

The following theorem says that these geometries **P**(V) are in fact projective spaces. In the next chapter we shall see that, conversely, if a projective space has dimension 3 or greater then it is essentially **P**(V) for some V. (There are, however, lots of projective planes that are not of this type.)

2.1.1 Theorem. *Let* V *be a vector space of dimension* $d+1 \geq 3$ *over a division ring* F. *Then* **P**(V) *is a projective space. We call* **P**(V) *the projective space* ***coordinatized*** *by* F.

Proof. We have to show that axioms 1 to 4 hold.

Axiom 1. Let P and Q be two distinct points of **P**(V). Then, by definition, there are vectors $v, w \in V$ different from the zero-vector o with $P = \langle v \rangle$ and $Q = \langle w \rangle$. (Here, $\langle\rangle$ denotes the span in the vector space.) Since $P \neq Q$ we have $\langle v \rangle \neq \langle w \rangle$. Hence the vectors v and w are linearly independent. Therefore $\langle v, w \rangle$ is a 2-dimensional subspace of V and thus, by definition, a line of **P**(V).

Since this subspace contains $P\ (= \langle v \rangle)$ as well as $Q\ (= \langle w \rangle)$ it follows that $\langle v, w \rangle$ is a line through the points $\langle v \rangle$ and $\langle w \rangle$. Since, on the other hand, any line through $\langle v\rangle$ and $\langle w\rangle$ must contain the subspace $\langle v, w\rangle$, the line through $\langle v\rangle$ and $\langle w\rangle$ is uniquely determined.

Hence $\langle v, w\rangle$ *is the uniquely determined line through* $\langle v\rangle$ *and* $\langle w\rangle$.

Axiom 2. Let $g_1 = P_0P_1$ and $g_2 = P_0P_2$ be lines through a common point P_0 that are intersected by the line $h = P_1P_2$. Moreover, let $Q \neq P_1, P_2$ be a point on h and $Q_1 \neq P_0, P_1$ a point on g_1. We have to show that QQ_1 intersects the line g_2 in some point. (Make yourself a picture.)

If $g_1 = g_2$ the assertion follows trivially. We suppose therefore $g_1 \neq g_2$ and choose the vectors $v_0, v_1, v_2 \in V$ in such a way that we have $P_0 = \langle v_0\rangle$, $P_1 = \langle v_1\rangle$ and $P_2 = \langle v_2\rangle$. Since the points P_0, P_1, and P_2 are noncollinear, the vectors v_0, v_1, and v_2 are linearly independent.

Consider the subspace V' of V that is generated by v_0, v_1, and v_2. Let w_1 be a vector with $Q_1 = \langle w_1\rangle$, then $w_1 \in \langle v_0, v_1\rangle \subseteq V'$. Moreover, $Q = \langle w\rangle \subseteq \langle v_0, v_1, v_2\rangle = V'$. In particular we have that the line QQ_1 is a 2-dimensional subspace of V'. Thus, from $V' \supseteq \langle g_2, QQ_1\rangle$ it follows that $V' = \langle g_2, QQ_1\rangle$.

Since g_2 is a also 2-dimensional subspace of V', by the dimension formula for vector spaces, we are able to compute the dimension $\dim_V(g_2 \cap QQ_1)$ of the intersection of g_2 and QQ_1:

$$\begin{aligned}\dim_V(g_2 \cap QQ_1) &= \dim_V(g_2) + \dim_V(QQ_1) - \dim_V(\langle g_2, QQ_1\rangle)\\ &= 2 + 2 - 3 = 1.\end{aligned}$$

Hence g_2 intersects the line QQ_1 in a 1-dimensional subspace of V, so in a point of **P**(V).

Axiom 3. An arbitrary line $\langle v_1, v_2\rangle$ contains at least the three different points $\langle v_1\rangle$, $\langle v_2\rangle$ and $\langle v_1 + v_2\rangle$.

Axiom 4. Since V has dimension at least 3, there exist three linearly independent vectors v_0, v_1, v_2 in V. Then $\langle v_0, v_1\rangle$ and $\langle v_0, v_2\rangle$ are two distinct lines of **P**(V). □

Remark. We used the hypothesis '$\dim_V(V) \geq 3$' only in order to show axiom 4.

2.1.2 Lemma. (a) *Let* V' *be a subspace of the vector space* V. *Then* **P**(V') *is a subspace of* **P**(V).
(b) *Let* **U** *be a subspace of the projective space* **P**(V). *Then there exists a vector subspace* V' *of* V *such that* **P**(V') = **U**.

Proof. (a) We have to show that the set of 1-dimensional subspaces contained in V' forms a linear set of **P**(V): Let $\langle v\rangle, \langle w\rangle \subseteq V'$. Then $v, w \in V'$, and so $\langle v, w\rangle \subseteq V'$.

(b) Let $\{P_0, P_1, \ldots, P_t\}$ be a basis of **U**. Then there are vectors $v_0, v_1, \ldots, v_t$ with

$$P_i = \langle v_i \rangle \quad (i = 0, 1, \ldots, t).$$

Let $v_0, v_1, \ldots, v_t$ generate the vector subspace V' of V.
We claim that $\mathbf{P}(V') = \mathbf{U}$.

We shall show this by induction on t. For $t = 0$ the assertion is trivial.

We consider also the case $t = 1$. Let $g = P_0P_1$ be a line of $\mathbf{P}(V)$ with $P_0 = \langle v_0 \rangle$, $P_1 = \langle v_1 \rangle$. Then the assertion follows, since by **2.1.1** $g = \langle v_0, v_1 \rangle$.

Assume now that the assertion is true for $t \geq 1$. Consider a $(t+1)$-dimensional subspace

$$\mathbf{U} = \langle P_0, P_1, \ldots, P_t, P_{t+1} \rangle$$

of $\mathbf{P}(V)$. By induction, there is a vector subspace V" of V such that $\mathbf{P}(V'') = \langle P_0, P_1, \ldots, P_t \rangle$. Thus

$$\mathbf{U} = \langle \langle P_0, P_1, \ldots, P_t \rangle, P_{t+1} \rangle = \langle \mathbf{P}(V''), P_{t+1} \rangle.$$

Let v_{t+1} be a vector with $P_{t+1} = \langle v_{t+1} \rangle$, and let V' be the vector subspace spanned by V" and v_{t+1}.

First we show the inclusion $\mathbf{P}(V') \subseteq \mathbf{U}$. Let $P = \langle v \rangle$ be a point of $\mathbf{P}(V')$. Then there is a vector $v'' \in V''$ such that

$$v = av'' + bv_{t+1} \quad (a, b \in F).$$

Therefore the point P is incident with the line of **U** that passes through the points $\langle v'' \rangle$ and $\langle v_{t+1} \rangle$ of **U**. Thus any point of $\mathbf{P}(V')$ lies in **U**.

Conversely, let Q be a point of **U**. Then, by **1.3.1**, Q is incident with a line XP_{t+1}, where X is a point of $\mathbf{P}(V'')$. If v" is a vector with $X = \langle v'' \rangle$, then Q is on the line through $\langle v'' \rangle$ and $\langle v_{t+1} \rangle$. So, by case $t = 1$, there exist $a, b \in F$ with

$$v = av'' + bv_{t+1} \in V'$$

and $Q = \langle v \rangle$. Hence Q is a point of $\mathbf{P}(V')$. □

2.1.3 Corollary. *The projective space* $\mathbf{P}(V)$ *has dimension* d.

Proof. If $\dim(\mathbf{P}(V)) \geq d + 1$ then there would exist a flag

$$\mathbf{U}_0 \subset \mathbf{U}_1 \subset \ldots \subset \mathbf{U}_d$$

of nontrivial subspaces of $\mathbf{P}(V)$. By **2.1.2** this corresponds to a chain of d nontrivial vector subspace of V that are properly contained in each other. This is impossible, since $\dim_V(V) = d + 1$.

Similarly one proves that $\dim(\mathbf{P}(V)) \geq d$: Since $\dim_V(V) = d+1$ there is a chain of $d-1$ nontrivial subspaces of V that are properly contained in each other. These subspaces correspond to a flag of $d-1$ nontrivial subspaces of $\mathbf{P}(V)$. □

2.1.4 Lemma. *The line of* $\mathbf{P}(V)$ *through the points* $\langle v \rangle$ *and* $\langle w \rangle$ *consists of the point* $\langle w \rangle$ *and the points* $\langle v + aw \rangle$ $(a \in F)$. *In particular we have: if* F *is a finite field with* q *elements,* $q \in \mathbf{N}$, *then any line of* $\mathbf{P}(V)$ *has exactly* $q+1$ *points. In this case* $\mathbf{P}(V)$ *has order* q.

Proof. Each point $\langle u \rangle$ of the line through $\langle v \rangle$ and $\langle w \rangle$ can be written as

$$\langle u \rangle = \langle av + bw \rangle \text{ with } a, b \in F.$$

If $a = 0$ then $b \neq 0$, hence $\langle u \rangle = \langle bw \rangle = \langle w \rangle$. If $a \neq 0$ then

$$\langle u \rangle = \langle av + bw \rangle = \langle v + ba^{-1}w \rangle.$$

Conversely, any point of the form $\langle v + aw \rangle$ is a point on the line through $\langle v \rangle$ and $\langle w \rangle$. □

Remarks. 1. Lemma **2.1.4** implies in particular that for any prime power q (that is for any integer $q = p^a$ where p is a prime number and $a \in \mathbf{N}$) and any positive integer $d \geq 2$ there is a projective space of order q and dimension d. (For any prime power q there is a finite field F with exactly q elements. See for instance [Lang65], VII,5.)
2. If V is a $(d+1)$-dimensional vector space over the division ring F then the projective space $\mathbf{P}(V)$ is denoted by $PG(d, F)$; one calls $\mathbf{P}(V)$ the d-dimensional projective space over the division ring F. If F is a finite field of order q then $\mathbf{P}(V)$ is also denoted by $PG(d, q)$. (This notion makes sense since up to isomorphism there is only one finite field with q elements.)

2.2 The theorems of Desargues and Pappus

We shall now prove two extremely important **configuration theorems**, namely the theorem of Desargues (Girard Desargues (1591–1661)) and the theorem of Pappus (Pappus of Alexandria, ca. 300 A.D.). The assertions of the theorems of Desargues and Pappus are called 'theorems' since they were studied by Desargues and Pappus in the real plane, where they truly hold. For us, these assertions are simply unproved statements that could hold in a projective space or not. We shall present examples for both phenomena.

Definition. Let $\mathbf{P}$ be a projective space. We say that in $\mathbf{P}$ the **theorem of Desargues** holds if the following statement is valid. For any choice $A_1, A_2, A_3, B_1, B_2, B_3$ of points with the properties
- A_i, B_i are collinear with a point C, $C \neq A_i \neq B_i \neq C$ $(i = 1, 2, 3)$,
- no three of the points C, A_1, A_2, A_3 and no three of the points C, B_1, B_2, B_3 are collinear,

we have that the points

$$P_{12} := A_1A_2 \cap B_1B_2,\ P_{23} := A_2A_3 \cap B_2B_3,\ P_{31} := A_3A_1 \cap B_3B_1$$

lie on a common line (see Figure 2.1).

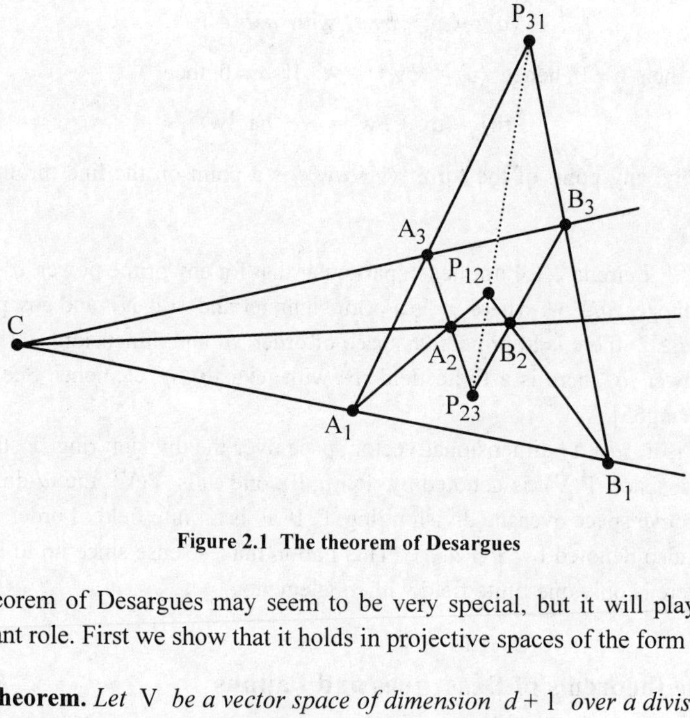

Figure 2.1 The theorem of Desargues

The theorem of Desargues may seem to be very special, but it will play a very important role. First we show that it holds in projective spaces of the form $\mathbf{P}(V)$.

2.2.1 Theorem. *Let* V *be a vector space of dimension* $d+1$ *over a division ring* F. *Then the theorem of Desargues holds in* $\mathbf{P}(V)$.

Proof. We suppose that the hypotheses of the theorem of Desargues are fulfilled.

Let v_1, v_2, v_3 be vectors such that $A_1 = \langle v_1 \rangle, A_2 = \langle v_2 \rangle, A_3 = \langle v_3 \rangle$. Since A_1, A_2, A_3 are not on a common line, v_1, v_2, v_3 are linearly independent. Therefore, they form a basis of the 3-dimensional vector space $V' := \langle v_1, v_2, v_3 \rangle$. We distinguish two cases.

Case 1. The point C lies in the plane that is spanned by A_1, A_2 and A_3.

Then there are $a_1, a_2, a_3 \in F$ with $C = \langle a_1v_1 + a_2v_2 + a_3v_3\rangle$. Since no three of the points A_1, A_2, A_3, C are collinear, we have $a_1, a_2, a_3 \neq 0$. Therefore, replacing v_i by a_iv_i, if necessary, we may assume w.l.o.g. that $C = \langle v_1 + v_2 + v_3\rangle$.

Since C, A_i, and B_i are collinear there are $a_1, a_2, a_3 \in F$ such that

$$B_1 = \langle v_1 + v_2 + v_3 + a_1v_1\rangle = \langle (a_1 + 1)v_1 + v_2 + v_3\rangle,$$
$$B_2 = \langle v_1 + (a_2 + 1)v_2 + v_3\rangle,$$
$$B_3 = \langle v_1 + v_2 + (a_3 + 1)v_3\rangle.$$

Now we can determine the vectors that represent the points P_{ij}:

$$\begin{aligned} P_{12} &= A_1A_2 \cap B_1B_2 \\ &= \langle v_1, v_2\rangle \cap \langle (a_1 + 1)v_1 + v_2 + v_3, v_1 + (a_2 + 1)v_2 + v_3\rangle \\ &= \langle a_1v_1 - a_2v_2\rangle. \end{aligned}$$

We get the last equality as follows. The vector $a_1v_1 - a_2v_2 \neq o$ is contained in $\langle v_1, v_2\rangle$ as well as in $\langle (a_1 + 1)v_1 + v_2 + v_3,\ v_1 + (a_2 + 1)v_2 + v_3\rangle$. Since these subspaces are distinct, they can intersect each other in a subspace whose dimension is at most 1. Therefore, the assertion holds. Similarly one gets

$$P_{23} = \langle a_2v_2 - a_3v_3\rangle$$

and

$$P_{31} = \langle a_3v_3 - a_1v_1\rangle.$$

This implies together that all three points P_{12}, P_{23}, and P_{31} lie on the line

$$\langle a_1v_1 - a_2v_2, a_3v_3 - a_1v_1\rangle.$$

Case 2. The points $C = \langle v\rangle$ and $A_1 = \langle v_1\rangle, A_2 = \langle v_2\rangle$ and $A_3 = \langle v_3\rangle$ do not lie in a common plane.

This case is easier. The vectors v, v_1, v_2, v_3 are linearly independent. Therefore we assume w.l.o.g. that

$$B_1 = \langle v + v_1\rangle, B_2 = \langle v + v_2\rangle, B_3 = \langle v + v_3\rangle.$$

From this it follows easily that

$$P_{12} = A_1A_2 \cap B_1B_2 = \langle v_1, v_2\rangle \cap \langle v + v_1, v + v_2\rangle = \langle v_1 - v_2\rangle,$$
$$P_{23} = \langle v_2 - v_3\rangle,$$
$$P_{31} = \langle v_3 - v_1\rangle.$$

Since the points P_{12}, P_{23}, P_{31} lie on the line $\langle v_1 - v_2, v_2 - v_3\rangle$ we have proved the theorem of Desargues. □

Remark. Observe how cleverly we have used the w.l.o.g.tool.

Definition. Let $\mathbf{P}$ be a projective space. We say that in $\mathbf{P}$ the **theorem of Pappus** holds if any two intersecting lines g and h with $g \neq h$ satisfy the following condition. If A_1, A_2, A_3 are distinct points on g and B_1, B_2, B_3 are distinct points on h all different from $g \cap h$ then the points

$$Q_{12} := A_1B_2 \cap B_1A_2,\ Q_{23} := A_2B_3 \cap B_2A_3 \text{ and } Q_{31} := A_3B_1 \cap B_3A_1$$

lie on a common line (see Figure 2.2).

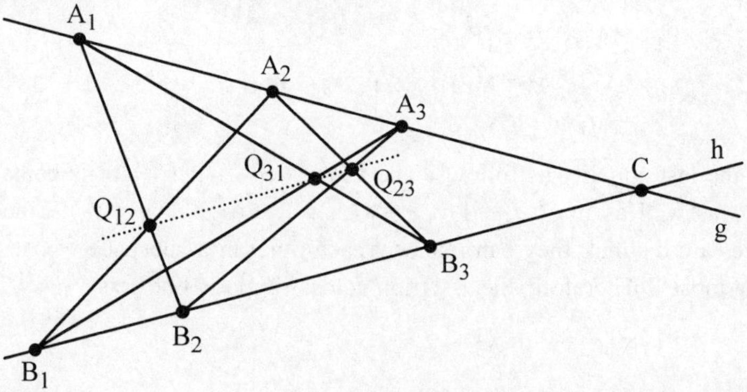

Figure 2.2 The theorem of Pappus

Remark. The theorem of Pappus always operates in a plane, namely in the plane spanned by g and h.

2.2.2 Theorem. *Let* V *be a vector space over the division ring* F. *Then the theorem of Pappus holds in* $\mathbf{P}(V)$ *if and only if* F *is commutative (in other words, if* F *is a field).*

Proof. First we consider only the points A_1, A_2, B_1, B_2 and the point $C := g \cap h = A_1A_2 \cap B_1B_2$.

Let u, v, w be vectors with $C = \langle u \rangle$, $A_1 = \langle v \rangle$, and $B_1 = \langle w \rangle$. Then, clearly, $A_2 = \langle u + av \rangle$ and $B_2 = \langle u + a'w \rangle$ with $a, a' \neq 0$. Therefore we may assume w.l.o.g. that

$$A_1 = \langle v \rangle,\ A_2 = \langle u + v \rangle,\ B_1 = \langle w \rangle, \text{ and } B_2 = \langle u + w \rangle.$$

Let a and b be arbitrary elements of F. Since for the question of commutativity only the elements $\neq 0, 1$ play a role (for 0 and 1 commute with any element) we may suppose that $a, b \neq 0, 1$. We define

$$A_3 := \langle u + av \rangle,\ B_3 := \langle u + bw \rangle.$$

Since $a, b \neq 0, 1$ we have $A_3 \neq A_1, A_2$ and $B_3 \neq B_1, B_2$.
We claim: the points Q_{12}, Q_{23}, Q_{31} *are collinear if and only if* $ab = ba$.

In order to show this we compute the coordinates of the points Q_{ij}. (Observe that in computing a spanning vector of Q_{ij} we will not use that F is commutative.)

$$Q_{12} = A_1B_2 \cap A_2B_1 = \langle v, u + w \rangle \cap \langle u + v, w \rangle = \langle u + v + w \rangle.$$

For obviously we have $u + v + w \in \langle v, u + w \rangle$ and $u + v + w \in \langle u + v, w \rangle$. Hence we have $\langle u + v + w \rangle \subseteq \langle v, u + w \rangle \cap \langle u + v, w \rangle$. A dimension argument shows that equality holds. Similarly we see that

$$Q_{31} = A_3B_1 \cap A_1B_3 = \langle u + av, w \rangle \cap \langle v, u + bw \rangle = \langle u + av + bw \rangle.$$

Finally, we get

$$\begin{aligned} Q_{23} = A_2B_3 \cap A_3B_2 &= \langle u + v, u + bw \rangle \cap \langle u + av, u + w \rangle \\ &= \langle (a + (a-1)(b-1)^{-1}) \cdot u + a \cdot v + (a-1)(b-1)^{-1} \cdot bw \rangle \end{aligned}$$

for we have

$$\begin{aligned} (a + (a-1)(b-1)^{-1}) \cdot u &+ a \cdot v + (a-1)(b-1)^{-1} \cdot bw \\ &= a \cdot (u + v) + (a-1)(b-1)^{-1} \cdot (u + bw) \\ &\in \langle u + v, u + bw \rangle, \end{aligned}$$

and

$$\begin{aligned} (a + (a-1)(b-1)^{-1}) \cdot u &+ a \cdot v + (a-1)(b-1)^{-1} \cdot bw \\ &= (a + (a-1)(b-1)^{-1} + (a-1)(b-1)^{-1}b - (a-1)(b-1)^{-1}b) \cdot u \\ &\qquad + a \cdot v + (a-1)(b-1)^{-1} \cdot bw \\ &= (a + (a-1)(b-1)^{-1}(1-b) + (a-1)(b-1)^{-1}b) \cdot u \\ &\qquad + a \cdot v + (a-1)(b-1)^{-1} \cdot bw \\ &= (a - (a-1) + (a-1)(b-1)^{-1}b) \cdot u + a \cdot v + (a-1)(b-1)^{-1} \cdot bw \\ &= (u + av) + (a-1)(b-1)^{-1}b \cdot (u + w) \\ &\in \langle u + av, u + w \rangle. \end{aligned}$$

Again a dimension argument proves the assertion.

Putting these together we see

$$Q_{23} \in Q_{31}Q_{12}$$

$$\Leftrightarrow \langle (a + (a-1)(b-1)^{-1}) \cdot \mathrm{u} + a \cdot \mathrm{v} + (a-1)(b-1)^{-1} \cdot b\mathrm{w} \rangle$$
$$\subseteq \langle \mathrm{u} + a\mathrm{v} + b\mathrm{w}, \mathrm{u} + \mathrm{v} + \mathrm{w} \rangle$$

$\Leftrightarrow$ there exist $x, y \in \mathrm{F}$ with
$$(a + (a-1)(b-1)^{-1}) \cdot \mathrm{u} + a \cdot \mathrm{v} + (a-1)(b-1)^{-1} \cdot b\mathrm{w}$$
$$= x\mathrm{u} + xa\mathrm{v} + xb\mathrm{w} + y\mathrm{u} + y\mathrm{v} + y\mathrm{w}$$

$\Leftrightarrow$ the following equations hold:

$$a + (a-1)(b-1)^{-1} = x + y,$$
$$a = xa + y,$$
$$(a-1)(b-1)^{-1}b = xb + y.$$

The first equation means $y = a + (a-1)(b-1)^{-1} - x$, and together with the second gives

$$x = (a-1)(b-1)^{-1}(1-a)^{-1},$$

thus

$$y = a + (a-1)(b-1)^{-1}(1-(1-a)^{-1}).$$

Substituting these results in the third equation we get by patient formula manipulating

$$\mathrm{Q}_{23} \in \mathrm{Q}_{31}\mathrm{Q}_{12}$$

$$\Leftrightarrow (a-1)(b-1)^{-1}b$$
$$= (a-1)(b-1)^{-1}(1-a)^{-1}b + a + (a-1)(b-1)^{-1}(1-(1-a)^{-1})$$

$$\Leftrightarrow b = (1-a)^{-1}b + (b-1)(a-1)^{-1}a + (1-(1-a)^{-1})$$

$$\Leftrightarrow (1-a)b = b + (1-a)(b-1)(a-1)^{-1}a + (1-a) - 1$$

$$\Leftrightarrow (1-a)b$$
$$= b + (1-a)(b-1)(a-1)^{-1}a$$
$$+ (1-a)(b-1)(a-1)^{-1} - (1-a)(b-1)(a-1)^{-1} - a$$

$$\Leftrightarrow (1-a)b$$
$$= b + (1-a)(b-1)(a-1)^{-1}(a-1) + (1-a)(b-1)(a-1)^{-1} - a$$

$$\Leftrightarrow (1-a)b = b + (1-a)(b-1) - a + (1-a)(b-1)(a-1)^{-1}$$

$$\Leftrightarrow \quad 0 = b - (1-a) - a + (1-a)(b-1)(a-1)^{-1}$$

$$\Leftrightarrow \quad 1 - b = (1-a)(b-1)(a-1)^{-1}$$

$$\Leftrightarrow \quad (1-b)(a-1) = (1-a)(b-1)$$

$$\Leftrightarrow \quad ba = ab. \qquad \square$$

Remark. The theorems of Desargues and Pappus are not independent, the theorem of Pappus is stronger. This is expressed in the following theorem.

2.2.3 Hessenberg's theorem. *Let* **P** *be an arbitrary projective space. If the theorem of Pappus holds in* **P** *then the theorem of Desargues is also true in* **P**.

We will not prove this theorem here. Proofs can be found in the literature, for instance in [Lin69], [Lenz65]. The most beautiful proof is due to A. Herzer [Her72], who proved Hessenberg's theorem using dualities.

2.3 Coordinates

Definition. We fix a basis $\{v_0, v_1, \ldots, v_d\}$ of V. Then we can express any vector

$$v = a_0v_0 + a_1v_1 + \ldots + a_dv_d \in V$$

uniquely by its **coordinates** $(a_0, a_1, \ldots, a_d)$.

We define an equivalence relation $\sim$ on the set of $(d+1)$-tuples different from $(0, \ldots, 0)$ whose elements are in F by

$$(a_0, a_1, \ldots, a_d) \sim (b_0, b_1, \ldots, b_d) :\Leftrightarrow$$
$$\text{there is an } a \in F\setminus\{0\} \text{ with } (a_0, a_1, \ldots, a_d) = a\cdot(b_0, b_1, \ldots, b_d).$$

We say that a point $\langle v\rangle$ of **P**(V) has **homogeneous coordinates** $(a_0, a_1, \ldots, a_d)$ if

$$\langle v\rangle = \langle a_0v_0 + a_1v_1 + \ldots + a_dv_d\rangle.$$

Clearly, two such $(d+1)$-tuples are equivalent if and only if they are homogeneous coordinates of the same point of **P**(V). Therefore, homogeneous coordinates are not uniquely determined by the point they represent. (Be careful when using the definite article.)

Therefore we proceed as follows. We denote the equivalence class containing $(a_0, a_1, \ldots, a_d)$ by $(a_0: a_1: \ldots : a_d)$ and then write

$$P = (a_0: a_1: \ldots : a_d),$$

if $P = \langle a_0v_0 + a_1v_1 + \ldots + a_dv_d\rangle$. We shall also call $(a_0: a_1: \ldots : a_d)$ the **homogeneous coordinates** of P.

Homogeneous coordinates have the advantage of being flexible. One important feature is that we can w.l.o.g. normalize the first (or last) nonzero entry to 1.

Example. The line through the points with homogeneous coordinates $(a_0: a_1: \ldots: a_d)$ and $(b_0: b_1: \ldots: b_d)$ consists of the points with the following coordinates:

$$(a_0: a_1: \ldots: a_d) \text{ and } (b_0: b_1: \ldots: b_d) + a\cdot(a_0: a_1: \ldots: a_d),\ a \in F.$$

In particular, the line through (1: 0, 0: . . .: 0) and (0: 1: 0: . . .: 0) consists of

$$(1: 0: 0: \ldots: 0) \text{ and } (a: 1: 0: \ldots: 0).$$

Next, we shall study how one can describe higher-dimensional subspaces by coordinates.

2.3.1 Theorem. *Let* $P_1, P_2, \ldots, P_t$ *be points of* $\mathbf{P}(V)$ *with homogeneous coordinates*

$$P_1 = (a_{10}: a_{11}: \ldots: a_{1d}),$$

$$\ldots,$$

$$P_t = (a_{t0}: a_{t1}: \ldots: a_{td}).$$

Then the points $P_1, P_2, \ldots, P_t$ *are independent if and only if the matrix*

$$\begin{pmatrix} a_{10} & a_{11} & \ldots & a_{1d} \\ \vdots & \vdots & & \vdots \\ a_{t0} & a_{t1} & \ldots & a_{td} \end{pmatrix}$$

has rank t.

In particular we have: $d+1$ *points form a basis if and only if the matrix whose rows are the homogeneous coordinates of the points is nonsingular.*

Proof. If the above matrix has rank t then the homogeneous coordinates of the points span a t-dimensional subspace of V. So, by **2.1.2,** the points P_1,

$P_2, \ldots, P_t$ span a $(t-1)$-dimensional subspace **P**(V); therefore they are independent.

Conversely we suppose that $P_1, P_2, \ldots, P_t$ are independent. If the rank of the above matrix were smaller than t, then the homogeneous coordinates of the points $P_1, P_2, \ldots, P_t$ would span a subspace of V with dimension smaller than t. Thus $P_1, P_2, \ldots, P_t$ would span a subspace of **P**(V) with dimension smaller than $t-1$; hence $P_1, P_2, \ldots, P_t$ would not be independent. □

2.3.2 Theorem. *Let* V *be a vector space of dimension* $d+1$ *over the division ring* F, *and let* **P**(V) *be the corresponding projective space. Let* **H** *be a hyperplane of* **P**(V). *Then the homogeneous coordinates of the points of* **H** *are the solutions of a homogeneous equation with coefficients in* F. *Conversely, any homogeneous equation that is different from the 'zero equation' describes a hyperplane of* **P**(V).

Proof. Let H' be the hyperplane of V that corresponds to **H** (see **2.1.2**(b)); we have, therefore, **P**(H') = **H**. The coordinates of the vectors in H' are solutions of a homogeneous equation. Conversely we know from linear algebra that the solutions of any homogeneous equation different from the zero equation are a subspace of dimension d. Hence the space of solutions is a hyperplane of V and therefore a hyperplane of **P**(V). □

2.3.3 Corollary. *Any* t*-dimensional subspace* **U** *of a projective space of dimension* d *given by homogeneous coordinates can be described by a homogeneous system of* $d-t$ *linear equations. More precisely: there exists a* $(d-t)\times(d+1)$ *matrix* H *such that a point* $P = (a_0: a_1: \ldots: a_d)$ *is a point of* **U** *if and only if*

$$(a_0: a_1: \ldots: a_d)\cdot H^T = 0.$$

Proof. By **1.3.10**, **U** is the intersection of $d-t$ hyperplanes. □

Now we can describe any hyperplane by a homogeneous equation. For instance, we shall speak of the 'hyperplane $x_0 = 0$' and we mean by this the hyperplane whose points have homogeneous coordinates $(0: a_1: \ldots: a_d)$.

We shall also describe a hyperplane by the coefficients of a corresponding homogeneous equation. For instance, the hyperplane with equation $x_0 = 0$ will be represented by $[1: 0: \ldots: 0]$. Note that these are also homogeneous coordinates.

2.3.4 Theorem. *Let* **P** = **P**(V) *be a projective space whose points are given by homogeneous coordinates* $(a_0: a_1: \ldots: a_d)$ *with* $a_i \in$ F, $(a_0, a_1, \ldots, a_d) \neq (0, 0, \ldots, 0)$. *Then any hyperplane of* **P**(V) *can be represented by* $[b_0: b_1: \ldots: b_d]$

with $b_i \in F$, $[b_0, b_1, \ldots, b_d] \neq [0, 0, \ldots, 0]$. *Conversely, to any such* $(d+1)$*-tuple* $[b_0: b_1: \ldots: b_d]$ *there belongs a hyperplane. Moreover, we have*

$$(a_0: a_1: \ldots: a_d) \mathrm{I} [b_0: b_1: \ldots: b_d] \Leftrightarrow a_0b_0 + a_1b_1 + \ldots + b_da_d = 0.$$

The *proof* follows directly from **2.3.2**, since any homogeneous linear equation in $d+1$ variables can be described by a $(d+1)$-tuple $[b_0: b_1: \ldots: b_d]$ of homogeneous coordinates. □

Now we consider the dual geometry of a projective space **P**. This means that we consider **P** as a rank 2 geometry consisting of points and hyperplanes. Then $\mathbf{P}^\Delta$ is the geometry whose points are the hyperplanes of **P**, and whose blocks are the points of **P** (cf. **1.2.2**).

2.3.5 Corollary. *Let* $\mathbf{P} = \mathbf{P}(V)$ *be a coordinatized projective space. Then* $\mathbf{P}^\Delta \cong \mathbf{P}$. *In particular,* $\mathbf{P}^\Delta$ *is coordinatized as well. Therefore the principle of duality holds for the class of all coordinatized projective spaces of fixed dimension* d. □

Now we shall introduce coordinates for affine spaces.

2.3.6 Theorem. *Let* $\mathbf{H}_\infty$ *be the hyperplane of* $\mathbf{P}(V)$ *with equation* $x_0 = 0$. *Then the affine space* $\mathbf{A} = \mathbf{P} \setminus \mathbf{H}_\infty$ *can be described as follows.*
– *The points of* **A** *are the vectors* $(a_1, \ldots, a_d)$ *of the* d*-dimensional vector space* F^d;
– *the lines of* **A** *are the* **cosets** *of the* 1*-dimensional subspaces of* F^d; *that is the set* $u + \langle v \rangle$, where $u, v \in F^d$, $v \neq o$.
– *incidence is set-theoretical containment.*

Proof. Since the homogeneous coordinates of the points on $\mathbf{H}_\infty$ have as first entry 0, any point P outside $\mathbf{H}_\infty$ has homogeneous coordinates $(a_0: a_1: \ldots: a_d)$ with $a_0 \neq 0$. Thus it has homogeneous coordinates $(1: a_1: \ldots: a_d)$ with uniquely determined $a_1, \ldots, a_d \in F$. Hence we can identify a point P of **A** with the *d*-tuple $(a_1, \ldots, a_d)$. One calls $(a_1, \ldots, a_d)$ the **inhomogeneous coordinates** of the point P.

Now we shall describe the lines of **A**: We know that any line g of **A** has precisely one point $(0: b_1: \ldots: b_d)$ in common with $\mathbf{H}_\infty$. If $(1: a_1: \ldots: a_d)$ is an arbitrary point of **A** on g, then the points of **A** on g have the following homogeneous coordinates:

$$(1: a_1: \ldots: a_d) + a \cdot (0: b_1: \ldots: b_d) \text{ with } a \in F.$$

Therefore, the inhomogeneous coordinates of the points on g are as follows:

$$(a_1, \ldots, a_d) + a \cdot (b_1, \ldots, b_d) \text{ with } a \in F.$$

In other words: the affine points on g are precisely the points of the coset $(a_1, \ldots, a_d) + \langle (b_1, \ldots, b_d) \rangle$ of the 1-dimensional subspace $\langle (b_1, \ldots, b_d) \rangle$ of F^d.

Conversely, let $(a_1, \ldots, a_d) + \langle (b_1, \ldots, b_d) \rangle$ be a coset of a 1-dimensional subspace $\langle (b_1, \ldots, b_d) \rangle$ of F^d. Then, by the above construction, this coset corresponds to the 2-dimensional subspace $\langle (1, a_1, \ldots, a_d), (0, b_1, \ldots, b_d) \rangle$ of F^{d+1}; it is therefore a line of **P**(V). Since this line intersects $\mathbf{H}_\infty$ just in the point $(0: b_1: \ldots: b_d)$, the coset in question is a line of **A**. □

Notation. If $\mathbf{H}_\infty$ is a hyperplane of $\mathbf{P} = PG(d, F)$ then we denote the affine space $\mathbf{A} = \mathbf{P} \setminus \mathbf{H}_\infty$ by $AG(d, F)$ and call it the affine space of dimension d coordinatized over F. If F has finite order q, then we also denote **A** by $AG(d, q)$.

Since any two hyperplanes of $PG(d, F)$ can be mapped onto each other by a collineation, the definition of $AG(d, F)$ is independent of the choice of the hyperplane $\mathbf{H}_\infty$ (see also exercise 14).

Remark. If the theorem of Desargues holds in **P**, it holds also in $\mathbf{A} = \mathbf{P} \setminus \mathbf{H}_\infty$. When formulating the theorem of Desargues in affine spaces one has to observe that two lines in a plane do not necessarily meet, but can also be parallel.

2.4 The hyperbolic quadric of PG(3, F)

We now present a theorem which plays a crucial role for the construction of geometrically important objects, namely quadrics (see Chapter 4).

Definition. Let **P** be a projective space.
(a) We call a set $\mathcal{S}$ of subspaces of **P** **skew** if no two distinct subspaces of $\mathcal{S}$ have a point in common. We also speak of **skew subspaces**.
(b) Let $\mathcal{S}$ be a set of skew subspaces. A line is called a **transversal** of $\mathcal{S}$ if it intersects each subspace of $\mathcal{S}$ in exactly one point.

2.4.1 Lemma. *Let* **P** *be a projective space. Let* g_1 *and* g_2 *be two skew lines, and denote by* P *a point outside* g_1 *and* g_2. *Then there is at most one transversal of* g_1 *and* g_2 *through* P. *If* **P** *is 3-dimensional then there is exactly one transversal of* g_1 *and* g_2 *through* P.

Proof. Assume that there are two transversals h_1 and h_2 through P. Then each of these transversals meets the lines g_1 and g_2 in different points. So h_1 and h_2 span a plane, which contains the two skew lines g_1 and g_2, a contradiction.

Now suppose that **P** is 3-dimensional. Then the plane $\langle P, g_1\rangle$ must intersect the line g_2 in some point Q. Therefore the line PQ intersects g_1 and g_2, hence it is a transversal of g_1 and g_2. □

The following theorem has been proved for the real projective space by G. Gallucci (see Coxeter [Cox69], p. 257); the theorem is known as the **theorem of Dandelin** (Germinal Pierre Dandelin, 1794–1847).

2.4.2 Theorem (16 point theorem). *Let* **P** *be a 3-dimensional projective space over the division ring* F. *Let* $\{g_1, g_2, g_3\}$ *and* $\{h_1, h_2, h_3\}$ *be sets of skew lines with the property that each line* g_i *meets each line* h_j. *Then the following is true:* F *is commutative (hence a field) if and only if each transversal* $g \notin \{g_1, g_2, g_3\}$ *of* $\{h_1, h_2, h_3\}$ *intersects each transversal* $h \notin \{h_1, h_2, h_3\}$ *of* $\{g_1, g_2, g_3\}$ *(see Figure 2.3).*

Proof. First we fix the following notations.

$$\langle v_1\rangle := g_1 \cap h_1,\quad \langle v_2\rangle := g_1 \cap h_2,$$
$$\langle v_3\rangle := g_2 \cap h_1,\quad \langle v_4\rangle := g_2 \cap h_2.$$

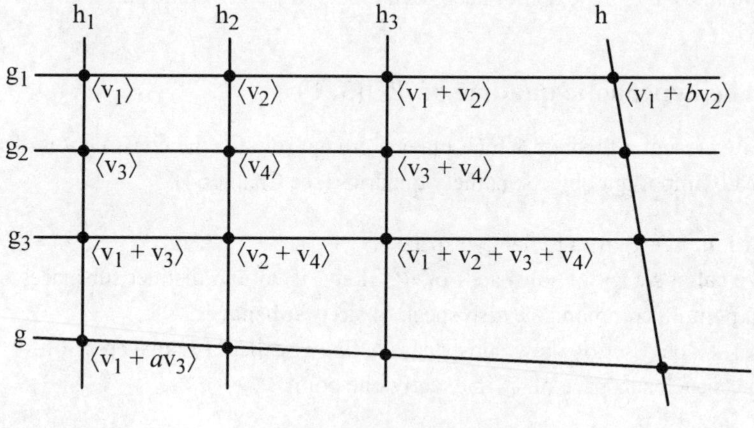

Figure 2.3 The 16 point theorem

Then $g_3 \cap h_1 = \langle av_1 + bv_3\rangle$ with $0 \neq a, b \in F$. We may assume w.l.o.g. that $b = 1$. Replacing v_1 by $v_1' = av_1$ we get $g_3 \cap h_1 = \langle v_1' + v_3\rangle$; since we can replace v_1' by v_1 we have

$$g_3 \cap h_1 = \langle v_1 + v_3\rangle.$$

Multiplying v_2 by a suitable factor we can similarly assume that

$$g_1 \cap h_3 = \langle v_1 + v_2 \rangle.$$

Multiplying v_4 suitably we can assume that

$$g_2 \cap h_3 = \langle v_3 + v_4 \rangle.$$

Finally we get that

$$g_3 \cap h_2 = \langle v_2 + a v_4 \rangle.$$

By hypothesis there is a unique point of intersection of g_3 and h_3. For this we see

$$g_3 \cap h_3 = \langle a_1(v_1 + v_3) + a_2(v_2 + a v_4) \rangle = \langle b_1(v_1 + v_2) + b_2(v_3 + v_4) \rangle.$$

Since the vectors v_i, $i = 1, 2, 3, 4$, are linearly independent we get $b_1 = b_2 = a_1 = a_2$ and $a = 1$, hence

$$g_3 \cap h_2 = \langle v_2 + v_4 \rangle$$

and

$$g_3 \cap h_3 = \langle v_1 + v_2 + v_3 + v_4 \rangle.$$

Let $g \notin \{g_1, g_2, g_3\}$ be a transversal of $\{h_1, h_2, h_3\}$, and let $h \notin \{h_1, h_2, h_3\}$ be a transversal of $\{g_1, g_2, g_3\}$. We define the elements a and b of F by

$$g \cap h_1 =: \langle v_1 + a v_3 \rangle,$$
$$h \cap g_1 =: \langle v_1 + b v_2 \rangle.$$

We claim: the lines g *and* h *intersect each other if and only if* $ab = ba$.

In order to determine the equation of the line g we use the fact that by **2.4.1** there is at most one line through the point $\langle v_1 + a v_3 \rangle$ that intersects h_2 as well as h_3; this line must be g. Now luckily we can convince ourselves easily that the points $\langle v_2 + a v_4 \rangle \in \langle v_2, v_4 \rangle = h_2$ and $\langle v_1 + v_2 + a(v_3 + v_4) \rangle \in \langle v_1 + v_2, v_3 + v_4 \rangle = h_3$ lie together with $\langle v_1 + a v_3 \rangle$ on a common line. Therefore, necessarily

$$g = \langle v_1 + a v_3, v_2 + a v_4 \rangle.$$

Similarly we get

$$h = \langle v_1 + b v_2, v_3 + b v_4 \rangle.$$

Now we simply compute the intersection of the subspaces g and h. For each common vector there are $x, y, z, w \in F$ such that

$$x(v_1 + a v_3) + y(v_2 + a v_4) = z(v_1 + b v_2) + w(v_3 + b v_4).$$

Since v_1, v_2, v_3, v_4 are linearly independent, this is the case if and only if

$$x = z, \ \ y = zb, \ \ xa = w \ \text{ and } \ ya = wb.$$

This, however, is true if and only if

$$xba = zba = ya = wb = xab.$$

If g and h have a common intersection then $x \neq 0$ (otherwise $x = 0$, $z = 0$, $y = 0$, and $w = 0$); this implies $ba = ab$. If conversely $ab = ba$ then

$$\langle \mathrm{v}_1 + a\mathrm{v}_3 + b\mathrm{v}_2 + ba\mathrm{v}_4 \rangle$$

is a common point of $\langle \mathrm{v}_1 + a\mathrm{v}_3, \mathrm{v}_2 + a\mathrm{v}_4 \rangle = \mathrm{g}$ and $\langle \mathrm{v}_1 + b\mathrm{v}_2, \mathrm{v}_3 + b\mathrm{v}_4 \rangle = \mathrm{h}$. □

Definition. Let **P** be a 3-dimensional projective space. A nonempty set $\mathfrak{R}$ of skew lines of **P** is called a **regulus** if the following are true:
- Through each point of each line of $\mathfrak{R}$ there is a transversal of $\mathfrak{R}$.
- Through each point of a transversal of $\mathfrak{R}$ there is a line of $\mathfrak{R}$.

It is clear that the set $\mathfrak{R}'$ of all transversals of a regulus $\mathfrak{R}$ again form a regulus; we call it the **opposite regulus** of $\mathfrak{R}$.

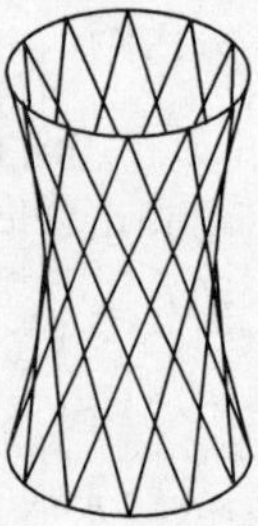

Figure 2.4 A regulus and its opposite regulus

If **P** has finite order q then any regulus consists of exactly $q+1$ lines.

2.4.3 Theorem. *Let* **P** *be a 3-dimensional projective space over the division ring* F. *Let* $\mathrm{g}_1, \mathrm{g}_2, \mathrm{g}_3$ *be three skew lines of* **P**. *Then the following assertions are true.*
(a) *There is at most one regulus containing* $\mathrm{g}_1, \mathrm{g}_2,$ *and* g_3.
(b) *If* F *is noncommutative then there is no regulus in* **P**.
(c) *If* F *is commutative then there is exactly one regulus through* $\mathrm{g}_1, \mathrm{g}_2,$ *and* g_3.

Proof. (a) By **2.4.1** each point of g_3 is on exactly one transversal of $\mathrm{g}_1, \mathrm{g}_2, \mathrm{g}_3$. Let $\mathfrak{R}'$ be the set of those transversals. These are exactly the transversals of any regulus containing $\mathrm{g}_1, \mathrm{g}_2, \mathrm{g}_3$. Since through each point of g_3 there is a line of $\mathfrak{R}'$ there exist at least three such transversals $\mathrm{h}_1, \mathrm{h}_2, \mathrm{h}_3$.

Let P be a point on h_1 that is not incident with $\mathrm{g}_1, \mathrm{g}_2,$ or g_3. Then any regulus $\mathfrak{R}$ through $\mathrm{g}_1, \mathrm{g}_2, \mathrm{g}_3$ has the property that the line of $\mathfrak{R}$ through P is necessarily the transversal of h_1, h_2, h_3 through P. Hence $\mathcal{R}$ is uniquely determined.

(b) follows directly from the above theorem.

(c) The existence of a regulus also follows from the 16 point theorem. □

Remark. Since by the theorem of Wedderburn any finite division ring is commutative, i.e. is a field (see for instance [Her64], Theorem 7.2.1), any *finite* 3-dimensional projective space **P**(V) has the property that through any three skew lines there is (exactly) one regulus.

We shall now show that the points on the lines of a regulus form a 'quadric', which means that they are the solutions of a quadratic equation.

2.4.4 Theorem. *Let* **P** *be a 3-dimensional projective space over the field* F *which is represented by homogeneous coordinates. Let*

$$g_1 = \langle(1: 0: 0: 0), (0: 1: 0: 0)\rangle,$$
$$g_2 = \langle(0: 0: 1: 0), (0: 0: 0: 1)\rangle,$$
$$g_3 = \langle(1: 0: 1: 0), (0: 1: 0: 1)\rangle$$

be three skew lines. Then the set $\mathcal{Q}$ *of points on the uniquely determined regulus* $\mathcal{R}$ *through* g_1, g_2, g_3 *can be described as*

$$\mathcal{Q} = \{(a_0: a_1: a_2: a_3) \mid a_0a_3 = a_1a_2 \text{ with } a_i \in F, \text{ not all } a_i = 0\};$$

therefore the coordinates of the points of $\mathcal{Q}$ *satisfy the quadratic equation*

$$x_0x_3 - x_1x_2 = 0.$$

Proof. We use the notation introduced in the proof of **2.4.2**. If $v_1 = (1: 0: 0: 0)$, $v_2 = (0: 1: 0: 0)$, $v_3 = (0: 0: 1: 0)$, and $v_4 = (0: 0: 0: 1)$ then the points $(a_0: a_1: a_2: a_3)$ on $\mathcal{Q}$ satisfy

$$(a_0: a_1: a_2: a_3) = k\cdot v_1 + k\cdot bv_2 + k\cdot av_3 + k\cdot abv_4 = (k: kb: ka: kab)$$

where $a, b \in F$ are arbitrary, and $k \in F\setminus\{0\}$. Hence the coordinates (a_0, a_1, a_2, a_3) of any point of $\mathcal{Q}$ fulfil the condition

$$0 = k\cdot kab - ka\cdot kb = a_0a_3 - a_1a_2.$$

Conversely, let $P = (a_0: a_1: a_2: a_3)$ denote a point with $a_0a_3 = a_1a_2$. If $a_0 \neq 0$ then this point can be written as

$$P = (a_0: a_1: a_2: a_1a_2/a_0),$$

and this is a point on $\mathfrak{Q}$. If $a_0 = 0$ then $a_1 = 0$ or $a_2 = 0$; w.l.o.g. $a_1 = 0$. Then P can be written as $P = (0: 0: a_2: a_3)$. W.l.o.g. we may assume $a_2 \neq 0$. Therefore $P = (0: 0: a: ab)$ with $a, b \in F$. From these results together it follows that $P \in \mathfrak{Q}$.□

Definition. One calls $\mathfrak{Q}$ the **hyperbolic quadric** of the 3-dimensional projective space $\mathbf{P} = PG(3, F)$.

In Chapter 4 we shall study quadrics in detail. Then the hyperbolic quadric will play a fundamental role.

2.5 Normal rational curves

For many reasons one is interested in large sets of points that are as independent as possible ('in general position'). One reason is the application of finite projective geometry in coding theory and cryptography. There we will often need such sets.

Definition. Let **P** be a projective space of dimension d. We say that a set $\mathcal{S}$ of at least $d+1$ points of **P** is in **general position** if any $d+1$ points of $\mathcal{S}$ form a basis of **P**.

Examples. (a) A set of at least three points in a projective plane is in general position if no three are on a common line.
(b) A set of at least four points of a 3-dimensional projective space is in general position if no four of them are in a common plane.

The natural candidates for large sets of points in general position are the 'normal rational curves'.

Definition. Let $\mathbf{P} = \mathbf{P}(V)$ be a d-dimensional projective space coordinatized over a field F. Let the points of **P** be described by homogeneous coordinates. Then the set $\mathcal{C}$ of points defined as follows is called a **normal rational curve** of **P**:

$$\mathcal{C} = \{(1: t: t^2: t^3: \ldots: t^d) \mid t \in F\} \cup \{(0: 0: \ldots: 0: 1)\}.$$

Example. In a projective plane, a normal rational curve consists of the points

$$(1: t: t^2) \text{ with } t \in F$$

together with the point (0: 0: 1). If we interpret this in the affine plane that is obtained by removing the line with equation $x_0 = 0$ we see that the affine points of the normal rational curve are the points (x_1, x_2) with $x_2 = x_1^2$. These form a parabola in the affine plane.

The following theorem says that the normal rational curves are those objects we are looking for.

2.5.1 Theorem. *The points of a normal rational curve are in general position.*

Proof. We choose $d+1$ points of a normal rational curve and consider the matrix whose rows are the coordinates of those points. The points under consideration are independent if and only if the determinant of the matrix is different from zero.
Case 1. Our $d+1$ points all have the form $P_{t_i} = (1, t_i, t_i^2, \ldots, t_i^d)$, $i = 1, \ldots, d+1$.

Then the determinant is

$$\begin{vmatrix} 1 & t_1 & t_1^2 & \cdots & t_1^d \\ 1 & t_2 & t_2^2 & \cdots & t_2^d \\ \vdots & \vdots & \vdots & & \vdots \\ 1 & t_d & t_d^2 & \cdots & t_d^d \\ 1 & t_{d+1} & t_{d+1}^2 & \cdots & t_{d+1}^d \end{vmatrix}.$$

This is a Vandermonde determinant. It is different from zero if and only if all t_i are different. Since the points P_{t_i} are different the determinant is different from zero; hence the points are independent.
Case 2. The point $(0, 0, \ldots, 0, 1)$ is contained in our set.

Let $P_{t_i} = (1, t_i, t_i^2, \ldots, t_i^d)$, $i = 1, \ldots, d$, be the other points. Then the determinant is

$$\begin{vmatrix} 1 & t_1 & t_1^2 & \cdots & t_1^d \\ 1 & t_2 & t_2^2 & \cdots & t_2^d \\ \vdots & \vdots & \vdots & & \vdots \\ 1 & t_d & t_d^2 & \cdots & t_d^d \\ 0 & 0 & 0 & \cdots & 1 \end{vmatrix}.$$

We develop the determinant with respect to the last row and get as in case 1 a Vandermonde determinant, which is different from zero since the points are distinct. □

2.5.2 Corollary. *Each hyperplane intersects a normal rational curve in at most d points.* □

2.6 The Moulton plane

In this section we shall define an affine plane in which the theorem of Desargues does not hold universally. This shows that not every affine (and hence not any projective plane) is of the form **P**(V). This plane will be constructed using the affine plane over the reals. More precisely one can say that this plane consists of two halves of the real plane.

Definition. We define the geometry **M** as follows (see Figure 2.5).

- The *points* of **M** are all the pairs (x, y) with $x, y \in \mathbf{R}$.
- The *lines* of **M** are described by the equations $x = c$ and $y = mx + b$ with $m, b, c \in \mathbf{R}$.
- The *incidence* is as follows.

Let (x_0, y_0) be a point. It lies on a line $x = c$ if and only if $x_0 = c$. If $x_0 \leq 0$ or $m \geq 0$ then (x_0, y_0) is incident with $y = mx + b$ if and only if $y_0 = mx_0 + b$. Finally, if $x_0 > 0$ and $m < 0$ then (x_0, y_0) is incident with $y = mx + b$ if and only if $\boldsymbol{y_0 = 2mx_0 + b}$.

This geometry **M** is called the **Moulton plane** (after the American mathematician F.R. Moulton who first studied this geometry in 1902; see [Mou02]).

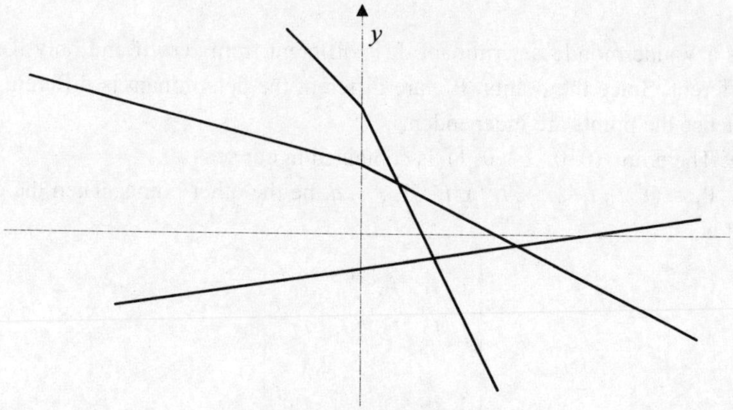

Figure 2.5 The Moulton plane

One can imagine that the Moulton plane is obtained from the real plane by bending all lines with negative slope at the y-axis by a factor of 2. (Obviously, one can take any factor $\neq 1$ instead of 2.)

2.6.1 Theorem. *The Moulton plane is an affine plane in which the theorem of Desargues is not true.*

Proof. First we shall show that **M** is an affine plane.

Axiom 1. Let (x_0, y_0) and (x_1, y_1) be two distinct points of **M**, w.l.o.g. $x_0 \le x_1$). In the following cases it is clear that these two points are joined by exactly one line of **M**:

$$x_0, x_1 \le 0;\ x_0, x_1 \ge 0;\ \text{or}\ \ y_0 \le y_1.$$

Suppose now that $x_0 < 0, x_1 > 0$ and $y_0 > y_1$. Then our points are certainly not on an 'unbent' line. Question: which bent lines are incident with (x_0, y_0) and (x_1, y_1)? In order to answer this question we have to search for all $m, b \in \mathbf{R}$ with

$$y_0 = mx_0 + b \text{ and } y_1 = 2mx_1 + b.$$

From this we get

$$y_0 - y_1 = mx_0 - 2mx_1 = m(x_0 - 2x_1),$$

hence

$$m = (y_0 - y_1) / (x_0 - 2x_1) < 0$$

and

$$b = (y_1x_0 - 2y_0x_1) / (x_0 - 2x_1).$$

Thus m and b are uniquely determined.

Axiom 2. Obviously, the lines $x = c$ as well as the lines with fixed slope $m \in \mathbf{R}$ form a parallel class of **M**. It is easy to see that this parallelism satisfies Playfair's parallel postulate (see exercise 18).

The nondegeneracy axioms are clear. Therefore **M** is an affine plane.

It is, however, conceivable that **M** is also the real plane, only very strangely disguised. But this is not the case, as is shown in the following assertion.

Claim: In **M** *the theorem of Desargues does not hold universally.*

The idea is to draw a Desargues figure in such a way that all but one of the crossing points lie in one half, and the last point of intersection is definitely distinct from the place where it would lie in the real plane. Since in the real plane the theorem of Desargues is true, it cannot be true in the Moulton plane. This will become clear from Figure 2.6.

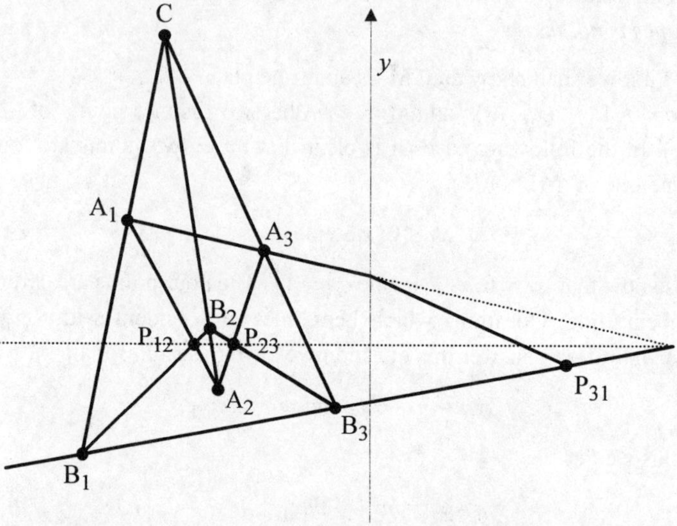

Figure 2.6 A 'Desargues figure' in the Moulton plane

□

2.7 Spatial geometries are Desarguesian

The following very important theorem says in particular that there is no 'spatial Moulton geometry'.

2.7.1 Theorem. *Let* **P** *be a projective space of dimension* d. *If* $d \geq 3$ *then the theorem of Desargues holds in* **P**.

Proof. Let $A_1, A_2, A_3, B_1, B_2, B_3, P_{12}, P_{23}, P_{31}$, and C be points that satisfy the hypotheses of the theorem of Desargues. We have to show that P_{12}, P_{23}, and P_{31} are collinear.

Case 1. The planes $\pi := \langle A_1, A_2, A_3 \rangle$ and $\psi := \langle B_1, B_2, B_3 \rangle$ are different (see Figure 2.7).

Since A_i and B_i are collinear with C we have $B_i \in \langle C, A_1, A_2, A_3 \rangle$ ($i = 1, 2, 3$). Therefore all points and lines in question are contained in the 3-dimensional subspace $\mathbf{U} := \langle C, A_1, A_2, A_3 \rangle$ of $\mathbf{P}$. The points P_{12}, P_{23}, P_{31} lie in $\pi \cap \psi$. Since any two distinct planes of $\mathbf{U}$ intersect in a line, the points P_{12}, P_{23}, P_{31} are collinear.

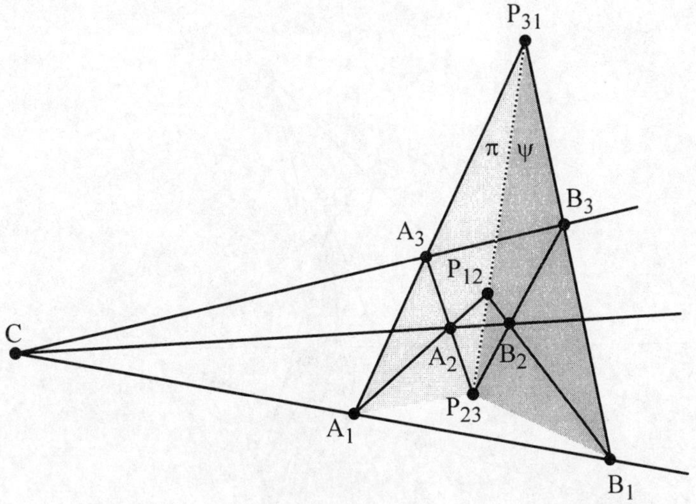

Figure 2.7

Case 2. The points $A_1, A_2, A_3, B_1, B_2, B_3$, and C lie in a common plane π.

We reduce this problem to case 1. In order to do this we construct three non-collinear points D_1, D_2, D_3 not in π and two points C', C'' such that for $C', D_1, D_2, D_3, A_1, A_2, A_3$ and $C'', D_1, D_2, D_3, B_1, B_2, B_3$ the hypotheses of case 1 are satisfied (see Figure 2.8).

Let C', C'' be two distinct points outside π such that the line $C'C''$ intersects the plane π in C. Since $C \in C'C'' \cap A_1B_1$ the lines $C'C''$ and A_1B_1 generate a plane. Therefore, the lines $C'A_1$ and $C''B_1$ intersect in a point D_1, which is outside π.

Similarly, there exist points D_2, D_3 with

$$D_2 = C'A_2 \cap C''B_2 \quad \text{and} \quad D_3 = C'A_3 \cap C''B_3.$$

We now show that $C', D_1, D_2, D_3, A_1, A_2, A_3$ and $C'', D_1, D_2, D_3, B_1, B_2, B_3$ satisfy the hypotheses of the theorem of Desargues.

If three of the points C', D_1, D_2, D_3 are collinear then $\dim(\langle D_1, D_2, D_3, C'\rangle) \le 2$. Since, by construction, A_i is on $C'D_i$, $i = 1, 2, 3$, we have that the points A_1, A_2, A_3 lie in $\pi \cap \langle D_1, D_2, D_3, C'\rangle$. Since the dimension of this intersection is at most 1, it is a point or a line; therefore A_1, A_2, A_3 would be collinear, a contradiction.

By construction of C' it is clear that no three of the points C', A_1, A_2, A_3 are collinear. Similarly it follows that $C'', D_1, D_2, D_3, B_1, B_2, B_3$ also satisfy the hypotheses of the theorem of Desargues.

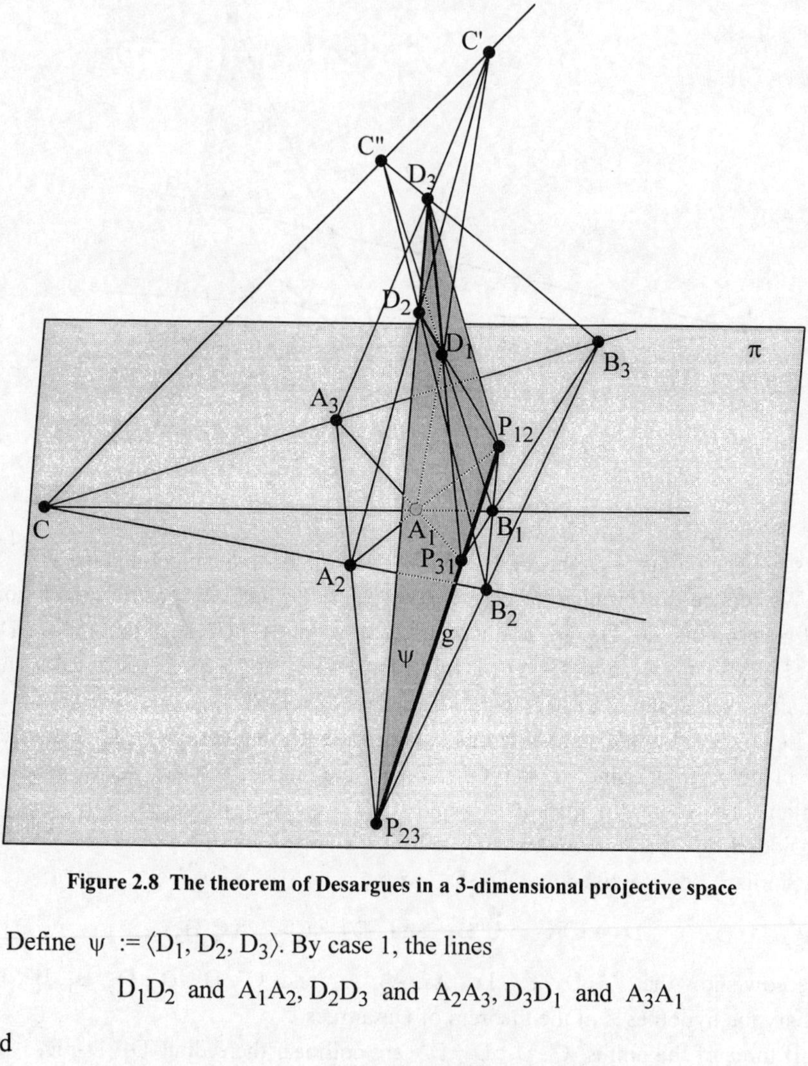

Figure 2.8 The theorem of Desargues in a 3-dimensional projective space

Define $\psi := \langle D_1, D_2, D_3 \rangle$. By case 1, the lines

$$D_1D_2 \text{ and } A_1A_2,\ D_2D_3 \text{ and } A_2A_3,\ D_3D_1 \text{ and } A_3A_1$$

and

$$D_1D_2 \text{ and } B_1B_2,\ D_2D_3 \text{ and } B_2B_3,\ D_3D_1 \text{ and } B_3B_1$$

intersect in points of the line $g := \pi \cap \psi$. In particular it follows that $A_iA_j \cap B_iB_j$ is also a point of g: If, for instance, $X := D_1D_2 \cap A_1A_2$ and $Y := D_1D_2 \cap B_1B_2$ then X is the intersection of D_1D_2 and g; similarly $Y = D_1D_2 \cap g$. Hence $X = Y$, and this point can also be described as $X = Y = A_1A_2 \cap B_1B_2 = P_{12}$.

Therefore the points P_{12}, P_{23}, P_{31} are all incident with the line g. □

2.7.2 Corollary. *Let* **P** *be a projective plane. Then the theorem of Desargues holds in* **P** *if and only if* **P** *can be embedded as a plane in a projective space of dimension* ≥ 3.

Proof. If **P** is a plane of a projective space whose dimension is at least 3 then, by the preceding theorem, the theorem of Desargues holds in **P**. □

Remark. In the next chapter we shall see that the converse of **2.7.2** is also true: the Desarguesian projective planes are precisely those projective planes that can be embedded in a higher-dimensional projective space.

2.8 Application: a communication problem

We suppose a set of users who would like to communicate with each other. An example is participants of a telephone system. It is impossible that two users get in direct contact, but they have to use a network with its switches. Each switch is responsible for a certain number of users and can connect any two of 'its' users. Any connection between two users needs at least one switch. For economic reasons, the number of switches should be as small as possible. Therefore, our first requirement is

– *any two users can be connected using just one switch.*

Since any switch should have some use, the second requirement is simple:

– *each switch connects at least two users.*

If all users were connected by only one switch then there would be a lot of mutual interference; therefore our third requirement reads (seemingly very modest):

– *there are at least two switches.*

Finally, it should be possible to produce the switches cheaply; therefore we require also

– *all switches look 'alike'.*

Clearly, this requirement has to be specified further.

After having formulated the requirements for a communication system, we translate this into geometric language. The crucial notion is that of a linear space.

Definition. A **linear space** is a geometry **L** consisting of points and lines such that the following three axioms hold.

(L1) Any two distinct points of **L** are incident with precisely one line.

(L2) On any line of **L** there are at least two points.

(L3) **L** has at least two lines.

Example. Figure 2.9 shows all linear spaces with at most five points. (By the way, lines having only two points are not drawn; they can be uniquely reconstructed.)

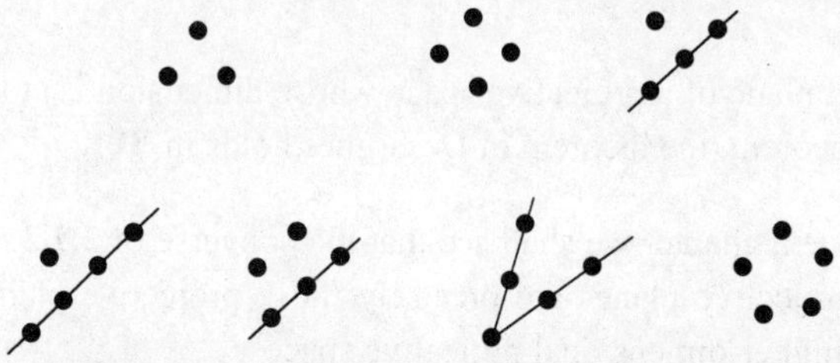

Figure 2.9 The linear spaces with at most five points

Using linear spaces we can translate the communication problem into geometric language (see [Hag71]). In order to do this, we call the users points and the switches lines. Then the first three above requirements translate into the axioms for a linear space. This means: in order to obtain a good communication system we have to find a linear space

- that has a number of lines that is a small as possible, and
- in which all lines look 'alike'.

We do not know yet what 'looking alike' means precisely, but it certainly means that any line has the same number of points (in the language of communication systems: that each switch connects the same number of users).

Now we can construct a first example of a communication system. As linear space we consider the projective plane of order 2 with the points 1, 2, 3, 4, 5, 6, 7 and the lines 123, 145, 167, 246, 257, 347, 356 (see Figure 2.10).

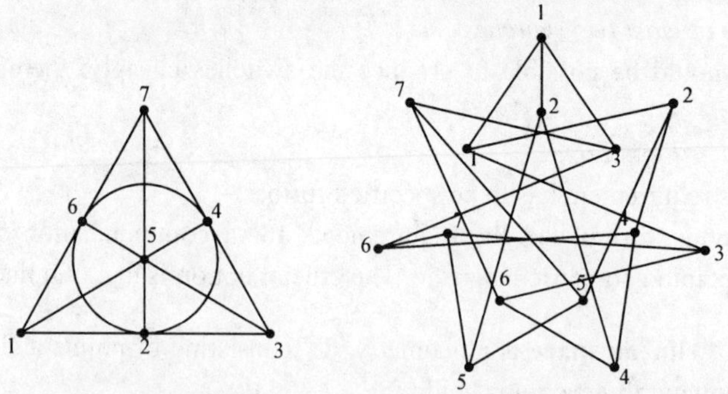

Figure 2.10 A communication system using the projective plane of order 2

It is easy to see that the drawing at the right-hand side of Figure 2.10 is a communication system: the 'outer' points represent the switches (lines), while the 'inner' points represent the users (points). On this first glance, this picture looks wild and complicated, but it really is a communication system, and, as we shall see, a communication system with very few switches.

We shall now analyse the two additional requirements for a linear space, namely that it should contain as few lines as possible and that all lines 'look alike'. In order to fulfil the second of these requirements, we can certainly suppose that all lines are incident with the same number of points. Given these hypotheses it is easy to answer the question, what is the minimum number of switches.

2.8.1 Theorem. *Let* **L** *be a finite linear space with the property that any line has the same number of points. Then we have: the number of lines is at least as big as the number of points; equality holds if and only if* **L** *is a projective plane or* **L** *has only three points.*

Proof. Let v be the number of points and denote by b the number of lines of **L**; moreover, we denote by k the fixed number of points per line.
Claim 1: Each point is on the same number r *of lines.*

Let P be an arbitrary point, and denote by r the number of lines through that point. Since any of the $v-1$ points different from P lies on exactly one of the r lines through P and any of these lines has exactly $k-1$ points different from P, we have

$$v-1 = r\cdot(k-1).$$

In particular, the value $r = (v-1)/(k-1)$ is independent of the choice of the point P.
Claim 2: We have $r \geq k$.

Consider an arbitrary line g. In view of axioms (**L2**) and (**L3**) there is at least one point P outside g. Since P is connected with each of the k points on g by a line and since all these lines are different, P is on at least k lines. Hence $r \geq k$.
Claim 3: We have $rv = bk$.

On both sides there is the number of incident point–line pairs.

Now we are ready to prove the assertion of the theorem. Since $r \geq k$ the third claim implies that $b = vr/k \geq vk/k = v$. Equality holds if and only if $r = k$. The proof of claim 2 shows that this is the case if and only if any line through a point outside a line g intersects this line. This is equivalent to the fact that any two lines meet. Therefore we have $b = v$ if and only if **L** is a projective plane or 'the triangle' (which one could also describe as the 'projective plane of order 1'). □

Remark. The above theorem also remains true without the hypothesis that all lines have the same number of points. This is the important theorem of de Bruijn and Erdös:

2.8.2 Theorem (de Bruijn, Erdös [BrEr48]**).** *Let* **L** *be a finite linear space with* v *points and* b *lines. Then we have* $b \geq v$ *with equality if and only if* **L** *is a projective plane or a 'near-pencil' (see Figure 2.11).*

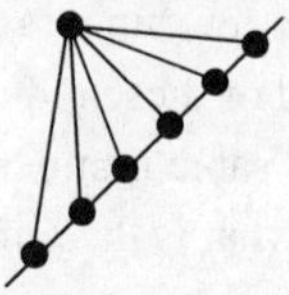

Figure 2.11 A near-pencil

In [BaBe93] the interested reader can find various proofs of this theorem and a general introduction to the theory of linear spaces.

Thus we have got an important partial result: the communication systems with the least number of switches are obtained from projective planes. Now we have to look for those projective planes with the property that all lines 'look alike'. For this we consider a different representation of the projective plane of order 2 and the corresponding communication system. As above, the points are the numbers 1, 2, 3, 4, 5, 6, 7; the lines are now the point sets 124, 235, 346, 457, 561, 672, 713. (These are exactly the sets $\{(1+i,\ 2+i,\ 4+i) \mid i \in \{1, 2, \ldots, 7\}\}$.) Thus the communication system has the convincing structure shown in Figure 2.12.

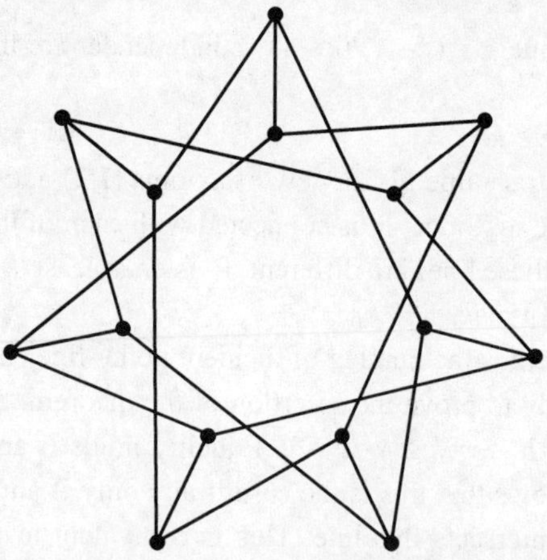

Figure 2.12 A communication system in which all switches look alike

Here, in fact, all switches look alike: each switch sees the participant connected with it in the same way. We can express this in a practical down-to-earth way: the actual connections at all switches (and at all participants) are alike; so one can use the same hardware for any switch.

Which projective planes correspond to those ideal communication systems? These are exactly those projective planes that can be constructed using difference sets.

Definition. Let n be a positive integer. A set $\mathfrak{D}$ of positive integers is called a **difference set** of **order** n, if

- $\mathfrak{D}$ has exactly $n+1$ elements,
- any integer from $\{1, 2, 3, \ldots, n^2+n\}$ can be written in a unique way as

$$d - d' \bmod n^2+n+1 \quad \text{with } d, d' \in \mathfrak{D}.$$

Remark. We use the symbol 'mod' in two ways. Firstly, $a \bmod n$ denotes the least nonnegative remainder when a is divided by n. Secondly, $a \equiv b \bmod n$ ('a is congruent to b modulo n') means that $a-b$ is divisible by n, or in other words, $a \bmod n = b \bmod n$.

Examples of difference sets are easy to find.

(a) The set $\mathfrak{D}_2 = \{1, 2, 4\}$ is a difference set of order 2. In order to verify this we compute all differences modulo 7 of distinct elements of $\mathfrak{D}_2$:

$$1-2 \bmod 7 = \mathbf{6},\ 1-4 \bmod 7 = \mathbf{4},\ 2-4 \bmod 7 = \mathbf{5},$$
$$2-1 \bmod 7 = \mathbf{1},\ 4-1 \bmod 7 = \mathbf{3},\ 4-2 \bmod 7 = \mathbf{2}.$$

Hence all elements of $\{1, 2, \ldots, 6\}$ occur exactly once as a difference.

(b) Using the same method one can check that the sets

$$\mathfrak{D}_3 = \{1, 2, 4, 10\} \quad \text{and} \quad \mathfrak{D}_4 = \{1, 2, 5, 15, 17\}$$

are difference sets of orders 3 and 4.

The importance of difference sets is based on the fact that one can easily construct projective planes with them.

2.8.3 Theorem. *Let $\mathfrak{D}$ be a difference set of order $n \geq 2$. Then the incidence structure $\mathbf{P}(\mathfrak{D})$ defined as follows is a projective plane of order n:*

- *the points of $\mathbf{P}(\mathfrak{D})$ are the integers $0, 1, 2, \ldots, n^2+n$;*
- *the lines of $\mathbf{P}(\mathfrak{D})$ are the sets $\mathfrak{D}+i$ ($i \in \{0, 1, 2, \ldots, n^2+n\}$). Here, $\mathfrak{D}+i$ is defined as follows: let $\mathfrak{D} = \{d_0, d_1, \ldots, d_n\}$, then*

$$\mathfrak{D}+i := \{d_0+i \bmod n^2+n+1, \ldots, d_n+i \bmod n^2+n+1\};$$

- *the incidence of $\mathbf{P}(\mathfrak{D})$ is set-theoretic inclusion.*

Before we present the proof, we illustrate the claim of the theorem by means of an *example*. Let $n = 3$, and let $\mathfrak{D} = \mathfrak{D}_3$. Then the sets $\mathfrak{D} + i$ are as follows:

$$\begin{aligned}
\mathfrak{D}+0 &= \{1,\ 2,\ 4,\ 10\},\\
\mathfrak{D}+1 &= \{2,\ 3,\ 5,\ 11\},\\
\mathfrak{D}+2 &= \{3,\ 4,\ 6,\ 12\},\\
\mathfrak{D}+3 &= \{4,\ 5,\ 7,\ 0\},\\
\mathfrak{D}+4 &= \{5,\ 6,\ 8,\ 1\},\\
\mathfrak{D}+5 &= \{6,\ 7,\ 9,\ 2\},\\
\mathfrak{D}+6 &= \{7,\ 8,\ 10,\ 3\},\\
\mathfrak{D}+7 &= \{8,\ 9,\ 11,\ 4\},\\
\mathfrak{D}+8 &= \{9,\ 10,\ 12,\ 5\},\\
\mathfrak{D}+9 &= \{10,\ 11,\ 0,\ 6\},\\
\mathfrak{D}+10 &= \{11,\ 12,\ 1,\ 7\},\\
\mathfrak{D}+11 &= \{12,\ 0,\ 2,\ 8\},\\
\mathfrak{D}+12 &= \{0,\ 1,\ 3,\ 9\}.
\end{aligned}$$

One can convince oneself without great difficulties that these sets are the lines of a projective plane of order 3.

Proof of Theorem **2.8.3**: First we show axiom 1 of a projective plane: Let x and x' be two distinct points of $\mathbf{P}(\mathfrak{D})$. Then we have

$$\begin{aligned}
&x, x' \in \mathfrak{D} + i\\
\Leftrightarrow\ &\text{there are } d, d' \in \mathfrak{D} \text{ with } x = d + i \bmod n^2 + n + 1, \text{ and}\\
&\quad x' = d' + i \bmod n^2 + n + 1\\
\Rightarrow\ &\text{there are } d, d' \in \mathfrak{D} \text{ with } x - x' \bmod n^2 + n + 1 = d - d' \bmod n^2 + n + 1.
\end{aligned}$$

Thus there is at least one set of the form $\mathfrak{D} + i$ containing x and x'. In order to obtain $\mathfrak{D} + i$ one first determines $d, d' \in \mathfrak{D}$ with

$$d - d' \bmod n^2 + n + 1 = x - x' \bmod n^2 + n + 1,$$

and then computes the integer i with

$$i = x - d \bmod n^2 + n + 1\ (= x' - d' \bmod n^2 + n + 1).$$

Since, by definition, there is exactly one pair $d, d' \in \mathfrak{D}$ with this property, it follows that $\mathfrak{D} + i$ is the uniquely determined line through x and x'.

Now we show axiom 2'. Let $\mathfrak{D} + i$ and $\mathfrak{D} + j$ be two distinct lines. A point x lies in $\mathfrak{D} + i$ and $\mathfrak{D} + j$ if and only if there exist d, d' in $\mathfrak{D}$ with

$$x = d + i \quad \text{and} \quad x = d' + j.$$

From this we see how to proceed in order to find x: First one determines the elements $d, d' \in D$ with $d - d' \bmod n^2 + n + 1 = j - i \bmod n^2 + n + 1$. Then it follows that

$$x := d + i \bmod n^2 + n + 1 \ (= d' + j \bmod n^2 + n + 1)$$

is a common point of $\mathcal{D} + i$ and $\mathcal{D} + j$.

Since $\mathbf{P}(\mathcal{D})$ has exactly $n^2 + n + 1$ points and any line is incident with precisely $n + 1 \geq 3$ points, the nondegeneracy axioms 3 and 4 are satisfied. □

The question of which projective planes can be constructed using difference sets is one of the most important and most difficult questions in finite geometry. It is known that any finite Desarguesian projective plane can be obtained from a difference set ([Sing38]). It is conjectured that only the Desarguesian planes stem from difference sets.

Inside geometry, difference sets are very important for two reasons. Firstly, difference sets provide an extremely economical representation of geometries: in order to describe a projective plane of order n, one needs only $n + 1$ integers. Secondly, difference sets are closely connected with certain interesting collineations. This will be explained in the following.

2.8.4 Theorem. *Let* $\mathbf{P} = \mathbf{P}(\mathcal{D})$ *be a projective plane that is constructed using a difference set* $\mathcal{D}$ *of order* n. *Then the map* σ *that is defined by*

$$\sigma: x \mapsto x + 1 \bmod n^2 + n + 1$$

is a collineation of order $n^2 + n + 1$ *of* $\mathbf{P}$. *(The* ***order*** *of a permutation* σ *is the smallest positive integer* t *such that* $\sigma^t = \text{id}$.*) The collineation* σ *cyclically permutes the points (and the lines) of* $\mathbf{P}$. *One calls the group of collineations generated by* σ *a* ***Singer cycle***.

Proof. Clearly, σ acts bijectively on the point set of $\mathbf{P}$. We show that σ maps any three collinear points x, y, z onto collinear points: Since x, y, z are collinear, there is a line $\mathcal{D} + i$ containing these three points. Then, clearly, the points $\sigma(x) = x + 1$, $\sigma(y) = y + 1$, $\sigma(z) = z + 1$ lie on the line $\mathcal{D} + (i + 1)$. Similarly it follows from the fact that $\sigma(x)$, $\sigma(y)$, and $\sigma(z)$ are collinear that x, y, z also lie on a common line.

Finally we show that σ acts bijectively on the set of lines: The line $\mathcal{D} + i$ has as preimage the line $\mathcal{D} + (i - 1)$; hence σ is surjective. Since any two lines $\mathcal{D} + i$ and $\mathcal{D} + j$ are mapped onto $\mathcal{D} + (i + 1)$ and $\mathcal{D} + (j + 1)$, σ is also injective, because from $\mathcal{D} + (i + 1) = \mathcal{D} + (j + 1)$ it follows that $i + 1 = j + 1$, hence $i = j$. □

Interestingly the converse of the above theorem is also true.

2.8.5 Theorem. *Let* **P** *be a finite projective plane of order* n *that has a collineation* σ *of order* n^2+n+1 *such that the group of collineations generated by* σ *permutes the points of* **P** *as well as the lines in a cyclic way. Then there exists a difference set* $\mathcal{D}$ *of order* n *such that* $\mathbf{P} = \mathbf{P}(\mathcal{D})$.

Proof. Let Q be an arbitrary point and g an arbitrary line of **P**. Since any point of **P** has the form $\sigma^i(\mathrm{Q})$, there are exactly $n+1$ integers d such that $\sigma^d(\mathrm{Q})$ is a point of g. We define the set $\mathcal{D}$ to consist of exactly those numbers:

$$\mathcal{D} = \{d \mid \sigma^d(\mathrm{Q}) \in \mathrm{g}\}.$$

We claim that $\mathcal{D}$ *is a difference set.* For this we have to show that for any positive integer z with $z \le n^2+n$ there is exactly one pair $d, d' \in \mathcal{D}$ with $z = d' - d$:

Since Q and $\mathrm{Q}' = \sigma^z(\mathrm{Q})$ are distinct points, there is exactly one line h through Q and Q'. Therefore, by hypothesis, there exists a $y \in \{0, 1, \ldots, n^2+n\}$ with

$$h = \sigma^y(\mathrm{g}).$$

Since Q, Q' are incident with h, there are points P, P' on g such that

$$\mathrm{Q} = \sigma^y(\mathrm{P}),\ \ \mathrm{Q}' = \sigma^y(\mathrm{P}').$$

By definition of $\mathcal{D}$ there are $d, d' \in \mathcal{D}$ with

$$\mathrm{P} = \sigma^d(\mathrm{Q}),\ \ \mathrm{P}' = \sigma^{d'}(\mathrm{Q}).$$

This implies on the one hand that

$$\mathrm{Q} = \sigma^y(\mathrm{P}) = \sigma^y(\sigma^d(\mathrm{Q})) = \sigma^{y+d}(\mathrm{Q}),$$

and therefore

$$y + d = n^2 + n + 1.$$

On the other hand it follows that

$$\mathrm{Q}' = \sigma^y(\mathrm{P}') = \sigma^y(\sigma^{d'}(\mathrm{Q})) = \sigma^{y+d'}(\mathrm{Q}) = \sigma^{n^2+n+1-d+d'}(\mathrm{Q}) = \sigma^{d'-d}(\mathrm{Q}).$$

Since $\mathrm{Q}' = \sigma^z(\mathrm{Q})$, we have that $z = d' - d$ with $d, d' \in \mathcal{D}$.

We leave it as an exercise for the reader to show that there is only one such pair (d, d') and that **P** is isomorphic to $\mathbf{P}(\mathcal{D})$ (see exercise 26). □

Remarks. (a) In Theorem **2.8.5** one hypothesis was that the group of collineations generated by σ acts transitively on the points and on the lines. One can weaken this hypothesis by showing that a group of collineations of a finite projective

plane acts transitively on the points if and only if it acts transitively on the lines (Dembowski–Hughes–Parker theorem, see for instance [HuPi73], p. 257).
(b) In Chapter 6 we shall prove the theorem of Singer that any finite Desarguesian projective plane has a collineation σ as described above (see **6.2.2** and exercise 1 of Chapter 6). Then it follows that any finite Desarguesian projective plane can be described using a difference set.

Exercises

1 Draw the theorem of Desargues. Is it possible that the point C lies on the line through the points P_{12}, P_{23}, and P_{31}?

2 Draw the theorem of Pappus.

3 Draw on a sheet of paper two lines g and h that are not parallel, but do not intersect on the sheet. For an arbitrary point X on your sheet, construct, using the theorem of Pappus, the line through X and $g \cap h$.

4 Is the projective plane constructed in exercise 4 of Chapter 1 of the form **P**(V)? [Hint: What would be the field F?]

5 The **theorem of Fano** (Gino Fano (1871–1952)) reads as follows: if the points P_1, P_2, P_3, P_4 of a projective plane form a quadrangle, then the points

$$Q_1 := P_1P_4 \cap P_2P_3,\ Q_2 := P_2P_4 \cap P_1P_3, \text{ and } Q_3 := P_3P_4 \cap P_1P_2$$

are on a common line.
Show: in **P**(V) the theorem of Fano is true if and only if in F the equation $1 + 1 = 0$ holds, i.e. if F has characteristic 2.

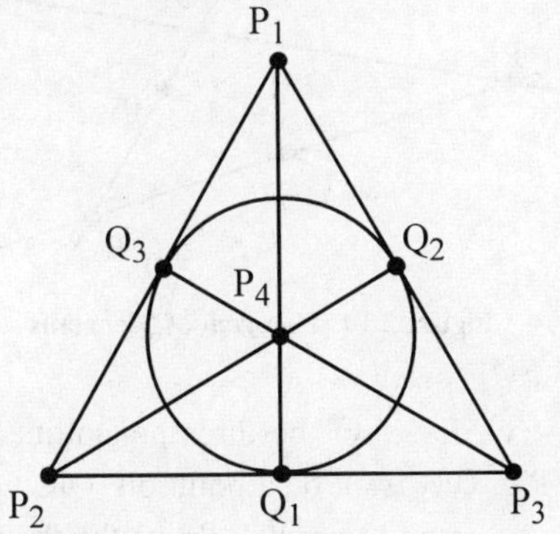

Figure 2.13 The theorem of Fano

[*Historical remark.* Fano studied 'foundations of geometry' – an axiomatic treatment of Euclidean geometry. In this context he of course required not the 'theorem of Fano', but its converse, namely that the points Q_1, Q_2, Q_3 are *not* collinear. His observation was that one cannot prove the corresponding statement, but has to require it axiomatically.]

6 Imagine how difficult it would be to prove the theorem of Pappus using inhomogeneous coordinates.

7 Consider a projective plane **P** represented by homogeneous coordinates. The points are therefore of the form $(a_1 : a_2 : a_3)$, where the elements a_i are taken from a field F. Express the fact that three given points are on a common line in term of the determinant of the 3×3 matrix whose rows are the coordinates of the points.

8 Generalize the preceding exercise to the 3-dimensional (*n*-dimensional) space.

9 In a projective plane described by homogeneous coordinates, by **2.3.4** any line can be represented by a triple $[b_1 : b_2 : b_3]$, $(b_1, b_2, b_3 \in \mathrm{F})$. Express the fact that three lines pass through a common point by using a determinant (cf. exercise 7). Prove your assertion.

10 Generalize the preceding exercise to the 3-dimensional (*n*-dimensional) space.

11 Let $P = \langle u \rangle$, $Q = \langle v \rangle$, $R = \langle w \rangle$ be three noncollinear points of a projective plane **P**(V). Let $P' = \langle v + aw \rangle$ be a point on QR, $Q' = \langle w + bu \rangle$ a point on RP, and $R' = \langle u + cv \rangle$ a point on PQ. Prove the **theorem of Menelaus** (Menelaus of Alexandria, about 100 B. C.):
the points P', Q', R' *are collinear if and only if* $abc = -1$ (see Figure 2.14).

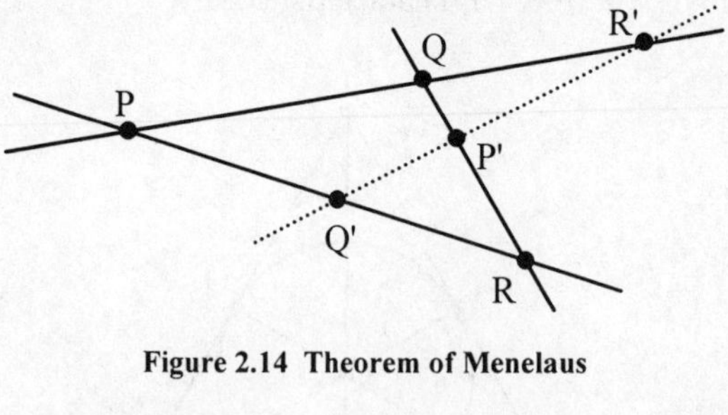

Figure 2.14 Theorem of Menelaus

12 Let $P = \langle u \rangle$, $Q = \langle v \rangle$, $R = \langle w \rangle$ be three noncollinear points of a projective plane **P**(V). Let $P' = \langle v + aw \rangle$ be a point on QR, $Q' = \langle w + bu \rangle$ a point on RP, and $R' = \langle u + cv \rangle$ a point on PQ. Prove the **theorem of Ceva** (Giovanni Ceva (1647 or 1648–1734)):

the lines PP', QQ', RR' *pass through a common point if and only if* $abc = 1$ (see Figure 2.15).

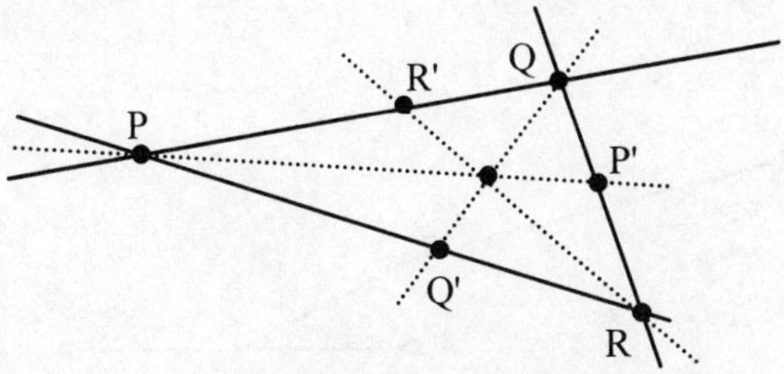

Figure 2.15 Theorem of Ceva

Remark. This theorem was already stated and proved by Al-Mu'taman (who died in 1085), an Islamic king of Zaragoza (in northeastern Spain), in his *Book of Perfection* (cf. [Hog94]).

13 Let V be a vector space over the division ring F.
(a) If $\langle u \rangle$ is a point on the line $\langle v, w \rangle$ of **P**(V) then there exists a $t \in F$ such that

$$\langle u \rangle = \langle tv + (1-t)w \rangle.$$

(b) If F is commutative then t is uniquely determined.

14 Let V be a $(d+1)$-dimensional vector space over the field F. Let $\mathbf{H}_1$ and $\mathbf{H}_2$ be two hyperplanes of the projective space $\mathbf{P} = \mathbf{P}(V) = PG(d, F)$, and let H_1, H_2 be the subspaces of V with $\mathbf{P}(H_i) = \mathbf{H}_i$. Show the following assertions:
(a) There exists a linear map of V that maps H_1 onto H_2.
(b) There exists a collineation of **P** that maps $\mathbf{H}_1$ onto $\mathbf{H}_2$.
(c) The affine spaces $\mathbf{P} \setminus \mathbf{H}_1$ and $\mathbf{P} \setminus \mathbf{H}_2$ are isomorphic.

15 Let U be a subspace of the vector space V. Show that the 1-dimensional subspace $\langle v + U \rangle$ of the factor space V/U consists of all vectors of the subspace $\langle v \rangle + U = \langle v, U \rangle$ of V.

16 Let V be a $(d+1)$-dimensional vector space, and let $Q = \langle w \rangle$ be a point of **P**(V). Then the quotient geometry $\mathbf{P}(V)/Q$ is isomorphic to $\mathbf{P}(V/\langle w \rangle)$, where $V/\langle w \rangle$ is the factor space of V modulo $\langle w \rangle$.

17 (a) Let P and Q be two points of the Moulton plane that are not connected by an 'unbent' line (then, w.l.o.g., P is on the left-hand side and Q on the right-hand side of the y-axis, and P is 'above' Q). Show that the line of the

Moulton plane connecting P and Q can be constructed as follows (see Figure 2.16):

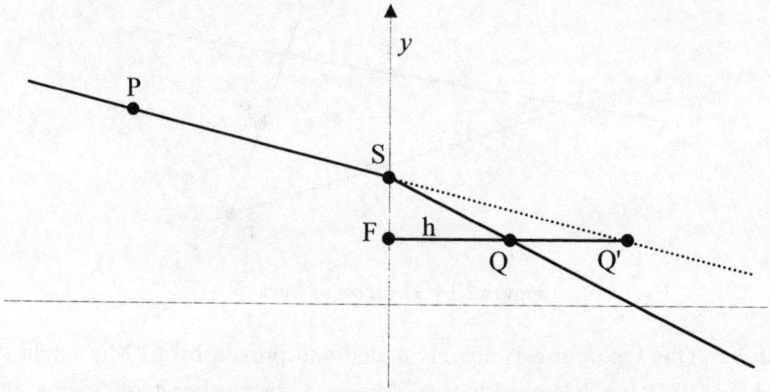

Figure 2.16 Construction of a Moulton line

Let h be the line through Q that is perpendicular to the y-axis, let F be the foot of that line, and let Q' the point with the property that Q is the midpoint of F and Q'. Finally, let S be the intersection of the line PQ' with the y-axis. Then the Moulton line through P and Q is the line that coincides left of the y-axis with PS and right of the y-axis with SQ.

(b) Use (a) in order to show that any two points of the Moulton plane are joined by exactly one line.

18 Show that the Moulton plane satisfies Playfair's parallel postulate.

19 Show that in the Moulton plane the theorem of Pappus does not hold universally.

20 Why does one, in constructing the Moulton plane, not bend all lines at their intersection with the y-axis?

21 What does not work if one tries to generalize the construction of the Moulton plane to the 3-dimensional real space? (One would bend all lines with 'negative slope' at the (y, z)-plane by a factor 2.)

22 Prove Theorem **2.8.2** (theorem of de Bruijn and Erdös).
[This exercise is difficult and needs some time, cf. [BaBe93].]

23 (a) Prove that there is no finite linear space with $b = v + 1$ in which all lines are incident with the same number of points.
(b) Determine all finite linear spaces with $b = v + 1$.

24 Construct examples of difference sets of orders 5, 7, and 8.

25 Draw in a way analogous to Figure 2.12 communication systems having 13, 21, and 31 participants, respectively.

26 Complete the proof of **2.8.5**.

True or false?

☐ Homogeneous coordinates determine a point uniquely.

☐ Inhomogeneous coordinates determine a point uniquely.

☐ Every projective space can be described as a space **P**(V).

☐ Every projective space that is not a projective plane can be described as a space **P**(V).

☐ If the theorem of Desargues holds in **P** it also holds in **P****H**.

☐ Three points in a projective plane that is described by homogeneous coordinates form a triangle if the matrix of its homogeneous coordinates is non-singular.

☐ Four points in a projective plane that is described by homogeneous coordinates form a quadrangle if the matrix of its homogeneous coordinates is non-singular.

☐ A hyperbolic quadric is uniquely determined by any three skew lines contained in it.

☐ In the Moulton plane there is no configuration of points for which the theorem of Desargues holds.

☐ One obtains an example of a non-Desarguesian affine plane, if one bends all lines with negative slope of the real affine plane at the x-axis by a certain factor.

☐ Every finite linear space has a line with just two points.

☐ For any pair (v, b) of positive integers there is a linear space with v points and b lines.

You should know the following notions

P(V), geometric dimension vs. vector space dimension, PG(d, K), PG(d, q), AG(d, K), AG(d, q), theorem of Desargues, theorem of Pappus, theorem of Fano, homogeneous coordinates, inhomogeneous coordinates, 16 point theorem, regulus, hyperbolic quadric, points in general position, normal rational curve, Moulton plane, linear space, difference set.

3 The representation theorems, or good descriptions of projective and affine spaces

This chapter provides the connection between synthetic and analytic geometry. In the second chapter we have introduced the spaces **P**(V) as examples of projective spaces. The question is whether these are nearly all the projective spaces or whether they provide only a small percentage of all projective spaces.

The aim of analytic geometry is to describe 'the geometry' using coordinates. So one has to prove that there are no other geometries. We shall show that this is true if and only if the projective space under consideration is Desarguesian, hence in particular if it is not only a plane.

This chapter belongs to the classical part of projective geometry. However, we will use nearly nothing of it in the next chapters. If you are a quick reader it is sufficient to look only at the representation theorems in Sections **3.4** and **3.5** and later on enjoy the beautiful parts of this chapter.

3.1 Central collineations

We always denote by **P** a projective space of dimension $d \geq 2$.

Our aim is to construct – with the help of the theorem of Desargues – a vector space V such that **P**(V) = **P**. For this we have to construct a division ring F. It will be composed of collineations (automorphisms) of **P**. Therefore our first aim must be to construct – using the theorem of Desargues – lots of collineations of **P**.

We recall: a **collineation** of **P** is a bijective map from the point set and line set of **P** onto themselves that preserves incidence.

Clearly, the set of all collineations of **P** forms a group with respect to composition of maps (see exercise 1).

3.1.1 Lemma. *Let* α *be a collineation of* **P**. *Then for any two distinct points* X, Y *of* **P** *we have*

$$\alpha(XY) = \alpha(X)\alpha(Y).$$

Proof. Since α is a collineation, all points incident with the line $g = XY$ are mapped onto the image g' of the line g. Since $\alpha(X)$ and $\alpha(Y)$ are on g', necessarily $g' = \alpha(X)\alpha(Y)$. □

The collineations that are important for our purposes are not just arbitrary collineations, but central collineations.

Definition. A collineation α of **P** is called a **central collineation** if there is a hyperplane **H** (the **axis** of α) and a point C (the **centre** of α) with the following properties:
- every point X of **H** is a **fixed point** of α (that is $\alpha(X) = X$);
- every line x on C is a **fixed line** of α (that is $\alpha(x) = x$).

We remark that the image of every point on a fixed line g lies on g; but this does not mean that every point on g is a fixed point.

Example. Each reflection in the real affine plane is a central collineation in the corresponding projective closure. The axis g of the reflection σ is also the axis of the corresponding central collineation since every point on g is fixed. What is the centre of σ? In order to see this we observe that each line orthogonal to g is fixed under σ. Hence the point C at infinity that is the intersection of all lines orthogonal to g is the centre of σ.

Similarly one can prove that each point reflection at the point C is a central collineation with centre C and the line at infinity as axis (see exercise 2).

3.1.2 Lemma. *Let* **H** *be a hyperplane and* C *a point of* **P**. *Then the set of central collineations with axis* **H** *and centre* C *is a group with respect to composition of maps.*

Proof. Let Γ be the set of all central collineations with axis **H** and centre C.

Clearly, Γ is nonempty, since the identity is contained in Γ.

Furthermore, Γ is closed under composition since the product of any two elements $\alpha, \beta \in \Gamma$ fixes every point of **H** and every line through C, hence is an element of Γ.

Finally, for each $\alpha \in \Gamma$, the inverse collineation α^{-1} is also an element of Γ: since $\alpha\alpha^{-1} = \text{id}$, α^{-1} must also fix every point on **H** and every line through C.□

3.1.3 Lemma. *Let* α *be a central collineation of* $\mathbf{P}$ *with axis* $\mathbf{H}$ *and centre* C. *Let* $P \neq C$ *be a point not on* $\mathbf{H}$, *and let* $P' = \alpha(P)$ *be the image of* P. *Then* α *is uniquely determined. In particular, the image of each point* X *that is neither on* $\mathbf{H}$ *nor on* PP' (= PC) *satisfies*

$$\alpha(X) = CX \cap FP',$$

where $F = PX \cap \mathbf{H}$ *(see Figure 3.1).*

Proof. In view of the definition of a central collineation the image $X' = \alpha(X)$ of a point X is determined by the following restrictions:
– On the one hand, the line CX is mapped onto itself (as is any line through C); since α is a collineation the point X' is on $\alpha(CX) = CX$.
– On the other hand we consider the point $F := PX \cap \mathbf{H}$. Being a point of the axis $\mathbf{H}$ it is fixed by α. Using **3.1.1** it follows that

$$X' = \alpha(X)\ \mathrm{I}\ \alpha(PX) = \alpha(FP) = \alpha(F)\alpha(P) = FP'.$$

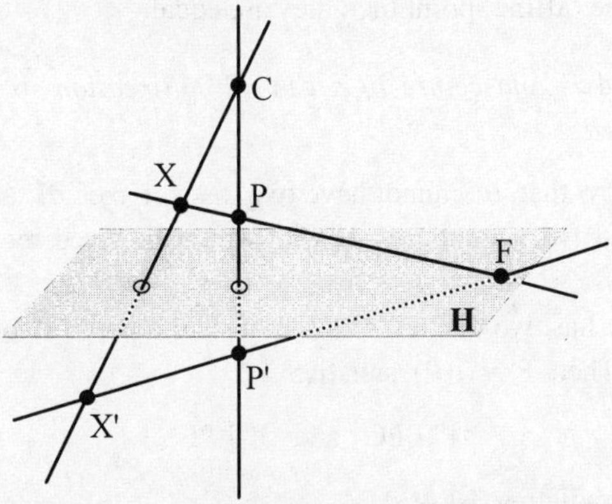

Figure 3.1 Uniqueness of a central collineation

Since X is not on PP', F is not on CX. Hence X' is the intersection of the two distinct lines CX and FP'. Thus X' is uniquely determined.

It now follows that the image of each point Y on CP is also uniquely determined: Replace (P, P') by a pair (R, R') with $R' \neq R$ and $R \notin CP$. Then it follows that $Y' = CY \cap F^*R'$ (with $F^* = RY \cap \mathbf{H}$) is uniquely determined. □

3.1.4 Corollary (uniqueness of central collineations). *Let* α *be a central collineation of* $\mathbf{P}$ *with axis* $\mathbf{H}$ *and centre* C *that is not the identity. Then we have:*
(a) *If* P *is a point* $\neq C$ *not on* $\mathbf{H}$ *then* P *is not fixed by* α.
(b) *The central collineation* α *is uniquely determined by one pair* $(P, \alpha(P))$ *with* $P \neq \alpha(P)$.

Proof. (a) Assume that the point P is fixed by α. We show that then each point X is fixed by α. First, let X not be on the line CP. Then, using the notation of **3.1.3** we have that $\alpha(X) = CX \cap FP' = CX \cap FP = X$ since X is on FP.

Using a (fixed) point X_0 outside PC we now see that each point on PC is also fixed by α. Hence α is the identity, contradicting the hypothesis.

(b) follows directly from **3.1.3**. □

Remark. The above corollary is extremely useful, and we shall often apply it.

In fact, we shall often have the following situation (see for instance **3.2**): We consider central collineations with the same axis **H** such that their respective centres lie on **H** (we shall call those central collineations 'translations'). In order to show that two such central collineations α and β are equal, we are not forced to show for every point X that $\alpha(X) = \beta(X)$, but it suffices to show this for just one point $X_0 \notin \mathbf{H}$. (If $X_0' := \alpha(X_0) = \beta(X_0)$ then the centre of α and β is the point $X_0X_0' \cap \mathbf{H}$, and the desired equality follows from **3.1.4**.) Shortly: if translations agree on one 'affine' point then they are equal.

3.1.5 Corollary. *Axis and centre of a central collineation* $\alpha \neq \mathrm{id}$ *of* **P** *are uniquely determined.*

Proof. First we show that α cannot have two distinct axes **H** and **H'**. Otherwise any of the (at least two) points of $\mathbf{H}'\setminus\mathbf{H}$ would be fixed by α, contradicting **3.1.4**(a).

Assume that α has two centres C, C'. Consider a point P outside the axis **H** and the line CC'. Then $P' = \alpha(P)$ satisfies

$$P' \mathrm{I}\, PC \quad \text{and} \quad P' \mathrm{I}\, PC',$$

hence $P' = P$ contradicting **3.1.4**(a). □

Definition. Let $\alpha \neq \mathrm{id}$ be a central collineation of a projective space **P**. We call α an **elation** if its centre is incident with its axis and a **homology** if centre and axis are not incident. The identity is considered as homology and as elation.

3.1.6 Lemma. *Let* **P** *be a projective space, and let* $\alpha \neq \mathrm{id}$ *be a central collineation with centre* C *and axis* **H**. *If* **U** *is a subspace of* **P** *with* $C \in \mathbf{U}$ *but* $\mathbf{U} \not\subset \mathbf{H}$ *then the restriction of* α *to* **U** *is a central collineation, which is not the identity.*

Proof. Since $C \in \mathbf{U}$, the subspace **U** is mapped by α onto itself. Hence the restriction α' of α to **U** is a collineation of **U**. Since α' has the point C as its centre and $\mathbf{U} \cap \mathbf{H}$ as its axis it is a central collineation. Since $\mathbf{U} \not\subset \mathbf{H}$, α' is not the identity. □

Remark. In the situation of Lemma **3.1.6** one says that α **induces** a central collineation of **U**. In **3.1.10** we shall see that each central collineation of a subspace is induced by a central collineation of the whole projective space.

For our purposes the theorem of Baer (see **3.1.8**) is crucial for it relates the theorem of Desargues to the existence of central collineations.

In order to show this theorem we need the following general lemma on the extendibility of collineations.

3.1.7 Lemma *Let* **P** *be a projective space of dimension of at least* 2, *and let* g_0 *be a line of* **P**. *Let* **P'** *be the rank* 2 *geometry consisting of the points* **P** *that are not on* g_0 *and the lines of* **P** *that are different from* g_0.

Let α *be a collineation of* **P'**. *Then* α *can be uniquely extended to a collineation* α^* *of* **P**. *This collineation* α^* *fixes the line* g_0.

Proof. We first show the following assertion, which will imply everything: *Let* g *and* h *be two lines that intersect* g_0 *in the same point* P. *Then the lines* g' = α(g) *and* h' = α(h) *also intersect each other in a point* P' *of* g_0.

This can be shown as follows. Since α preserves the incidence of **P'** the lines g' and h' also have no point of **P'** in common. We have to distinguish two cases.
Case 1. The order of **P** is bigger than 2.

Then there are a point Q in **P'** and two lines m and n through Q that intersect each of the lines g and h in two distinct points.

The collineation α maps m and n onto two lines m' and n' that pass through a common point Q' of **P'** and intersect each of the lines g' and h' in distinct points. So g' and h' lie in a common plane of **P** (namely in the plane spanned by m' and n'), hence these lines intersect each other in a point **P**. Since they have no point of **P'** in common, they must intersect in a point g_0.
Case 2. The order of **P** is 2.

The critical situation is when g_0, g, and h lie in a common plane; for then these lines cover all points of that plane, and we cannot find a point Q as above.

However, if g, h, and g_0 are not in a common plane then there is a point Q and there exist two lines m, n through Q that intersect each of the lines g and h in distinct points. We may proceed as in the first case.

But if g, h, and g_0 lie in a common plane then we again have to distinguish two cases. If **P** is only a plane then one can directly verify the assertion (see exercise 13). If **P** is not only a plane then there exists a line l through the point P $= g \cap h \cap g_0$ that is not contained in the plane spanned by g and h. Then we apply our results obtained so far to the pairs (l, g) and (l, h): Since l, g, and g_0 are not in a common plane, the lines l' and g' pass through a common point of

g_0. Similarly, l' and h' contain a common point of g_0. These points coincide since both are equal to the intersection of l and g_0. Therefore, g' and h' also pass through the same point of g_0.

Thus we have proved our claim.

From this we infer the following: For a point P on g_0 let $\mathcal{G}_P$ be the set of lines of **P'** through P. Then there is a point P' on g_0 such that all images of the lines of $\mathcal{G}_P$ pass through P'. In other words, $\alpha(\mathcal{G}_P) = (\mathcal{G}_{P'})$.

Now we define α^* in such a way that the restriction of α^* to **P'** is α, that $\alpha(g_0) = g_0$, and that for each point P on g_0 we have $\alpha^*(P) = P'$. Then the above considerations imply not only that α^* is the only possible extension of α to **P**, but also that the map α^* is in fact a collineation. □

Now we are ready to prove the theorem of Baer.

3.1.8 Theorem (existence of central collineation, [Baer42]**).** *If in the projective space* **P** *the theorem of Desargues is valid we have: if* **H** *is a hyperplane and* C, P, P' *are distinct collinear points of* **P** *with* P, P' $\notin$ **H**, *then there is precisely one central collineation of* **P** *with axis* **H** *and centre* C *mapping* P *onto* P'.

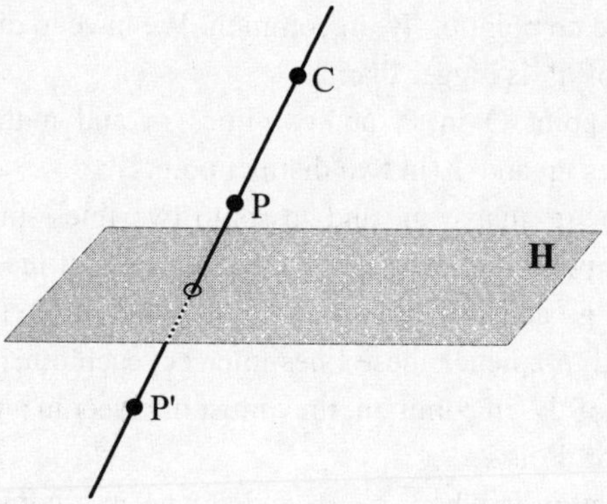

Figure 3.2 Existence of a central collineation

Proof. The *uniqueness* of the central collineation has already been proved in **3.1.4**(b). Thus, the central part is the *existence* of the central collineation. In essence, the theorem of Baer reads as follows: *in a Desarguesian projective space there exist 'all possible' central collineations.*

In order to show the existence of the central collineation in question we proceed as follows. We consider the rank 2 geometry **P'** that consists of the points of **P** not on CP and the lines of **P** that are different from CP. We shall define a map α and show that this is a collineation of **P'**. Then Lemma **3.1.7** implies that α can be extended to a collineation of **P**.

Definition of the map α: Each point of **H** is fixed by α. For a point X outside **H** define $X' := \alpha(X) := CX \cap FP'$, where $F = XP \cap$ **H**.

The main part of the proof consists in showing that α is a collineation of **P'**. For this we first show that α is a bijective map of the point set of **P'** onto itself.

To show that two different points X_1 and X_2 have different images we may suppose that they are not on **H**. If X_1, X_2, and C are collinear then the points $F_1 := X_1P \cap$ **H** and $F_2 := X_2P \cap$ **H** are distinct. Hence $\alpha(X_1)$ and $\alpha(X_2)$ are two distinct points on the line through C, X_1, X_2. If X_1, X_2, and C are not collinear then $\alpha(X_1)$ and $\alpha(X_2)$ are distinct since $\alpha(X_1)$ is on CX_1 and $\alpha(X_2)$ is on CX_2. Hence α is injective.

The map α is surjective since the point $Y_0 := CY \cap FP$, where $F := YP' \cap$ **H** is a preimage of Y.

We now show that any three collinear points are mapped onto collinear points; then α is a collineation (see exercise 26 in Chapter 1). For this we first prove the following claim:

for any two points X *and* Y *of* **P'**, *the lines* XY *and* X'Y' *intersect each other in a point of the axis* **H**.

We may suppose that X, Y $\notin$ **H**. In order to verify the claim, we consider once more the construction of X' and Y':

First we deal with the 'trivial' case that P, X, Y are collinear (see Figure 3.3). Then X' and Y' are also on the line through $F := PX \cap$ **H** and P'; hence P', X', Y' are collinear. In particular, XY = PX and X'Y' = X'P' intersect in the point F of **H**.

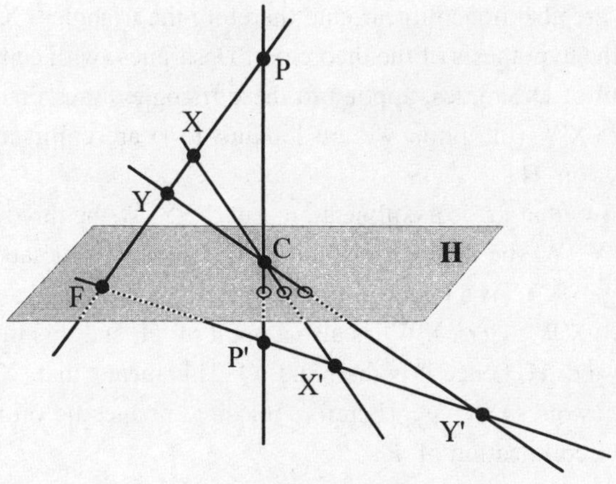

Figure 3.3 Construction of a central collineation

Now we study the 'general' case where X, Y, and P are not collinear (see Figure 3.4). Then the points $F_1 := PX \cap \mathbf{H}$ and $F_2 := PY \cap \mathbf{H}$ are distinct. Since P' is not a fixed point it does not lie in **H**, thus the lines F_1P' and F_2P' are distinct.

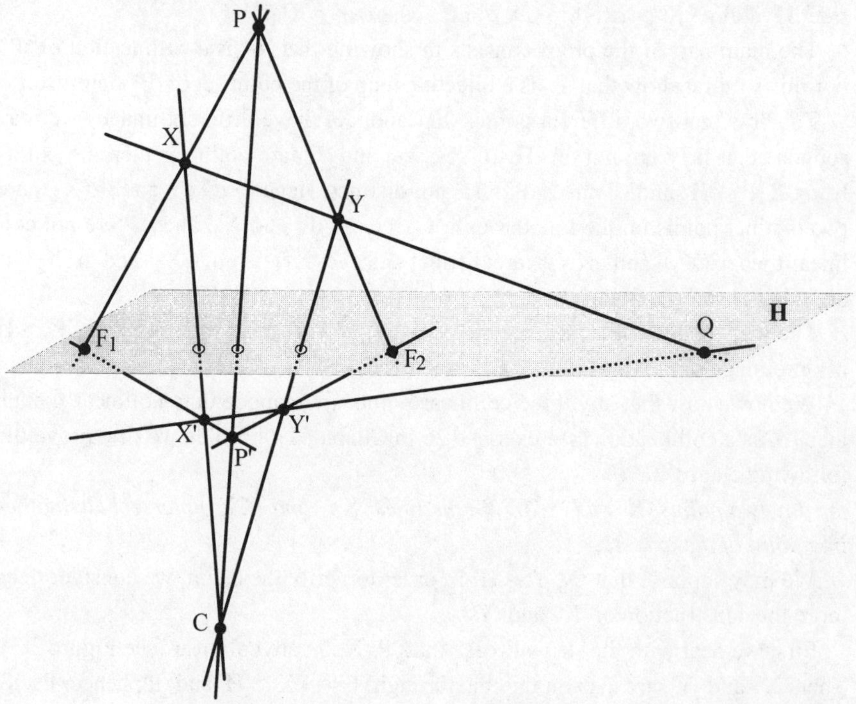

Figure 3.4 Application of the theorem of Desargues

So X', Y', P' are also noncollinear, and therefore the triangles X, Y, P and X', Y', P' satisfy the hypothesis of the theorem of Desargues (with centre C).

The theorem of Desargues, applied to these triangles, says that the points F_1, F_2, and $XY \cap X'Y'$ (the point we are looking for!) are collinear. In particular, $XY \cap X'Y'$ lies on **H**.

Now it follows that α is a collineation: Let X, Y, W be three points on a line g, and let X', Y', W' be their images under α. In view of the above claim, $Q := XY \cap X'Y' = g \cap X'Y'$ is a point of **H**. Moreover, Y' lies on $g' := QX'$.

Now $XW \cap X'W' = g \cap X'W'$ is also a point of **H**, and this must be the intersection of g and **H**, hence it is the point Q. This means that $X'W' = X'Q = g'$. So W' also lies on $QX' = g'$. Therefore all three images lie on a common line, and thus α is a collineation of **P'**.

By Lemma **3.1.7** we can extend α to a collineation α^* of **P**. It remains to show that α^* is a central collineation. By construction, **H** is an axis of α^*. Why is C the centre of α^*? Must α^* necessarily have a centre? The next lemma answers these questions. □

Remark. The above theorem goes back in its general form to R. Baer [Baer46], but was in essence known earlier; for instance, the case $d = 3$ can be found in Veblen and Young ([VeYo16], §29, Theorem 11).

3.1.9 Lemma. *Let* α *be a collineation of* **P** *such that there is a hyperplane* **H** *with the property that each point of* **H** *is fixed by* α. *Then there exists a point* C *of* **P** *such that each line through* C *is fixed by* α. *Shortly: each* ***axial*** *collineation is* ***central***.

Proof. If there is a point $C \notin \mathbf{H}$ with $\alpha(C) = C$ then C is a centre: for each line through C can be written as CP with $P \in \mathbf{H}$, and it follows that

$$\alpha(CP) = \alpha(C)\alpha(P) = CP.$$

Now we consider the case that no point outside **H** is fixed by α. In order to find the centre we consider an arbitrary point $P \notin \mathbf{H}$. *Then the line* $P\alpha(P)$ *is fixed by* α. For let $C := P\alpha(P) \cap \mathbf{H}$. Then

$$\alpha(P\alpha(P)) = \alpha(PC) = \alpha(P)\alpha(C) = \alpha(P)C = \alpha(P)P.$$

We claim that the point C considered above is the centre of α. For this, we have to show that each line g through C is fixed by α. We may suppose that g is not in **H**.

Claim: For each point $Q \notin \mathbf{H}$, $Q \notin P\alpha(P)$ *the line* $Q\alpha(Q)$ *passes through the point* $C := P\alpha(P) \cap \mathbf{H}$.

For let $S := PQ \cap \mathbf{H}$. Then

$$S = \alpha(S)\ \mathrm{I}\ \alpha(PQ) = \alpha(P)\alpha(Q).$$

Hence the points S, P, Q, $\alpha(P)$, $\alpha(Q)$ are contained in a common plane π. Therefore, the lines $Q\alpha(Q)$ and $P\alpha(P)$ intersect in some point X of π. Since the lines $Q\alpha(Q)$ and $P\alpha(P)$ are fixed by α, the point X satisfies

$$\alpha(X) = \alpha(P\alpha(P)) \cap \alpha(Q\alpha(Q)) = P\alpha(P) \cap Q\alpha(Q) = X.$$

Therefore X lies in **H**, thus it must coincide with the point $P\alpha(P) \cap \mathbf{H} = C$.

Hence all lines of the form $Q\alpha(Q)$ pass through C. Thus each line through C is fixed by α. □

A simple application of the theorem of Baer is the converse of Lemma **3.1.6**.

3.1.10 Theorem. *Let* **P** *be a Desarguesian projective space of dimension* $d \geq 2$. *Then each central collineation* α^* *of a subspace* **U** *of* **P** *is induced by a central collineation of* **P**. *More precisely, for each point* $P \notin \mathbf{U}$ *there is a central collineation of* **P** *whose axis passes through* P *and induces* α^*.

Proof. Let C be the centre and $\mathbf{H}^*$ the axis of α^*. Let **H** be a hyperplane of **P** through P such that $\mathbf{H} \cap \mathbf{U} = \mathbf{H}^*$.

Consider an arbitrary point Q of **U** with $Q \neq C$ and $Q \notin \mathbf{H}^*$; let $Q' = \alpha^*(Q)$. By the theorem of Baer there is a central collineation α of **P** with centre C and axis **H** that maps Q onto Q'.

Let α' be the central collineation of **U** that is induced by α (cf. Lemma **3.1.6**). Since α' and α^* share the same centre and the same axis and map Q onto Q' it follows that $\alpha' = \alpha^*$. Hence α^* $(= \alpha')$ is induced by the central collineation α of **P**. □

3.2 The group of translations

In this section we are carefully approaching the vector space that coordinatizes **P**. First we shall study a group which later on will turn out to be the additive group of the vector space we aim for.

One basic idea is to consider only an affine space. The main work will be to show that any Desarguesian affine space $\mathbf{A} = \mathbf{P}\backslash\mathbf{H}$ can be coordinatized by a vector space. From this, the coordinatization of projective spaces will follow relatively easily.

The following theorem says that one does not lose much information if one considers only affine spaces.

3.2.1 Theorem. *Let* α *be a collineation of the affine space* $\mathbf{A} = \mathbf{P}\backslash\mathbf{H}$. *If the order of* **P** *is greater than* 2 *then there is precisely one collineation of* **P** *such that its restriction to* **A** *is* α. *Shortly,* α *has precisely one projective extension.*

Proof. First we show that α has at most one extension to a collineation of **P**: Let α^* and α^+ be extensions of α. Then $\beta := \alpha^+\alpha^{*-1}$ is a collineation of **P**, which fixes each point and each line of **A**. Since each line of **A** and the hyperplane at infinity are fixed, β also fixes each point at infinity. It follows that $\beta = \mathrm{id}$, hence $\alpha^* = \alpha^+$.

The existence is more difficult to show. We first prove the following
Claim: If g *and* h *are parallel lines then* α(g) *and* α(h) *are also parallel.*

For we may suppose that $g \neq h$. Then the images of g and h also have no point in common – but they could be skew! Therefore we have to argue a bit more subtly. Since g and h are parallel they are contained in some plane π. We have to show that the images of g and h also lie in a common plane.

Since the order of **P** is greater than 2 there is a point P in π that is neither on g nor on h. Let l and m be two lines in π through P that are not parallel to g. Then the lines l and m both intersect g and h.

By α the lines l and m are mapped onto two lines $\alpha(l)$ and $\alpha(m)$ that intersect each other and intersect the images $\alpha(g)$ and $\alpha(h)$ of g and h. Thus $\alpha(g)$ and $\alpha(h)$ lie in the plane $\langle \alpha(l), \alpha(m) \rangle$ and are therefore parallel.

We now define an extension α^* of α in **P** as follows. For $P \in \mathbf{A} = \mathbf{P} \setminus \mathbf{H}$ define $\alpha^*(P) = \alpha(P)$. For $P \in \mathbf{H}$ let g be a line through P that is not in **H**. We define $\alpha^*(P) = \alpha(g) \cap \mathbf{H}$. Since g is an affine line $\alpha(g)$ is welldefined.

We have to show that α^* is welldefined: If g and h are two lines through P that are not in **H** then they are parallel as lines of **A**. By our claim, $\alpha(g)$ and $\alpha(h)$ are also parallel, hence they intersect each other in a point of **H**, namely in $\alpha^*(P)$. Thus α^* is welldefined.

The map α^* acts bijectively on the point set of **H** since α is bijective on the line set of **A**.

In order to show that α^* is a collineation it is sufficient to show that α^* maps collinear points onto collinear points. This is clear for points of **A**, and also if only one of the points lies in **H**. Let P_1, P_2, P_3 be three collinear points in **H**. Consider an arbitrary point $X \in \mathbf{A}$. Then $\pi = \langle X, P_1, P_2, P_3 \rangle$ is a plane. By definition of α^*, $\alpha^*(\pi)$ is a plane that intersects **H** in a line containing the points $\alpha^*(P_1), \alpha^*(P_2), \alpha^*(P_3)$. Therefore these points are collinear. □

Remark. The existence part of **3.2.1** does not remain true if the order of **P** is 2; see exercises 13 and 14. However, if a collineation also preserves the planes of **A**, it can be extended to a collineation of **P** (see exercise 15).

For the remainder of this chapter we suppose that **P** is Desarguesian.

Furthermore, we distinguish a hyperplane **H**. Then the theorem of Desargues also holds in the affine space $\mathbf{A} = \mathbf{P} \setminus \mathbf{H}$.

Definition. We denote the set of all central collineations with axis **H** and centre on **H** by $T(\mathbf{H})$. (In the affine space $\mathbf{A} = \mathbf{P} \setminus \mathbf{H}$ the elements of $T(\mathbf{H})$ are the **translations**; this is the reason for the letter T.)

3.2.2 Theorem. *For each hyperplane* **H** *of* **P**, T(**H**) *is an abelian group, which acts* ***sharply transitively (regularly)*** *on the points of* **P****H**. *This means: for any two points* P, Q *of* **P****H** *there is precisely one* $\alpha \in$ T(**H**) *such that* $\alpha(P) = Q$. *In particular, the identity is the only element of* T(**H**) *that fixes a point of* **P****H**.

Proof. First we show that T(**H**) acts transitively on the points of **P****H**. Let P and Q be two points of **P****H**, w.l.o.g. $P \neq Q$. If P is mapped by a translation onto Q then the centre of this translation is the point $C := PQ \cap$ **H**. Since by the theorem of Baer (**3.1.8**) there is precisely one central collineation with axis **H** and centre C mapping P onto Q, the set T(**H**) even acts sharply transitively on the points of **P****H**.

Now we show that T(**H**) is a group. We know that the set of all collineations of **P** is a group with respect to composition of maps (see exercise 1). Thus we have only to show that T(**H**) is a subgroup of the group of all collineations.

Since T(**H**) is nonempty, we only have to show that for any two elements α and β of T(**H**) the collineations α^{-1} and $\alpha\beta$ are also in T(**H**). Let α and β be arbitrary elements of T(**H**). Then each point of **H** is fixed by α^{-1} and by $\alpha\beta$. So, by **3.1.9**, α^{-1} and $\alpha\beta$ are central collineations. W.l.o.g. we may suppose that $\alpha \neq \mathrm{id}$. Then α fixes no point of **P****H**, hence α^{-1} also fixes no point of **P****H**. In particular, the centre of α^{-1} is in **H**, so $\alpha^{-1} \in$ T(**H**). We may suppose that $\alpha\beta \neq \mathrm{id}$. We have to show that the centre of $\alpha\beta$ is on **H**: Assume that some point $P \notin$ **H** is fixed by $\alpha\beta$. Then $\beta(P) = \alpha^{-1}(P)$, hence by **3.1.4**(b) also $\beta = \alpha^{-1}$, and so $\alpha\beta = \mathrm{id}$.

Finally we show that T(**H**) is abelian. Let α_1, α_2 be arbitrary elements of T(**H**). We have to show that $\alpha_1\alpha_2 = \alpha_2\alpha_1$. Observe that in view of **3.1.4**(b) we have to verify for only one point $X \notin$ **H** that $\alpha_1\alpha_2(X) = \alpha_2\alpha_1(X)$.

We may suppose w.l.o.g. that α_1 and α_2 are not the identity. Let C_1 and C_2 be the centres of α_1 and α_2.

Case 1. $C_1 \neq C_2$.

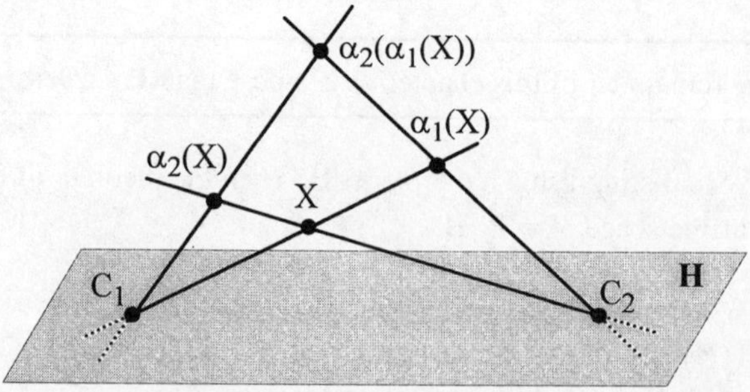

Figure 3.5 Commutativity of translations

Let X be an arbitrary point of **P****H**. Consider the points X, $\alpha_2(X)$, $\alpha_1(X)$, and $\alpha_2(\alpha_1(X))$. Since $\alpha_2(\alpha_1(X))$ is the image of $\alpha_1(X)$ under α_2 the point $\alpha_2(\alpha_1(X))$ lies on the line $C_2\alpha_1(X)$. Moreover, we have

$$C_1 = \alpha_2(C_1)\ I\ \alpha_2(X\alpha_1(X)) = \alpha_2(X)\alpha_2(\alpha_1(X)).$$

Thus the point $\alpha_2(\alpha_1(X))$ is also incident with $C_1\alpha_2(X)$ (see Figure 3.5). Similarly one sees that the point $\alpha_1(\alpha_2(X))$ is incident with $C_2\alpha_1(X)$ as well as with $C_1\alpha_2(X)$. Therefore, $\alpha_1\alpha_2(X) = \alpha_1(\alpha_2(X)) = \alpha_2(\alpha_1(X)) = \alpha_2\alpha_1(X)$. As we have observed above, this implies that $\alpha_1\alpha_2 = \alpha_2\alpha_1$.

Case 2. $C_1 = C_2$.

Consider a point $C_3 \neq C_1$ and an $\alpha_3 \in T(\mathbf{H})$, $\alpha_3 \neq id$, with centre C_3. Since the translations with centre C_1 form a group $\alpha_1\alpha_3$ also has centre $\neq C_1$. By case 1 we get

$$\alpha_2\alpha_3 = \alpha_3\alpha_2 \text{ and } \alpha_2(\alpha_1\alpha_3) = (\alpha_1\alpha_3)\alpha_2.$$

From these results together it follows that

$$(\alpha_1\alpha_2)\alpha_3 = \alpha_1(\alpha_2\alpha_3) = \alpha_1(\alpha_3\alpha_2) = (\alpha_1\alpha_3)\alpha_2 = \alpha_2(\alpha_1\alpha_3) = (\alpha_2\alpha_1)\alpha_3,$$

hence $\alpha_1\alpha_2 = \alpha_2\alpha_1$. □

Definition. We now arbitrarily distinguish a point O of **P****H**. In view of **3.2.2** we can identify each point P of **P****H** with a translation of T(**H**), more precisely with the element $\tau_P \in T(\mathbf{H})$ that maps O onto P. In other words, τ_P is the uniquely determined translation that maps O onto P. For instance, one has $\tau_O = id$ and $\tau_P(O) = P$ for each point P.

This definition enables us to 'add' two points of **P****H**; we define

$$P + Q := \tau_P(Q) = \tau_P(\tau_Q(O)).$$

3.2.3 Theorem. *The set* $\mathcal{P}^*$ *of the points of the affine space* **P****H** *is, with the above defined addition* +, *a group, which is isomorphic to* (T(**H**), ∘).

Proof. We define the map f: $\mathcal{P}^* \to T(\mathbf{H})$ by $f(P) := \tau_P$. In view of **3.2.2**, f is a bijective map. It remains to show that $f(P + Q) = f(P)\circ f(Q)$ for all points $P, Q \in \mathcal{P}^*$. By definition of addition of points we have

$$f(P + Q) = \tau_{P+Q} = \tau_{\tau_P(Q)}.$$

So $f(P + Q)$ is the translation that maps O onto the point $\tau_P(Q)$.

What is the image of O under $f(P)\circ f(Q) = \tau_P\tau_Q$? Well, τ_Q maps O onto Q, and then this point is mapped onto $\tau_P(Q)$. Thus, the translations $f(P + Q)$ and f(P)∘f(Q) map O onto the same point. Hence these translations are equal (cf. the remark after **3.1.4**). In other words,

$$f(P + Q) = f(P) \circ f(Q). \qquad \square$$

In particular we note the useful equation

$$\tau_{P+Q} = \tau_P \tau_Q.$$

By **3.2.3** we see also that each point P has a 'negative point' –P, namely the point $\tau_P^{-1}(O)$. Since τ_P^{-1} is (like τ_P) a translation with centre P* := OP ∩ **H** the point –P lies on the line OP* = OP.

Remark. One can obtain the sum of two points in the following geometric way: For a point P ≠ O of **P****H** we denote by P* the point PO ∩ **H**. Then, if the points P, Q, and O are not collinear, the sum P + Q can be obtained in the following extremely simple way (see Figure 3.6):

$$P + Q := P^*Q \cap PQ^*.$$

This can be seen as follows. Obviously P* is the centre of τ_P, hence P + Q = $\tau_P(Q)$ lies on the line QP*. On the other hand, τ_P maps the line OQ* onto PQ* since Q* is fixed by τ_P. Hence $\tau_P(Q)$ is also incident with PQ*.

The case that O, P, Q are on a common line will be handled in exercise 19.

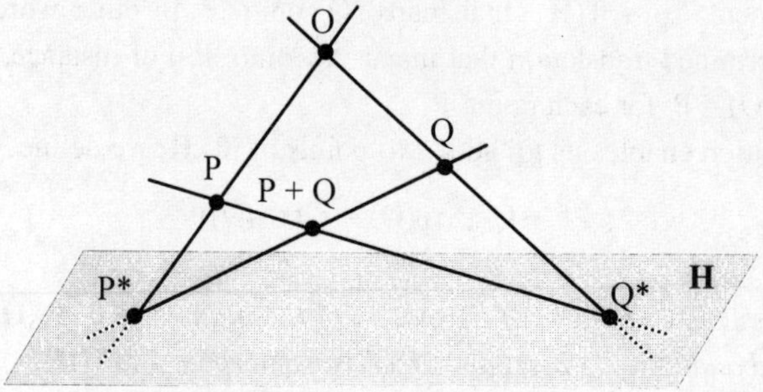

Figure 3.6 Sum of two points

From an affine point of view, the lines OP and Q(P + Q), as well as the lines OQ and P(P + Q), are parallel; thus the points O, P, Q, P + Q form a 'parallelogram'. Therefore we can express the above remark as follows. P + Q is the fourth point that completes O, P, Q to a parallelogram.

Definition. For a point $P \in \mathbf{H}$ let $T(P, \mathbf{H})$ be the set of all central collineations with axis $\mathbf{H}$ and centre P.

Clearly, for each $P \in \mathbf{H}$ the set $T(P, \mathbf{H})$ is an (abelian) subgroup of $T(\mathbf{H})$. Furthermore, according to the above identification we may identify the elements of $T(P, \mathbf{H})$ with the points of $\mathbf{P} \setminus \mathbf{H}$ on the line OP. This means that *the lines through* O *are in a* 1*-to-*1 *correspondence to the subgroups* $T(P, \mathbf{H})$ *of* $T(\mathbf{H})$.

With the help of $T(\mathbf{H})$ the affine space $\mathbf{A} = \mathbf{P} \setminus \mathbf{H}$ can already be described very well. If we knew that $T(\mathbf{H})$ is not only an abelian group, but also a vector space, we would have already the usual description of an affine space. This is the content of the following theorem.

3.2.4 Theorem. *The points of* $\mathbf{A}$ *form an abelian group, which we denote by* T. *The lines through* O *are certain normal subgroups of* T.
The other lines are the cosets with respect to these subgroups on O.

Proof. The first assertion follows from theorem **3.2.3**. In order to describe exactly what we have to prove, we introduce some notation. Let g be a line of $\mathbf{A}$; we define

$$T(g) = \{\tau_P \mid P \mathrm{I} g\}.$$

We have to show the following assertions.
(a) *If* g *is a line through* O *then we have* $T(g) = T(C, \mathbf{H})$, *where* C *is the point at infinity of* g. *Conversely, for any subgroup of* $T(\mathbf{H})$ *of the form* $T(C, \mathbf{H})$ *there is a line* g *through* O *such that* $T(g) = T(C, \mathbf{H})$.
(b) *If* g *is a line not through* O *then*

$$T(g) = \tau_P \circ T(C, \mathbf{H}),$$

where P *is an arbitrary point of* $\mathbf{A}$ *on* g *and* C *is the point at infinity of* g. *Conversely, for each point* P *and each point* C *at infinity there is a line* g *such that*

$$T(g) = \tau_P \circ T(C, \mathbf{H}).$$

These assertions are not difficult to prove. (a) Since the translations that fix g are exactly the translations with centre C, it follows that $T(g) = T(C, \mathbf{H})$.
(b) Let g be a line not through O, let P be a point of $\mathbf{A}$ on g, and let $C = g \cap \mathbf{H}$ (see Figure 3.7).
Claim:

$$g = \{P + X \mid X \text{ is an affine point of } OC\} =: P + OC.$$

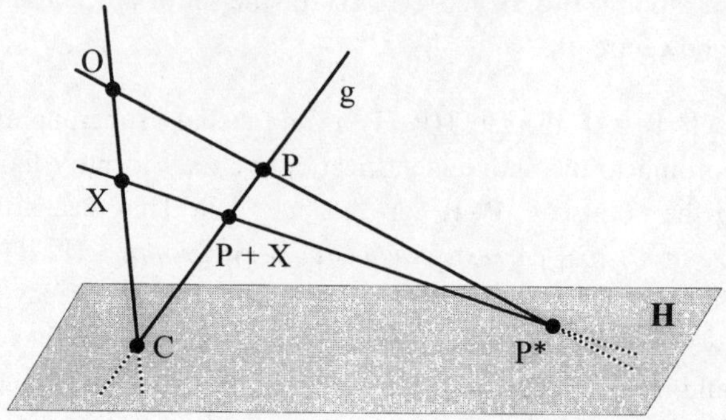

Figure 3.7 $\mathbf{g = P + OC}$

For on the one hand each point of the form $P + X$ $(X \in OC)$ is a point of g since in view of the remark after **3.2.3** $X^* = C$ implies that

$$P + X = P^*X \cap PX^* = P^*X \cap PC = P^*X \cap g \in g.$$

On the other hand, let Q be a point of $\mathbf{A}$ on g. Then the point $X := OC \cap P^*Q$ is a preimage of Q under τ_P.

From this claim it follows by (a) that

$$T(g) = \{\tau_{P+X} \mid X \text{ I } OC\} = \{\tau_P \circ \tau_X \mid X \text{ I } OC\} = \tau_P \circ T(C, \mathbf{H}).$$

If, conversely, P is a point of $\mathbf{A}$ and C a point at infinity then for $g = PC$ we have $T(g) = \tau_P \circ T(C, \mathbf{H})$. □

Remark. In the following we shall use the notation $g = P + OC$, where C is the intersection of g and $\mathbf{H}$.

3.3 The division ring

The aim of this section is to construct from a projective or affine space a division ring that will coordinatize the geometry. So far, we have used only the elations. Now homologies will play a key role. First we define the set that will form the division ring.

Definition. By D_O we denote the set of all central collineations of a projective space $\mathbf{P}$ with axis $\mathbf{H}$ and centre O. The elements of D_O are also called **dilatations** with centre O; hence the notation.

Remarks. 1. It follows directly from the theorem of Baer, **3.1.8,** that D_O is a group, which acts sharply transitively on the points of each line of **P\H** through O that are different from O.
2. What is the image of a line $g = P + OC$ of **P\H** under $\sigma \in D_O$? Since O and C are fixed by σ and the image must contain $\sigma(P)$ we get the following assertion:

$$\sigma(g) = \sigma(P) + OC$$

(see also exercise 23).

First we consider the element of D_O that will play the role of -1.

3.3.1 Lemma. *Let* $\mu: \mathcal{P}^* \to \mathcal{P}^*$ *be the map that is defined by*

$$\mu: X \mapsto -X.$$

This means that μ *maps an arbitrary affine point* X *onto the point* $-X = \tau_X^{-1}(O)$. *Then* μ *(more precisely: its projective extension) is an element of* D_O *(cf.* ***3.2.1****).*

Proof. Clearly, μ is a bijective map. Furthermore, μ fixes the point O. By simple computation one obtains that μ maps the line $g = P + OC$ onto the line $-P + OC$. For for each point X on OC the point $-X$ is also on $OX = OC$, therefore $\mu(P + X) = -(P + X) = -P + (-X)$ is a point of $-P + OC$. Hence μ is a collineation of **A**. By **3.2.1** μ has a projective extension, which we shall also denote by μ. (If the order of **P** is 2, then μ is the identity.)

Since along with P the point $-P$ is also on OC, μ maps any line through O onto itself. Hence, by exercise 8, μ is a central collineation with centre O. Since μ has no affine fixed point except O, the axis of μ must be the hyperplane at infinity. So $\mu \in D_O$. □

Remark. Every point X satisfies

$$X + -X = O.$$

For we have

$$X + -X = \tau_X(-X) = \tau_X(\tau_X^{-1}(O)) = O.$$

The elements of D_O act on the point set $\mathcal{P}^*$ of **A**. Thus the elements of D_O also act on the elements of the group $(\mathcal{P}^*, +)$. The question is, how they act on this group.

3.3.2 Lemma. *Each element of* D_O *is an automorphism of* $(\mathcal{P}^*, +)$. *Moreover, we have that* $\sigma \circ \mu = \mu \circ \sigma$, *that is* $\sigma(-X) = -\sigma(X)$ *for all* $\sigma \in D_O$ *and for all points* X.

Proof. Let σ be an arbitrary element of D_O. Since it is clear that σ acts bijectively on $\mathcal{P}^*$ we must show only that two arbitrary points X and Y satisfy $\sigma(X + Y) = \sigma(X) + \sigma(Y)$ (see Figure 3.8).

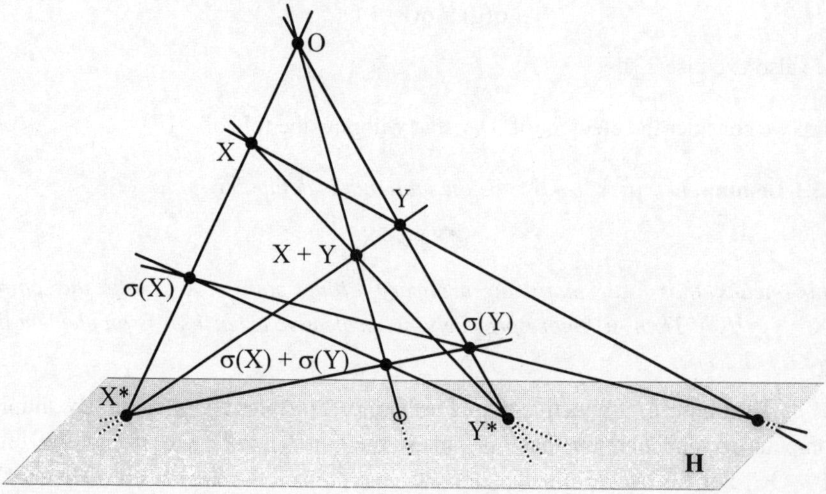

Figure 3.8 The automorphism σ

For this, we may assume that X, Y $\neq$ O.

First we suppose that X, Y, and O are not collinear. Since $X + Y = XY^* \cap X^*Y$ and $\sigma(X) + \sigma(Y) = \sigma(X)\sigma(Y)^* \cap \sigma(X)^*\sigma(Y)$ we must show that

$$\sigma(XY^* \cap X^*Y) = \sigma(X)\sigma(Y)^* \cap \sigma(X)^*\sigma(Y),$$

which, since σ is a collineation, means nothing else than

$$\sigma(X)\sigma(Y^*) \cap \sigma(X^*)\sigma(Y) = \sigma(X)\sigma(Y)^* \cap \sigma(X)^*\sigma(Y).$$

We know that $\sigma(X^*) = X^*$ since X^* is a point of **H**. Similarly we get that $\sigma(Y^*) = Y^*$. Moreover, $\sigma(X)^* = X^*$, since σ moves the point X on the line OX^*. Hence OX and $O\sigma(X)$ have the same point at infinity; in other words, $X^* = \sigma(X)^*$. Since $Y^* = \sigma(Y)^*$ follows similarly, the last equation is true, and therefore the assertion holds.

Next we show that $\sigma(-X) = -\sigma(X)$ for each point $X \neq O$. For this, let P be a point not on the line OX. Since O, X, and $-X$ are collinear $P + X$ is also not on $O(-X)$. Using what we have already proved it follows that

$$\begin{aligned}\sigma(P) &= \sigma(P + X + (-X))\\ &= \sigma(P + X) + \sigma(-X)\\ &= \sigma(P) + \sigma(X) + \sigma(-X),\end{aligned}$$

so

$$O = \sigma(X) + \sigma(-X),$$

thus $\sigma(-X) = -\sigma(X)$.

Now we are also able to show that $\sigma(X + Y) = \sigma(X) + \sigma(Y)$ holds for points X, Y with $O \in XY$. If $Y = -X$ then this is the above statement. So we assume that $Y \neq -X$. Consider a point $P \notin XY$. Then also $-P \notin XY$. Moreover, since $Y \neq X$ we have that O, $X + P$, and $Y - P$ are not collinear (see exercise 20). Using the first two parts of the proof it now follows that

$$\begin{aligned}\sigma(X + Y) &= \sigma(X + P - P + Y)\\ &= \sigma(X + P) + \sigma(Y - P)\\ &= \sigma(X) + \sigma(P) + \sigma(Y) + \sigma(-P)\\ &= \sigma(X) + \sigma(P) + \sigma(Y) - \sigma(P)\\ &= \sigma(X) + \sigma(Y).\end{aligned}$$ □

Now we define the addition in D_O. The main work consists in showing that this operation is closed. This will be done in the following lemma.

3.3.3 Main lemma. *Let* $\sigma_1, \sigma_2 \in D_O$. *We define the map* $\sigma_1 + \sigma_2$ *of* $\mathcal{P}^*$ *in* $\mathcal{P}^*$ *by*

$$(\sigma_1 + \sigma_2)(X) := \sigma_1(X) + \sigma_2(X).$$

If $\sigma_1 + \sigma_2$ *is not the zero map (the map that maps each point of* **A** *onto* O) *then the projective extension of* $\sigma_1 + \sigma_2$ *is an element of* D_O.

Proof. We define $\sigma := \sigma_1 + \sigma_2$. The beginning is easy.

Step 1. We have $\sigma(O) = O$ *and* $\sigma(X + Y) = \sigma(X) + \sigma(Y)$ *for all* $X, Y \in \mathcal{P}^*$.

For $\sigma(O) = (\sigma_1 + \sigma_2)(O) = \sigma_1(O) + \sigma_2(O) = O + O = O$.

By **3.3.2** we have that $\sigma_i(X + Y) = \sigma_i(X) + \sigma_i(Y)$ $(i = 1, 2)$; hence

$$\begin{aligned}\sigma(X + Y) &= (\sigma_1 + \sigma_2)(X + Y) = \sigma_1(X + Y) + \sigma_2(X + Y)\\ &= \sigma_1(X) + \sigma_1(Y) + \sigma_2(X) + \sigma_2(Y)\\ &= (\sigma_1 + \sigma_2)(X) + (\sigma_1 + \sigma_2)(Y)\\ &= \sigma(X) + \sigma(Y).\end{aligned}$$

Step 2. If $g = P + OC$ *denotes a line, where* C *is a point on* **H**, *then*

$$\sigma(g) \subseteq \sigma(P) + OC.$$

For suppose first that $P \text{ I } OC$. Then

$$\sigma(P) = \sigma_1(P) + \sigma_2(P) \in OC,$$

since $\sigma_1(P)$ and $\sigma_2(P)$ are points on OC. Hence each point of OC is mapped onto a point of OC.

Now let $g = P + OC$ be an arbitrary line, where P is not on OC. Then, by Step 1,

$$\begin{aligned}\sigma(g) &= \{\sigma(P + X) \mid X \text{ I } OC\} = \{\sigma(P) + \sigma(X) \mid X \text{ I } OC\} \\ &\subseteq \{\sigma(P) + Y \mid Y \text{ I } OC\}.\end{aligned}$$

Step 3. If σ *is not injective then* σ *is the zero map.*

Assume that σ is not injective. Then there are distinct points X and Y such that $\sigma(X) = \sigma(Y)$. This means that

$$(\sigma_1 + \sigma_2)(X) = (\sigma_1 + \sigma_2)(Y),$$

so

$$\sigma_1(X) + \sigma_2(X) = \sigma_1(Y) + \sigma_2(Y).$$

Since $\mu \in D_O$ and since all elements of D_O are automorphisms of $(\mathcal{P}^*, +)$ it follows that

$$\begin{aligned}\sigma_1(X - Y) &= \sigma_1(X) - \sigma_1(Y) = \sigma_2(Y) - \sigma_2(X) = \sigma_2(Y - X) \\ &= \sigma_2 \circ \mu(X - Y) = \mu \circ \sigma_2(X - Y),\end{aligned}$$

since by **3.3.2** the automorphism μ commutes with σ_2. Since $X - Y \neq O$ this implies $\sigma_1 = \mu \circ \sigma_2$, hence

$$\sigma_1 + \sigma_2 = \mu \circ \sigma_2 + \sigma_2 = 0,$$

where 0 denotes the zero map.

Step 4. If σ *is injective then* $\sigma \in D_O$.

Since σ is injective we have $\sigma(X) \neq \sigma(O)$ for each point $X \neq O$. We consider an arbitrary point X_0 with $X_0 \neq O$.

Since $\sigma(X_0) \neq O$ Step 2 implies that $\sigma(X_0) \in OX_0$. Hence by the theorem of Baer there is a $\sigma' \in D_O$ such that $\sigma'(X_0) = \sigma(X_0)$.

Claim: We have $\sigma = \sigma'$, *so in particular* $\sigma \in D_O$.

For this we consider an arbitrary point $Y \in \mathcal{P}^*$, where we first suppose that Y is not on OX_0. Then

$$Y = OY \cap [X_0 + (Y - X_0)O],$$

since $Y - X_0 \text{ I } (Y - X_0)O$. Therefore we have that

$$\sigma(Y) \subseteq \sigma(OY) \cap \sigma(X_0 + (Y - X_0)O) \subseteq OY \cap [\sigma(X_0) + (Y - X_0)O].$$

Since, on the other hand, in view of $\sigma'(X_0) = \sigma(X_0)$ it follows that

$$\sigma'(Y) = OY \cap [\sigma'(X_0) + (Y - X_0)O] = OY \cap [\sigma(X_0) + (Y - X_0)O],$$

we get $\sigma(Y) = \sigma'(Y)$.

From this we also obtain the assertion for points on OX_0, if a point Y_0 outside OX_0 plays the role of X_0.

Thus the lemma is proved. □

3.3.4 Theorem. *Let* 0 *be the zero map on* $\mathcal{P}^*$. *On the set* $F = D_O \cup \{0\}$ *we define as above an addition by*

$$(\sigma_1 + \sigma_2)(X) = \sigma_1(X) + \sigma_2(X)$$

and a multiplication as follows:

$$\sigma_1 \cdot \sigma_2 := \begin{cases} \sigma_1 \circ \sigma_2, & \text{if } \sigma_1, \sigma_2 \in D_O, \\ 0, & \text{if } \sigma_1 = 0 \text{ or } \sigma_2 = 0. \end{cases}$$

Then $(F, +, \cdot)$ *is a division ring.*

Proof. First we deal with the *addition.* The fact that F is closed under addition follows from **3.3.3**. The associativity and commutativity of $(F, +)$ can be reduced to the associativity and commutativity of $(\mathcal{P}^*, +)$. The neutral element is 0, and $\mu \circ \sigma$ is the inverse – the negative element – of σ. We shall simply write $-\sigma$ instead of $\mu \circ \sigma$.

Multiplication: Since $(D_O, \circ)$ is a group $(F \setminus \{0\}, \cdot)$ is also a group.

Finally we have to show that the *distributive laws* hold. Let $\sigma_1, \sigma_2, \sigma_3 \in F$. Then for each $X \in \mathcal{P}^*$ we have

$$\begin{aligned}\sigma_1(\sigma_2 + \sigma_3)(X) &= \sigma_1((\sigma_2 + \sigma_3)(X)) = \sigma_1(\sigma_2(X) + \sigma_3(X)) \\ &= \sigma_1\sigma_2(X) + \sigma_1\sigma_3(X) = (\sigma_1\sigma_2 + \sigma_1\sigma_3)(X),\end{aligned}$$

and

$$\begin{aligned}(\sigma_1 + \sigma_2)\sigma_3(X) &= (\sigma_1 + \sigma_2)(\sigma_3(X)) = \sigma_1(\sigma_3(X)) + \sigma_2(\sigma_3(X)) \\ &= \sigma_1\sigma_3(X) + \sigma_2\sigma_3(X).\end{aligned}$$

□

3.3.5 Corollary. *We define a multiplication with scalars on* $\mathcal{P}^*$ *by*

$$\sigma \cdot X := \sigma(X) \text{ for } \sigma \in F, X \in \mathcal{P}^*.$$

Then $\mathcal{P}^*$ *is a vector space over the division ring* F.

Proof. We already know from **3.2.3** that $(\mathcal{P}^*, +)$ is a commutative group. Moreover, F is a division ring, and, by **3.3.2**, $\sigma(O) = O$ and $\sigma(X + Y) = \sigma(X) + \sigma(Y)$ for all $\sigma \in D_O$ and all $X, Y \in \mathcal{P}^*$. Finally, the equation $(\sigma_1 + \sigma_2)(X) = \sigma_1(X) + \sigma_2(X)$ follows from the definition of addition in D_O. □

3.4 The representation theorems

We can already describe the structure of $\mathbf{A} = \mathbf{P}\backslash\mathbf{H}$ very well. We sum up what we have proved in the previous sections:

– The set D_O of all dilatations with centre O is, together with the zero map, a division ring F.

– The set $\mathcal{P}^*$ of the points of **A** is a vector space over F.

3.4.1 First representation theorem for affine spaces. *Let* $\mathbf{A} = \mathbf{P}\backslash\mathbf{H}$ *be an affine space. If the theorem of Desargues is valid in* **A** *then there are a division ring* F *and a vector space* V* *over* F *such that*

– *the points of* **A** *are the elements of* V*, *and*

– *the lines of* **A** *are the cosets of the* 1*-dimensional vector subspaces of* V*.

Proof. Let F be as in **3.3.4**; we define $V^* := \mathcal{P}^*$.

Each line g through O is of the form OP. Since, by the theorem of Baer, D_O acts transitively on the points of g that are different from O and $g \cap \mathbf{H}$, the multiplicative group D_O of the division ring F transforms each element of g different from O onto each other element of g that is different from O. Hence g is the 1-dimensional subspace of V* spanned by P.

Each line g not through O is of the form $g = P + OX$. Since OX is the 1-dimensional subspace $\langle X \rangle$, the line g is the coset $P + \langle X \rangle$.

Conversely, let $\langle X \rangle$ be a 1-dimensional subspace of V*. Then we have $\langle X \rangle = \{\sigma(X) \mid \sigma \in K\} = (OX)$. So the coset $P + \langle X \rangle$ is the line $P + OX$. □

Our original aim was the algebraic description of projective spaces. Now, this offers only small technical difficulties.

3.4.2 First representation theorem for projective spaces. *Let* $\mathbf{P} = (\mathcal{P}, \mathcal{L}, \mathrm{I})$ *be a projective space of dimension at least* 2. *If* $\mathbf{P}$ *is Desarguesian, then there is a vector space* V *over a division ring* F *such that* $\mathbf{P}$ *is isomorphic to* $\mathbf{P}(\mathrm{V})$.

In short, the Desarguesian projective spaces are precisely the geometries $\mathbf{P}(\mathrm{V})$.

Proof. We fix an arbitrary hyperplane H of P. Let F and V* be as in 3.4.1. We define

$$\mathrm{V} := \mathrm{F} \times \mathrm{V}^*.$$

Then V is a vector space over F with elements (a, v) $(a \in \mathrm{F}, \mathrm{v} \in \mathrm{V}^*)$.

We shall define a map α that assigns to each point X of $\mathbf{P}$ a 1-dimensional subspace of V:

– If X is a point of $\mathbf{P}\backslash\mathbf{H}$ then, by **3.4.1**, X is a vector of V*. In this case we define

$$\alpha(\mathrm{X}) := \langle(1, \mathrm{X})\rangle.$$

– If X is a point of $\mathbf{H}$ then we consider the line OX of $\mathbf{P}\backslash\mathbf{H}$. By **3.4.1** this line is a 1-dimensional subspace $\langle \mathrm{v} \rangle$ of V*. We define

$$\alpha(\mathrm{X}) := \langle(0, \mathrm{v})\rangle.$$

Claim 1: α *is bijective.*

Clearly, α is injective.

In order to show that α is also surjective we consider an arbitrary 1-dimensional subspace $\langle(a, \mathrm{v})\rangle$ of V. If $a \neq 0$ then the vector v/a of V* (or the corresponding point of $\mathbf{P}\backslash\mathbf{H}$) is a preimage of $\langle(a, \mathrm{v})\rangle$. If $a = 0$ then the point of $\mathbf{H}$ on the line $\langle \mathrm{v} \rangle$ of $\mathbf{P}\backslash\mathbf{H}$ is a preimage of $\langle(a, \mathrm{v})\rangle$.

Claim 2: The map α *transforms lines of* $\mathbf{P}$ *onto lines of* $\mathbf{P}(\mathrm{V})$.

Let g be an arbitrary line of $\mathbf{P}$.

Case 1. g is a line of $\mathbf{P}\backslash\mathbf{H}$.

Then we have $\mathrm{g} = \mathrm{u} + \langle \mathrm{v} \rangle$ with $\mathrm{u}, \mathrm{v} \in \mathrm{V}^*$, and it follows that

$$\alpha(\mathrm{g}) = \langle(1, \mathrm{u}), (0, \mathrm{v})\rangle.$$

Case 2. g is a line of $\mathbf{H}$.

Let $\mathrm{C}_1, \mathrm{C}_2$ be two points on g. Let $\mathrm{v}_1, \mathrm{v}_2$ be the vectors of V* such that

$$\mathrm{OC}_1 = \langle \mathrm{v}_1 \rangle, \mathrm{OC}_2 = \langle \mathrm{v}_2 \rangle.$$

Then $\langle(0,\mathrm{v}_1), (0,\mathrm{v}_2)\rangle$ is the line $\alpha(\mathrm{g})$.

It follows that α is an isomorphism from $\mathbf{P}$ onto $\mathbf{P}(\mathrm{V})$ (see exercise 27).

Thus the theorem is completely proved. □

The importance of the above theorem can also be seen from the next corollary.

3.4.3 Corollary. *If* $\dim(\mathbf{P}) \geq 3$ *then* $\mathbf{P} = \mathbf{P}(V)$.
In other words, projective spaces that are not only planes are coordinatizable by a division ring.

Proof. **2.7.1** and **3.4.2**. □

3.5 The representation theorems for collineations

In this section we shall determine all collineations of Desarguesian affine and projective spaces. More precisely, we shall describe the collineations in terms of the underlying vector space. As in the first representation theorems we shall begin with affine spaces.

Let $\mathbf{A} = \mathbf{P} \setminus \mathbf{H}$ be an affine space of dimension $d \geq 2$ in which the theorem of Desargues holds. We fix a point O of **A**. By $T = T(\mathbf{H})$ we denote the group of all translations of **A**, and by Γ the set of all collineations of **A**; these are those collineations of **P** that fix the hyperplane **H** as a whole. The set

$$\Gamma_O = \{\alpha \in \Gamma \mid \alpha(O) = O\}$$

of those collineations that fix the point O will play an important role. (Note that in Γ_O there are not only the elements of D_O.)

By the first representation theorem there are a division ring F and a vector space V* such that
– the points of **A** are the elements of V*, and
– the lines of **A** are the cosets of the 1-dimensional subspaces of V*.

We shall use these notations throughout.

3.5.1 Lemma. (a) Γ *is a group (with respect to composition of maps).*
(b) Γ_O *is a subgroup of* Γ.
(c) T *is a normal subgroup of* Γ.
(d) *Each* $\alpha \in \Gamma$ *can be uniquely written as*

$$\alpha = \tau\sigma \quad \textit{with} \quad \tau \in T, \sigma \in \Gamma_O.$$

Proof. (a) We show that Γ is a subgroup of the group of all permutations of the point set of **A**: The identity lies in Γ and the product of any two elements of Γ is again contained in Γ. Hence one has only to show that along with α the collineation α^{-1} also lies in Γ. This will be done in exercise 7.
(b) is obvious.

(c) Let $\tau \in T$ and $\alpha \in \Gamma$ be arbitrary elements. We have to show that $\alpha\tau\alpha^{-1} \in T$. Since τ fixes all points of **H**, for each point P of **H** we have that

$$\tau\alpha^{-1}(P) = \alpha^{-1}(P),$$

so

$$\alpha\tau\alpha^{-1}(P) = \alpha\alpha^{-1}(P) = P.$$

Thus the collineation $\alpha\tau\alpha^{-1}$ has axis **H**. If $\alpha\tau\alpha^{-1}$ fixes a point $Q \notin \mathbf{H}$ then

$$\alpha\tau\alpha^{-1}(Q) = Q, \text{ also } \tau(\alpha^{-1}(Q)) = \alpha^{-1}(Q),$$

so τ (and therefore $\alpha\tau\alpha^{-1}$) is the identity and trivially contained in T. If $\alpha\tau\alpha^{-1}$ fixes no point outside **H** then the centre of $\alpha\tau\alpha^{-1}$ is on **H**, and $\alpha\tau\alpha^{-1}$ is a translation.

(d) *Existence:* Let $\tau = \tau_{\alpha(O)}$ be the translation mapping O onto $\alpha(O)$. Putting $\sigma := \tau^{-1}\alpha$ it follows that

$$\alpha = \tau\sigma,$$

where $\tau \in T$, and $\sigma(O) = \tau^{-1}\alpha(O) = O$, hence $\sigma \in \Gamma_O$.

Uniqueness: Suppose that also $\alpha = \tau'\sigma'$ with $\tau' \in T, \sigma' \in \Gamma_O$. Then

$$\tau'^{-1}\tau = \sigma'\sigma^{-1} \in T \cap \Gamma_O = \{id\};$$

hence $\tau'^{-1}\tau = id = \sigma'\sigma^{-1}$, and so $\tau' = \tau$ and $\sigma' = \sigma$. □

Remark. **3.5.1**(d) is extremely useful since it means that the problem of describing all collineations of **A** splits into two smaller problems, namely describing the elements of T and the elements of Γ_O.

First we shall solve the simpler problem, namely to describe the translations.

3.5.2 Lemma. *Let* $\tau \in T$ *be an arbitrary translation. We choose an arbitrary point* P *of* **A** *and put* $P' := \tau(P)$. *If we consider the points* P *and* P' *as vectors of* V* *then we may describe* τ *as follows:*

$$\tau(X) = X + P' - P \quad \textit{for all points } X \textit{ of } \mathbf{A}.$$

Proof. In view of **3.2.2** (see also exercise 22) we know that the map τ' defined by

$$\tau'(X) := X + P' - P$$

is a translation. Thus τ and τ' are translations which both map the point P onto P'. This implies that $\tau' = \tau$. Hence, the definition of τ' applies to τ. This is the assertion. □

After having described the translations in a satisfactory way we now turn to the elements of Γ_O. We shall show that the collineations in Γ_O are 'semilinear' maps of the vector spaces V^*.

Definition. Let V be a vector space over the division ring F, and let λ be an automorphism of F. A map γ of V into itself is called a **semilinear map** with **accompanying automorphism** λ if for all $v, w \in V$ and for all $a \in F$ we have that

$$\gamma(v + w) = \gamma(v) + \gamma(w),$$

and

$$\gamma(a \cdot v) = \lambda(a) \cdot \gamma(v).$$

Examples. (a) The semilinear maps that have as accompanying automorphism the identity are precisely the linear maps.
(b) Let λ be an automorphism of F. For a basis $\{v_1, \ldots, v_d\}$ of the vector space V we define the map γ_λ of V into itself by

$$\gamma_\lambda(a_1 v_1 + \ldots + a_d v_d) := \lambda(a_1)v_1 + \ldots + \lambda(a_d)v_d.$$

By the simplest checking one can verify that γ_λ can be uniquely extended to a semilinear map with accompanying automorphism λ.

Now we are able to formulate and prove a description of the elements of Γ_O. The first step is the following lemma, which shows the effect of the elements of Γ_O on the sum of two vectors. Since the elements of V^* are the points of $\mathbf{A}$, the collineations of $\mathbf{A}$ also act on V^*.

For the remainder of this section we shall suppose that the order of $\mathbf{A}$ is at least 3.

3.5.3 Lemma. *Let* σ *be an arbitrary element of* Γ_O. *Then for all* $v, w \in V^*$ *we have that*

$$\sigma(v + w) = \sigma(v) + \sigma(w).$$

Proof. W.l.o.g. we may assume that $v, w \neq o$.
Case 1. $\langle v \rangle \neq \langle w \rangle$.

From Section **3.2** we know that

$$v + w = vw^* \cap v^*w.$$

Since σ may be extended to a collineation of the projective space **P** (see **3.2.1**), we have

$$\sigma(v + w) = \sigma(v)\sigma(w^*) \cap \sigma(v^*)\sigma(w).$$

By applying our knowledge to $\sigma(v)$ and $\sigma(w)$ we see that

$$\sigma(v) + \sigma(w) = \sigma(v)\sigma(w)^* \cap \sigma(v)^*\sigma(w).$$

Since σ is a collineation that fixes O and **H** we conclude that

$$\sigma(v)^* = O\sigma(v) \cap \mathbf{H} = \sigma(Ov \cap \mathbf{H}) = \sigma(Ov) \cap \sigma(\mathbf{H}) = \sigma(v^*),$$

and similarly

$$\sigma(w)^* = \sigma(w^*).$$

Putting these together we get

$$\sigma(v + w) = \sigma(v)\sigma(w^*) \cap \sigma(v^*)\sigma(w) = \sigma(v)\sigma(w)^* \cap \sigma(v)^*\sigma(w) = \sigma(v) + \sigma(w).$$

Since $\langle v - w\rangle \neq \langle w\rangle$ we also get

$$\sigma(v) = \sigma(v - w + w) = \sigma(v - w) + \sigma(w),$$

and hence

$$\sigma(v - w) = \sigma(v) - \sigma(w).$$

On the other hand we also have

$$\sigma(v - w) = \sigma(v) + \sigma(-w),$$

and so in particular $\sigma(-w) = -\sigma(w)$. This is true for each vector $w \in V^*$ since there is always a vector $v \in V^*$ with $\langle v\rangle \neq \langle w\rangle$.

Case 2. $\langle v\rangle = \langle w\rangle$.

If $v + w = o$ then

$$\sigma(v + w) = \sigma(o) = o = \sigma(v) - \sigma(v) = \sigma(v) + \sigma(-v) = \sigma(v) + \sigma(w).$$

Thus we may assume that $v + w \neq o$. Since $\dim(\mathbf{P}) \geq 2$ there is a $u \in V^*$ such that $\langle u\rangle \neq \langle v\rangle$. By the first case we now get

$$\begin{aligned}
\sigma(v + w) &= \sigma((v + w + u) - u) \\
&= \sigma(v + w + u) - \sigma(u) && (\text{since } \langle u\rangle \neq \langle v + w + u\rangle) \\
&= \sigma(v) + \sigma(w + u) - \sigma(u) && (\text{since } \langle v\rangle \neq \langle w + u\rangle) \\
&= \sigma(v) + \sigma(w) + \sigma(u) - \sigma(u) && (\text{since } \langle w\rangle \neq \langle u\rangle) \\
&= \sigma(v) + \sigma(w).
\end{aligned}$$

□

3.5.4 Theorem. *Each collineation* $\sigma \in \Gamma_O$ *is a semilinear map of the vector spaces* V^*.

This theorem implies immediately

3.5.5 Corollary *Suppose that* $F = \mathbf{Q}, \mathbf{R}$ *or* $\mathbf{Z}_p$ *(p a prime). Then each* $\sigma \in \Gamma_O$ *is a linear map of* V^* *in itself.*

Proof. The only automorphism of $\mathbf{Q}, \mathbf{R}$, or $\mathbf{Z}_p$ is the identity. □

Proof of Theorem **3.5.4**. This proof contains the main work of this section. Be prepared for a rather long argument.

Lemma **3.5.3** already shows that each element of Γ_O is additive. We have to show that there exists an automorphism λ of F such that for each collineation $\sigma \in \Gamma_O$ and for all $a \in F$ and $X \in V^*$ we have that

$$\sigma(a \cdot X) = \lambda(a) \cdot \sigma(X).$$

For $a \in F^*$ ($:= F \setminus \{0\}$) and a point $X \neq O$ of $\mathbf{P} \setminus \mathbf{H}$ the points O, X, and $a \cdot X$ are collinear. Since σ is a collineation with $\sigma(O) = O$ the points O, $\sigma(X)$, and $\sigma(a \cdot X)$ are also collinear. Hence $\sigma(a \cdot X)$ is a multiple of $\sigma(X)$. Let $\lambda_X(a)$ be the corresponding element of F, this means that

$$\sigma(a \cdot X) = \lambda_X(a) \cdot \sigma(X) \quad (a \in F^*, X \neq O).$$

Claim 1: For all $a \in F^*$ *and all* $X, Y \neq O$ *we have that*

$$\lambda_X(a) = \lambda_Y(a).$$

In order to see this we distinguish two cases.

Case 1. The points O, X, and Y are not collinear.

Then, by definition of λ we have on the one hand

$$\begin{aligned}\sigma(a \cdot (X + Y)) &= \lambda_{X+Y}(a) \cdot \sigma(X + Y) = \lambda_{X+Y}(a) \cdot (\sigma(X) + \sigma(Y)) \\ &= \lambda_{X+Y}(a) \cdot \sigma(X) + \lambda_{X+Y}(a) \cdot \sigma(Y)\end{aligned}$$

and on the other hand

$$\sigma(a \cdot (X + Y)) = \sigma(a \cdot X + a \cdot Y) = \sigma(a \cdot X) + \sigma(a \cdot Y) = \lambda_X(a) \cdot \sigma(X) + \lambda_Y(a) \cdot \sigma(Y).$$

Since σ is a collineation the points O ($= \sigma(O)$), $\sigma(X)$ and $\sigma(Y)$ are also non-collinear; hence $\sigma(X)$ and $\sigma(Y)$ are – considered as vectors – linearly independent. Thus we have

$$\lambda_X(a) = \lambda_{X+Y}(a) = \lambda_Y(a).$$

Case 2. The points O, X, and Y are collinear.

We consider a point Z of $\mathbf{P}\setminus\mathbf{H}$ not on OX. Then, by case 1 we have that

$$\lambda_X(a) = \lambda_Z(a) = \lambda_Y(a).$$

If we define $\lambda_O(a) := 0$ then we have got a map λ of F into itself defined by

$$\lambda(a) := \lambda_X(a) \quad \text{for any point } X \neq O \text{ of } \mathbf{P}\setminus\mathbf{H},$$

which satisfies

$$\sigma(a\cdot X) = \lambda(a)\cdot\sigma(X) \quad \text{for all } a \in F \text{ and all points X of } \mathbf{P}\setminus\mathbf{H}.$$

Of course, λ is our candidate for the accompanying automorphism.

Claim 2: The map λ *is an automorphism of* F.

First we show that λ is a homomorphism of F. For $a, b \in F$ and any point X we have

$$\begin{aligned}\lambda(a+b)\cdot\sigma(X) &= \sigma((a+b)\cdot X) = \sigma(a\cdot X + b\cdot X)\\ &= \sigma(a\cdot X) + \sigma(b\cdot X) = \lambda(a)\cdot(X) + \lambda(b)\cdot\sigma(X)\\ &= (\lambda(a) + \lambda(b))\cdot\sigma(X),\end{aligned}$$

so

$$\lambda(a+b) = \lambda(a) + \lambda(b).$$

Moreover we have that

$$\lambda(ab)\cdot\sigma(X) = \sigma(ab\cdot X) = \lambda(a)\cdot\sigma(b\cdot X) = \lambda(a)\lambda(b)\cdot\sigma(X),$$

and so

$$\lambda(ab) = \lambda(a)\lambda(b).$$

In order to show that λ is *injective* we suppose that $\lambda(a) = \lambda(b)$. Since σ acts bijectively on V* it follows that

$$\sigma(a\cdot X) = \lambda(a)\cdot\sigma(X) = \lambda(b)\cdot\sigma(X) = \sigma(b\cdot X),$$

hence $a\cdot X = b\cdot X$.

Surjectivity: In order to determine a $b \in F$ such that $\lambda(b) = a$ we determine for an arbitrary point $X \neq O$ the preimage Y of $a\cdot\sigma(X)$ under the map σ. Since $a\cdot\sigma(X)$ is a point of the line through O and $\sigma(X)$, the preimage Y must be a point of the line through O and X. Thus there is a b such that $Y = b\cdot X$. Therefore,

$$\sigma(b\cdot X) = \sigma(Y) = a\cdot\sigma(X).$$

This implies

$$a \cdot \sigma(X) = \sigma(b \cdot X) = \lambda(b) \cdot \sigma(X),$$

hence $\lambda(b) = a$.

Hence λ is an automorphism of F. □

3.5.6 Second representation theorem for affine spaces. *Let* $\mathbf{A} = \mathbf{P} \setminus \mathbf{H}$ *be a Desarguesian affine space of dimension* $d \geq 2$ *that is represented by the vector space* V* *over the division ring* F. *Then the following assertions are true.*
(a) *If* τ *is a translation and* σ *is an invertible semilinear map of* V* *then* $\tau\sigma$ *is a collineation of* **A**.
(b) *Each collineation* α *of* **A** *can be represented as*

$$\alpha = \tau\sigma,$$

where τ *is a translation and* σ *is an invertible semilinear map of* V*.

Proof. (a) In exercise 29 you are invited to show that σ (and so $\tau\sigma$) is a collineation.
(b) follows by the preceding theorems. □

Now we turn to the projective case. First we deal with the 'trivial' direction.

3.5.7 Theorem. *Let* V *be a vector space over the division ring* F. *If* γ *is a bijective semilinear map of* V *then* γ *induces a collineation of* **P**(V).

Proof. Let λ be the accompanying automorphism of γ. We define

$$\alpha(\langle v \rangle) := \langle \gamma(v) \rangle.$$

Then α is well defined since

$$\alpha(\langle a \cdot v \rangle) = \langle \gamma(a \cdot v) \rangle = \langle \lambda(a) \cdot \gamma(v) \rangle = \langle \gamma(v) \rangle$$

for $a \neq 0$. Moreover, α maps lines onto lines:

$$\begin{aligned}\alpha(\langle v, w \rangle) &= \alpha(\{\langle a \cdot v + b \cdot w \rangle \mid a, b \in F\}) \\ &= \{\langle \gamma(a \cdot v + b \cdot w) \rangle \mid a, b \in F\} \\ &= \{\langle \lambda(a) \cdot \gamma(v) + \lambda(b) \cdot \gamma(w) \rangle \mid a, b \in F\} \\ &= \langle \gamma(v), \gamma(w) \rangle.\end{aligned}$$

Since γ is bijective α is also bijective, and thus everything is shown. □

3.5.8 Second representation theorem for projective spaces. *Let* **P** *be a Desarguesian projective space of dimension* $d \geq 2$, *and let* V *be a vector space such* *that* **P** = P(V). *Then for each collineation* α *of* **P** *there is a bijective semilinear map* γ *of* V *that induces* α.

3.5.9 Corollary. *Let* F *be the division ring belonging to* **P**. *If* F = **Q**, **R**, *or* $\mathbf{Z}_p$, *(p a prime) then any collineation of* **P** *is induced by a bijective linear map of* V. □

Proof of **3.5.8**. Let **H**, O, V*, and V be as in the proof of **3.4.2**. Then the point O is represented by the subspace $\langle(1, O)\rangle$ of V. Let $\mathcal{B}^* = \{v_1, \ldots, v_d\}$ be a basis of V*. Then

$$\mathcal{B}_V = \{u_i := (0, v_i) \mid i = 1, \ldots, d\} \cup \{u_0 := (1, O)\}$$

is a basis of V and

$$\mathcal{B} = \{\langle(0, v_i)\rangle \mid i = 1, \ldots, d\} \cup \{\langle(1, O)\rangle\}$$

is a basis of **P**. Hence $\mathcal{B}\backslash\{O\} = (\mathcal{B}\backslash\{\langle(1, O)\rangle\})$ is a set of d independent points of **H**, hence a basis of **H**. Since α is a collineation $\alpha(\mathcal{B})$ is also a basis of **P**.

Let $w_0, w_1, \ldots, w_d$ be vectors of V such that

$$\langle w_i\rangle := \alpha(\langle(0, v_i)\rangle) \quad (i = 1, \ldots, d),$$

and

$$\langle w_0\rangle := \alpha(\langle(1, O)\rangle).$$

Then $\{w_0, w_1, \ldots, w_d\}$ is a basis of V. We define the map γ of V into itself in such a way that γ maps a vector $x = k_0u_0 + k_1u_1 + \ldots + k_du_d$ onto

$$\gamma(x) = k_0w_0 + k_1w_1 + \ldots + k_dw_d.$$

Then γ is a bijective linear map of V onto itself. By **3.5.7** γ induces a collineation β of **P**. It follows that

$$\sigma := \beta^{-1}\alpha$$

is a collineation of **P**, which fixes all points of $\mathcal{B}$. In particular, σ fixes the point O and a basis of **H**, hence it also fixes **H** (as a whole). Thus we may also consider σ as a collineation of **A** = **P****H**. Therefore, the 'affine part' of σ – the restriction of σ to **A** – is by **3.5.6** a semilinear map of V* into itself with accompanying automorphism λ.

Hence the map ρ of V into itself that is defined by

$$\rho(a, v) := (\lambda(a), \sigma(v)) \quad (a \in F, v \in V^*)$$

is a semilinear map of V into itself with accompanying automorphism λ.

Since σ coincides with the collineation of $\mathbf{P}\backslash\mathbf{H}$ induced by ρ and since by **3.2.1** the extensions of these collineations to $\mathbf{P}$ are unique ρ is the collineation induced by σ on $\mathbf{P}$.

Thus $\gamma' := \gamma\rho$ is the desired semilinear map. □

The importance of the second representation theorem consists, among other things, in the fact that the identity is the only collineation that fixes 'many points' in 'general position'. We will discuss this in the next section.

3.6 Projective collineations

Definition. A collineation of a projective space $\mathbf{P}(V)$ is called **projective** if it is induced by a *linear* map of V.

Clearly, the set of projective collineations of $\mathbf{P}(V)$ forms a group. The question is how big this group is. Can one describe this group particularly nicely?

First we show that all collineations we have considered so far in detail are in fact projective.

3.6.1 Theorem. *Each central collineation of* $\mathbf{P}(V)$ *is a projective collineation.*

Proof. Let $\mathbf{H}$ be an arbitrary hyperplane of $\mathbf{P} = \mathbf{P}(V)$. It is sufficient to show that all central collineations with axis $\mathbf{H}$ are projective.

Let the projective space $\mathbf{P}$ be coordinatized in such a way that O and V have the same meaning as in the proof of **3.5.8**. It is sufficient to show that all translations with axis $\mathbf{H}$ and all elements of D_O are projective collineations, since each central collineation with axis $\mathbf{H}$ is the product of a translation and an element of D_O (cf. **3.5.6**).

First we show that each element of D_O is projective. For this we consider the line g through the point $O = \langle(1, O)\rangle$ and a point $C = \langle(0, u_1)\rangle$ on $\mathbf{H}$. Then the points $\neq O, C$ on g have the form $\langle(a, u_1)\rangle$ with $a \neq 0$. Let $P = \langle(a, u_1)\rangle$ and $Q = \langle(b, u_1)\rangle$ be two distinct points $\neq O, C$ on g. We show that there is a projective collineation α in D_O mapping P onto Q. Then α is *the* element of D_O mapping P onto Q. It then follows that all elements of D_O are projective.

To prove this we extend $\langle(0, u_1)\rangle$ to a basis $\{\langle(0, u_1)\rangle, \langle(0, u_2)\rangle, \ldots, \langle(0, u_d)\rangle\}$ of $\mathbf{H}$. Then the linear map γ that is defined by

$$\gamma(1, O) := (b/a, O) \text{ and } \gamma(0, u_i) := (0, u_i) \quad (i = 1, \ldots, d)$$

induces a collineation α. Since γ fixes the vectors $(0, u_i)$ it fixes each vector in the span of these vectors. So α fixes each point of $\mathbf{H}$; hence α has axis $\mathbf{H}$.

Since O is also fixed, α has centre O. Finally, since $P = \langle a\cdot(1, O) + (0, u_1)\rangle$ is mapped onto the point $\langle b\cdot(1, O) + (0, u_1)\rangle = Q$, α is the projective collineation of D_O mapping P onto Q.

It remains to show that each translation is projective. Let τ be a translation with axis **H** mapping the point O onto a point $P \neq O$. In the language of **P**(V) this means: Let $\{\langle(0, u_1)\rangle, \ldots, \langle(0, u_d)\rangle\}$ be a basis of **H**. Then τ maps the points $\langle(0, u_1)\rangle, \ldots, \langle(0, u_d)\rangle$ onto themselves and each point $\langle(1, X)\rangle$ onto $\langle(1, X + P)\rangle$. Then the linear map γ that is defined by

$$\gamma(0, u_i) := (0, u_i) \text{ and } \gamma(1, O) := (1, P)$$

induces the translation τ. This may be seen as follows. The collineation induced by γ has **H** as its axis and fixes no point outside **H**, since $P \neq O$. Hence γ induces a translation. Since this translation maps O onto P, γ induces τ. □

Definition. A set of $d+2$ points of **P** in general position is called a **frame**. In other words, a frame is a set $\mathfrak{R}$ of $d+2$ points of **P** such that for each point P of $\mathfrak{R}$ the set $\mathfrak{R}\setminus\{P\}$ is a basis of **P**. An **ordered frame** is a sequence $(P_0, P_1, \ldots, P_d, P_{d+1})$ of $d+2$ points of **P** such that the set $\{P_0, P_1, \ldots, P_d, P_{d+1}\}$ is a frame.

Examples. In a projective plane a set of points is a frame if it consists of four points no three of which are collinear; a set of points in a 3-dimensional projective space is a frame if it has exactly five points any four of which span the whole space.

Observation. Let $\{\langle v_0\rangle, \langle v_1\rangle, \ldots, \langle v_d\rangle, \langle v_{d+1}\rangle\}$ be a frame of $\mathbf{P} = \mathbf{P}(V)$. Then we can w.l.o.g. assume that $v_{d+1} = v_0 + v_1 + \ldots + v_d$. For we have

$$v_{d+1} = a_0v_0 + a_1v_1 + \ldots + a_dv_d$$

with $a_i \neq 0$ $(i = 0, \ldots, d)$. Replace v_i by a_iv_i $(i = 0, \ldots, d)$.

In a certain contrast to **3.6.1** the next theorem says that there are only 'few' projective collineations.

3.6.2 Theorem. *If a projective collineation of* $\mathbf{P} = \mathbf{P}(V)$ *fixes each point of a frame then it is the identity.*

Proof. Let α be a projective collineation of **P**(V) that fixes each point of the frame $\mathfrak{R} = \{\langle v_0\rangle, \langle v_1\rangle, \ldots, \langle v_d\rangle, \langle v_0 + v_1 + \ldots + v_d\rangle\}$. Since α is projective there is a linear map γ of V that induces α. Since each point of $\mathfrak{R}$ is fixed it follows that

$$\gamma(v_i) = a_i \cdot v_i \quad (a_i \in F^* = F\backslash\{0\})$$

and

$$\gamma(v_0 + v_1 + \ldots + v_d) = a\cdot(v_0 + v_1 + \ldots + v_d) \quad (a \in F^*).$$

This implies

$$\begin{aligned} a\cdot(v_0 + v_1 + \ldots + v_d) &= \gamma(v_0 + v_1 + \ldots + v_d) \\ &= \gamma(v_0) + \gamma(v_1) + \ldots + \gamma(v_d) \\ &= a_0 v_0 + a_1 v_1 + \ldots + a_d v_d, \end{aligned}$$

also $a_0 = a_1 = \ldots = a_d = a$.

Hence γ maps each vector v onto a·v; thus α acts on the set of subspaces of V as the identity. □

3.6.3 Corollary. *Let* $\mathfrak{R} = \{P_0, P_1, \ldots, P_d, P_{d+1}\}$ *and* $\mathfrak{R}' = \{P'_0, P'_1, \ldots, P'_d, P'_{d+1}\}$ *be frames of* **P**(V). *Then there is exactly one projective collineation* α *of* **P**(V) *such that* $\alpha(P_i) = P'_i$ $(i = 0, 1, \ldots, d+1)$. *In other words, the group of projective collineations of* **P**(V) *acts sharply transitively on the set of ordered frames of* **P**(V).

Proof. First we show the *existence* of α: Let $P_i =: \langle v_i \rangle$, $P'_i =: \langle v'_i \rangle$ $(i = 0, 1, \ldots, d)$, and

$$P_{d+1} = \langle v_0 + v_1 + \ldots + v_d \rangle, \; P'_{d+1} = \langle v'_0 + v'_1 + \ldots + v'_d \rangle.$$

Then the linear map γ defined by

$$\gamma(v_i) := v'_i \quad (i = 0, 1, \ldots, d)$$

induces a projective collineation, which maps P_i onto P'_i $(i = 0, 1, \ldots, d, d+1)$.

Now we show the *uniqueness:* Let α and β be projective collineations that map P_i onto P'_i $(i = 0, 1, \ldots, d, d+1)$. Then $\alpha^{-1}\beta$ is a projective collineation, which fixes each point of the frame $\mathfrak{R}$. Hence, by **3.6.2**, $\alpha^{-1}\beta$ is the identity. Thus $\alpha = \beta$. □

An important theorem in projective geometry says that the projective collineations are exactly the products of central collineations. In order to prove this theorem we need some preparations.

3.6.4 Lemma. *Let* $\{P_0, P_1, \ldots, P_d\}$ *and* $\{Q_0, Q_1, \ldots, Q_d\}$ *be bases of* **P**. *Then there is a product* β *of at most* $d+1$ *central collineations such that*

$$\beta(P_i) = Q_i \quad \textit{for } i = 0, 1, \ldots, d.$$

Proof. We shall show the following assertion by induction on s: *for each* $s \in \{0, 1, \ldots, d\}$ *there is a product* β_s *of at most* $s+1$ *central collineations such that*

$$\beta_s(P_i) = Q_i \quad \textit{for} \quad i = 0, 1, \ldots, s.$$

If $s = 0$ we choose any central collineation α_0 with $\alpha_0(P_0) = Q_0$. Define $\beta_0 = \alpha_0$.

Assume that the assertion is true for $s - 1 \geq 0$. Let $\beta_{s-1}(P_s) = P_s'$. We construct a central collineation α_s fixing $Q_0, \ldots, Q_{s-1}$ and mapping P_s' onto Q_s: Since $s - 1 \leq d - 1$ we have that $\dim\langle Q_0, Q_1, \ldots, Q_{s-1}\rangle = s - 1 \leq d - 1$. Since $\{Q_0, \ldots, Q_{s-1}, Q_s\}$ and $\{Q_0, \ldots, Q_{s-1}, P_s'\} = \{\beta_{s-1}(P_0), \ldots, \beta_{s-1}(P_{s-1}), \beta_{s-1}(P_s)\}$ are independent sets, neither Q_s nor P_s' is a point of $\langle Q_0, Q_1, \ldots, Q_{s-1}\rangle$.

We shall show that there is a hyperplane **H** through $\langle Q_0, Q_1, \ldots, Q_{s-1}\rangle$ that contains neither Q_s nor P_s'. Then there is a central collineation α_s with axis **H** that maps P_s' onto Q_s. The assertion follows with $\beta_s = \alpha_s\beta_{s-1}$.

Claim: There is a hyperplane through $\langle Q_0, Q_1, \ldots, Q_{s-1}\rangle$ *containing neither* Q_s *nor* P_s'.

For if $Q_s = P_s'$ then $\langle Q_0, Q_1, \ldots, Q_{s-1}, Q_{s+1}, \ldots, Q_d\rangle$ is the desired hyperplane.

Thus, we may assume that $Q_s \neq P_s'$. Let P be an arbitrary point on the line Q_sP_s' that is different from Q_s and P_s'. Then there is a subset $\mathcal{B}$ of $\{Q_0, Q_1, \ldots, Q_d\}$ such that P is contained in $\langle \mathcal{B}\rangle$ but there is no proper subset $\mathcal{B}'$ of $\mathcal{B}$ such that P is contained in $\langle \mathcal{B}'\rangle$. By construction of P there is an element $Q_i \neq Q_s$ in $\mathcal{B}$. In view of the exchange lemma, $\{Q_0, Q_1, \ldots, Q_d\} \setminus Q_i \cup P$ is a basis of **P**. Hence $\langle\{Q_0, Q_1, \ldots, Q_d\} \setminus \{Q_i, Q_s\} \cup P\rangle$ is a hyperplane that contains P but not Q_s, hence it also does not contain P_s'. □

3.6.5 Lemma. *Let* $\{P_0, P_1, \ldots, P_d, P\}$ *and* $\{P_0, P_1, \ldots, P_d, Q\}$ *be frames of* **P**. *Then there is a product* γ *of at most* d *central collineations such that*

$$\gamma(P_i) = P_i \quad \textit{for} \quad i = 0, 1, \ldots, d$$

and

$$\gamma(P) = Q.$$

Proof. We show the following stronger statement by induction on d.

Let P_j *be an arbitrary point of* $\{P_0, P_1, \ldots, P_d\}$. *Then there is a product* γ *of at most* d *central collineations with* $\gamma(P_i) = P_i$ *for* $i = 0, 1, \ldots, d$ *and* $\gamma(P) = Q$, *where* P_j *is contained in the axis of each of these central collineations.*

We may assume w.l.o.g. that $P_j = P_0$. First we suppose $d = 2$ (see Figure 3.9).

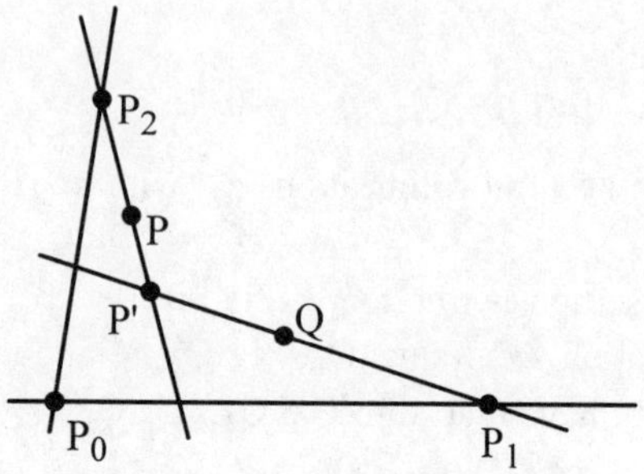

Figure 3.9 Product of central collineations

By a central collineation α_2 with centre P_2 and axis P_0P_1 the point P can be mapped onto the point $P' = PP_2 \cap P_1Q$. Then one can map the point P' using a central collineation α_1 with centre P_1 and axis P_0P_2 onto Q. Since α_1 and α_2 fix the points P_0, P_1, P_2 this is also true for $\gamma = \alpha_1\alpha_2$.

Suppose now that $d > 2$ and assume that the claim is true for $d-1$. We consider the hyperplane $\mathbf{H} = \langle P_1, \ldots, P_{d-1}, P\rangle$. First we construct two frames of **H** for which we may apply the induction hypothesis. Let $P'' = P_0P_d \cap \mathbf{H}$ and $P' = QP_d \cap \mathbf{H}$.

Claim 1: $\{P_1, \ldots, P_{d-1}, P'', P\}$ *is a frame of* **H**.

For this we have to show that any d of these points are independent. Since $\{P_1, \ldots, P_{d-1}, P\}$ is a subset of a frame of **P**, this set is independent. *Assume* that P'' is dependent on a set $\mathcal{B}$ of $d-1$ points of $\{P_1, \ldots, P_{d-1}, P\}$. Then $P'' \in \langle\mathcal{B}\rangle$ and therefore $P_0 \in P''P_d \subseteq \langle\mathcal{B}, P_d\rangle$, a contradiction, since $\{P_0, \ldots, P_d, P\}$ is a frame.

Claim 2: $\{P_1, \ldots, P_{d-1}, P'', P'\}$ *is a frame of* **H**.

By the first claim the points $\{P_1, \ldots, P_{d-1}, P''\}$ are independent. *Assume* that P' is dependent on $\{P_1, \ldots, P_{d-1}\}$. Then we have $Q \in P'P_d \subseteq \langle P_1, \ldots, P_d\rangle$, a contradiction, since $\{P_0, \ldots, P_d, Q\}$ is a frame. Assume that P' is dependent on $\{P_1, \ldots, P_{d-1}, P''\} \setminus P_i$ $(1 \le i \le d-1)$. W.l.o.g. $i = 1$. Then it follows that $Q \in P'P_d \subseteq \langle P_2, \ldots, P_d, P''\rangle = \langle P_2, \ldots, P_d, P_0\rangle$, a contradiction since $\{P_0, \ldots, P_d, Q\}$ is a frame.

Since $\{P_1, \ldots, P_{d-1}, P'', P\}$ and $\{P_1, \ldots, P_{d-1}, P'', P'\}$ are frames of **H**, by induction, there is a product $\gamma_{d-1}{}^*$ of $d-1$ central collineations $\alpha_1{}^*, \ldots, \alpha_{d-1}{}^*$ of **H** such that

$$\gamma_{d-1}{}^*(P_i) = P_i \quad \text{for } i = 1, \ldots, d-1,$$
$$\gamma_{d-1}{}^*(P) = P',$$

and with the property that P'' is contained in the axis of $\alpha_i{}^*$ $(i = 1, \ldots, d-1)$.

By **3.1.10** each central collineation $\alpha_i{}^*$ of **H** is induced by a central collineation α_i of **P** whose axis passes through P_0. Hence the line $P''P_0$, and therefore also the point $P_d \in P''P_0$, is contained in the axis of α_i $(i = 1, \ldots, d-1)$. In particular we have for $\gamma_{d-1} = \alpha_1 \circ \ldots \circ \alpha_{d-1}$

$$\gamma_{d-1}(P_d) = P_d.$$

We now have to find a central collineation α_d that maps P' onto Q and fixes the points P_i $(i = 0, \ldots, d)$, such that P_0 is contained in the axis. This is not difficult. We define α_d as the central collineation with axis $\langle P_0, P_1, \ldots, P_{d-1}\rangle$ and centre P_d that maps P' onto Q. We have to show that Q and P' are not contained in the axis of α_d: The point Q does not lie in the axis since $\{P_0, \ldots, P_d, Q\}$ is a frame. Assume that P' is contained in the axis of α_d. Then, by construction of P' we have

$$P' \in \langle P_0, \ldots, P_{d-1}\rangle \cap \langle P_1, \ldots, P_{d-1}, P\rangle = \langle P_1, \ldots, P_{d-1}\rangle,$$

and therefore

$$Q \in P'P_d \subseteq \langle P_1, \ldots, P_{d-1}, P_d\rangle,$$

a contradiction, since $\{P_0, \ldots, P_d, Q\}$ is a frame.

Thus the collineation

$$\gamma_d = \alpha_d \circ \alpha_{d-1} \circ \ldots \circ \alpha_1$$

has the following properties:

$$\gamma_d(P_i) = P_i \quad \text{for } i = 0, \ldots, d,$$
$$\gamma_d(P) = \alpha_d \circ \gamma_{d-1}(P) = \ldots = \alpha_d(Q') = Q.$$

Hence P_0 is contained in the axis of α_i $(i = 1, \ldots, d)$.

Thus we have shown the assertion. □

3.6.6 Corollary. *Let* $\{P_0, P_1, \ldots, P_{d+1}\}$ *and* $\{Q_0, Q_1, \ldots, Q_{d+1}\}$ *be arbitrary frames of* **P**. *Then there is a product* δ *of (at most* $2d+1$*) central collineations of* **P** *such that*

$$\delta(P_i) = Q_i \quad \textit{for } i = 0, 1, \ldots, d+1.$$

Proof. By Lemma **3.6.4** we can map $P_0, P_1, \ldots, P_d$ onto $Q_0, Q_1, \ldots, Q_d$ by a product β of at most $d+1$ central collineations. By Lemma **3.6.5** we can map the point $\beta(P_{d+1})$ onto Q_{d+1} by a product γ of at most d central collineations.

The assertion follows by putting $\delta = \gamma\beta$. □

Remark. One can show that $d+2$ central collineations are sufficient to map one ordered frame onto another (see for instance [Ped63]).

As another corollary we have the following theorem.

3.6.7 Theorem. *The projective collineations of* **P**(V) *are precisely the products of central collineations.*

Proof. Since by **3.6.1** each central collineation is projective and since the product of projective collineations is again projective, one direction easily follows.

Using our preparations, the other direction is also not difficult. Let α be an arbitrary projective collineation of **P**(V). Then α maps a frame $\mathfrak{R} = \{P_0, P_1, \ldots, P_d, P_{d+1}\}$ onto another frame $\mathfrak{R}' = \{P'_0, P'_1, \ldots, P'_d, P'_{d+1}\}$. By Corollary **3.6.6** there is a product δ of central collineations such that

$$\delta(P_i) = P'_i \quad (i = 0, 1, \ldots, d, d+1).$$

Then $\delta(P_i) = \alpha(P_i)$ for all i, hence, by **3.6.3** we have $\alpha = \delta$. Thus α is also a product of central collineations. □

3.6.8 Corollary. *Let* Σ *be the set of collineations of* **P**(V) *that fix each point of a frame. Then* Σ *is a group, which is isomorphic to the group* Aut(F) *of automorphisms of* F.

Proof. Let $\mathfrak{R} = \{\langle v_0\rangle, \langle v_1\rangle, \ldots, \langle v_d\rangle, \langle v_0 + v_1 + \ldots + v_d\rangle\}$ be a frame that is fixed pointwise by each element of Σ. Let $\sigma \in \Sigma$, and denote by γ the semilinear map of V with accompanying automorphism λ that induces σ. As in the proof of **3.6.2** one sees that σ maps each v_i onto av_i. W.l.o.g. we assume that $a = 1$. Therefore we have

$$\gamma(a_0v_0 + \ldots + a_dv_d) = \lambda(a_0)v_0 + \ldots + \lambda(a_d)v_d.$$

We denote this map more accurately by γ_λ. Thus we have found a map $\Sigma \to$ Aut(F), more precisely the map that maps σ onto λ. One easily checks (see exercise 33) that this map is a homomorphism. Since for $\lambda \neq \lambda'$ the collineations induced by γ_λ and $\gamma_{\lambda'}$ are distinct, the assertion follows. □

Exercises

1 Show that the set of all collineations of a projective space forms a group with respect to composition of maps.

2 Show that the following maps of the Euclidean plane are central collineations in the projective closure: point reflection, translation, rotation.

3 Interpret reflections, point reflections, and translations of the 3-dimensional Euclidean space as central collineations in the projective closure.

4 Let α be a collineation of a projective space $\mathbf{P}$. Show: if g and h are two intersecting lines of $\mathbf{P}$ then $\alpha(g)$ and $\alpha(h)$ also intersect each other, and we have

$$\alpha(g \cap h) = \alpha(g) \cap \alpha(h).$$

5 Show that a collineation α of a projective space $\mathbf{P}$ induces a collineation in a subspace $\mathbf{U}$ of $\mathbf{P}$, if α fixes the subspace $\mathbf{U}$ as a whole.

6 Convince yourself that a collineation α of a projective space $\mathbf{P}$ is also a collineation of the affine space $\mathbf{P}\backslash\mathbf{H}$ if α fixes the hyperplane $\mathbf{H}$ as a whole.

7 Show that each $\alpha \in \Gamma$ (that is each collineation α of the affine space $\mathbf{P}\backslash\mathbf{H}$) has the property that α^{-1} also lies in Γ.

8 Let α be a collineation of a projective space $\mathbf{P}$ such that there is a point C with the property that each line through C is fixed by α. Show that there exists a hyperplane $\mathbf{H}$ such that each point of $\mathbf{H}$ is fixed by α. Shortly, every central collineation is axial.

9 (a) Show: if $\mathbf{U}$ is a subspace of a hyperplane $\mathbf{H}$ of $\mathbf{P}$ then the central collineations with axis $\mathbf{H}$ and centre in $\mathbf{U}$ form a group.
(b) Does this assertion remain true if $\mathbf{U}$ is not a subspace of $\mathbf{H}$?

10 Determine all central collineations in the projective plane of order 2 having a fixed line g_0 as axis.

11 Let α be a central collineation of a projective space $\mathbf{P}$ with centre C and axis $\mathbf{H}$. Show that for each point Q of $\mathbf{P}$ with $Q \neq C$ and $Q \in \mathbf{H}$, α induces a central collineation in the quotient geometry $\mathbf{P}/Q$.

12 Let Q be a point of a Desarguesian projective space $\mathbf{P}$. Show that each central collineation of $\mathbf{P}/Q$ is induced by a central collineation of $\mathbf{P}$.

13 (a) Let **A** be the affine plane of order 2. Show that each permutation of the four points of **A** can be extended to a collineation of the corresponding projective plane of order 2.
(b) Compute the number of all collineations of the projective plane of order 2.

14 Show that not all collineations of AG(3, 2) can be extended to collineations of PG(3, 2). [Hint: There is a collineation of AG(3, 2) fixing all points but two.]

15 Let α be a collineation of an affine space **A** of order 2 with the additional property that α maps planes of **A** onto planes. Show that α can be uniquely extended to a collineation of the projective closure of **A**.

16 Determine the group T(**H**) of the projective plane of order 3. [Let **H** be the line at infinity.]

17 Let **H** be a hyperplane of a finite projective space **P** = PG(d, q). What is the number of elements of T(**H**)? How many elements are in T(P, **H**)?

18 Let g be a line through the point O of a Desarguesian projective space. Show that

$$T(g) = \{\tau \in T(\mathbf{H}) \mid \tau(g) = g\}.$$

19 Describe geometrically (as in Figure 3.6) the sum of two points P, Q on a common line through O.

20 Show that in the situation of the proof of **3.3.2** the following is true: if $Y \neq -X$, then O, X + P, and Y – P are not collinear.

21 Compare the addition of points in a Desarguesian affine space to the addition of vectors, which you have learned in school.

22 Let V* be the vector space the affine space **A** is constructed with. Let P be an arbitrary point. Then show that the map defined by

$$\tau(X) := X + P$$

is a translation of **A**.

23 Show that for each $\sigma \in D_O$ and each line $g = P + OC$ of **P****H** we have that

$$\sigma(g) = \sigma(P + OC) = \sigma(P) + OC.$$

24 Prove in detail that each element of D_O is an automorphism of $(\mathcal{P}^*, +)$ (cf. Lemma **3.3.2**).

25 Describe those central collineations that are collineations of $\mathbf{A} = \mathbf{P}\setminus\mathbf{H}$ and fix point O of **A**.

26 Prove **3.3.1** ('μ is an element of D_O').

27 Show that the map α defined in the proof of the first representation theorem (**3.4.2**) is also bijective on the set of lines.

28 Let γ_λ be the semilinear map of the vector space V corresponding to the automorphism $\lambda \in \mathrm{Aut}(F)$ (cf. the example before **3.5.3**). Show that if $\lambda \neq \lambda'$ then the collineations of **P**(V) induced by γ_λ and $\gamma_{\lambda'}$ are distinct.

29 Complete the proof of **3.5.6**. Show: if σ is a semilinear map of V* then σ is a collineation of **A**.

30 Show that for any two bases $\{P_0, P_1, \ldots, P_d\}$ and $\{Q_0, Q_1, \ldots, Q_d\}$ of a Desarguesian projective space **P** there is a product β of elations with

$$\beta(P_i) = Q_i \text{ for } i = 0, 1, \ldots, d.$$

31 Let $\{P_0, P_1, P_2, P_3\}$ and $\{P_0, P_1, P_2, Q\}$ be quadrangles of a projective plane **P** (so they are frames of **P**). Then show that, in general, there is no product γ of elations with

$$\gamma(P_i) = P_i \quad (i = 0, 1, 2) \quad \text{and} \quad \gamma(P) = Q.$$

32 In **3.6.6** we have shown that each projective collineation is a product of at most $2d + 1$ central collineations. Improve this bound.

33 Show that the map $\Sigma \rightarrow \mathrm{Aut}(F)$ defined in the proof of **3.6.8** is a homomorphism.

34 Is each projective collineation a product of elations (central collineations with centre on axis)?

35 Show that the number of ordered quadrangles (P_1, P_2, P_3, P_4) in a projective plane of order n is $(n^2 + n + 1)(n^2 + n)n^2(n - 1)^2$.

36 Determine the number of projective collineations of a Desarguesian projective plane.

37 Compute the number of all collineations of a Desarguesian projective plane of prime order.
[Use the fact that the field $\mathbf{Z}_p$ has only the identity as automorphism.]
Determine this number precisely for the orders 2, 3, and 5.

True or false?

☐ The set of all central collineations of **P** forms a group.

☐ The set of all central collineations of **P** with common centre forms a group.

☐ The set of all central collineations of **P** with a common fixed point Q forms a group.

☐ Each fixed point of a collineation α is a centre of α.

☐ Each fixed point of a central collineation α is a centre of α.

☐ In order to show that collineations α, β are equal one has to show $\sigma(P) = \beta(P)$ for just one point P.

☐ In order to show that central collineations α, β are equal one has to show $\sigma(P) = \beta(P)$ for just one point P.

☐ If a collineation α of **P** fixes a hyperplane **H** as a whole then each point of **H** is fixed by α.

☐ The identity of **P**(V) is induced only by the identity of V.

☐ Each projective space contains a frame.

☐ Each basis may be extended to a frame.

☐ Each basis may be uniquely extended to a frame.

You should know the following notions

Central collineation, fixed point, centre, axis, translation, T(**H**), τ_P, addition of points, dilatation, semilinear map, accompanying automorphism, collineation induced by a bijective semilinear map, frame, projective collineation.

4 Quadratic sets

In the preceding chapters we described all 'linear' subspaces – all sets of points in projective spaces that can be described by one or more *linear* equations. In this chapter we shall study sets of points in projective spaces **P**(V) that satisfy a *quadratic* equation. Those sets of points are called quadrics. They play a central and extremely important role in projective geometry. Quadrics were studied synthetically only in the seventies. F. Buekenhout coined the notion of a *quadratic set*, which is the synthetic counterpart of a quadric. In many situations it is sufficient to consider only quadratic sets.

Let $\mathbf{P} = (\mathcal{P}, \mathcal{L}, \mathrm{I})$ be a projective space of finite dimension.

4.1 Fundamental definitions

The first definition of a tangent already comes as a little surprise.

Definition. Let $\mathcal{Q}$ be a set of points of the projective space **P**.
(a) We call a line g a **tangent** of $\mathcal{Q}$ if either g has just one point in common with $\mathcal{Q}$ or each point of g is contained in $\mathcal{Q}$. If a tangent g has just one point P in common with $\mathcal{Q}$, then one calls g a tangent of $\mathcal{Q}$ **at** the point P.

A line g with the property that each point of g lies in $\mathcal{Q}$ is also called a $\mathcal{Q}$-line. In general, we call a subspace **U** a $\mathcal{Q}$-subspace, if each point of **U** lies in $\mathcal{Q}$.

(b) For each point P of $\mathcal{Q}$ let the set $\mathcal{Q}_P$ consist of the point P and all points $X \neq P$ of **P** such that the line XP is a tangent of $\mathcal{Q}$. One calls $\mathcal{Q}_P$ the **tangent space** of $\mathcal{Q}$ at the point P.
(c) We call the set $\mathcal{Q}$ a **quadratic set** of **P** if it satisfies the following conditions:
(i) **If-three-then-all axiom.** Any line g that contains at least three points of $\mathcal{Q}$ is totally contained in $\mathcal{Q}$ (which means that each point of g lies in $\mathcal{Q}$). In other words, any line has 0, 1, 2, or all points in common with $\mathcal{Q}$.
(ii) **Tangent-space axiom.** For any point $P \in \mathcal{Q}$, its tangent space $\mathcal{Q}_P$ is the set of points in a hyperplane or the set of all points of **P**.

One can also express the important tangent-space axiom as follows. For any point P of $\mathcal{Q}$ one has the following alternatives:
– *either* $\mathcal{Q}_P$ is a hyperplane, that means any line through P in $\mathcal{Q}_P$ is a tangent, and no line through P outside $\mathcal{Q}_P$ is a tangent, each such line contains exactly one further point of $\mathcal{Q}$,
– *or* $\mathcal{Q}_P$ is equal to **P**, and so each line through P is a tangent.

Examples. (1) The empty set, any set consisting of just one point, the set of points on a line, . . ., in short: the set of all points of a subspace is a quadratic set. These quadratic sets $\mathcal{Q}$ have the property that for each point $P \in \mathcal{Q}$ the set $\mathcal{Q}_P$ equals **P**.
(2) If $\mathcal{Q}$ consists of the points in the union of two hyperplanes $\mathbf{H}_1$, $\mathbf{H}_2$ of **P** then $\mathcal{Q}$ is a quadratic set. For each point $P \in \mathbf{H}_1 \cap \mathbf{H}_2$ we have that $\mathcal{Q}_P$ consists of all points of **P** while the points $P \in \mathbf{H}_i \backslash (\mathbf{H}_1 \cap \mathbf{H}_2)$ satisfy $\mathcal{Q}_P = \mathbf{H}_i$ $(i = 1, 2)$.
(3) From elementary geometry we immediately get examples. Each circle (and each ellipse) in the Euclidean plane is a quadratic set. (Here the tangent spaces are just lines.) Also the sphere in 3-dimensional (or d-dimensional) real affine space is a quadratic set.

In real 3-dimensional space there exist two other, quite different quadratic sets. These are, first of all, the cones; here, there is a point V, the vertex of the cone, such that $\mathcal{Q}_V$ is the whole space. Finally, there is the hyperbolic quadric (see Figure 4.1), which we shall later discuss in detail (cf. also Section **2.4**).

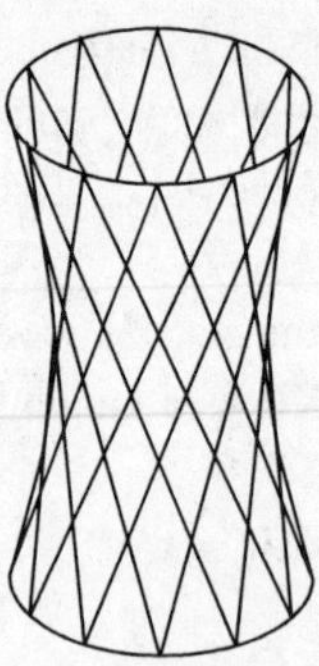

Figure 4.1 A hyperbolic quadric

Our aim is to study quadratic sets in detail. The $\mathcal{Q}$-subspaces of maximum dimension will play a particular role. This will lead to many results in classical geometry, in particular we will investigate the Klein quadric.

4.1.1 Lemma. *Let* $\mathcal{Q}$ *be a quadratic set of* **P**, *and let* **U** *be a subspace of* **P**. *Then the set* $\mathcal{Q}' := \mathcal{Q} \cap \mathbf{U}$ *of the points of* $\mathcal{Q}$ *in* **U** *is a quadratic set of* **U**. *Moreover, we have*

$$\mathcal{Q}'_P = \mathcal{Q}_P \cap \mathbf{U}$$

for all points $P \in \mathcal{Q}'$.

Proof. It is clear that $\mathcal{Q}'$ also satisfies condition (i). In order to show the tangent space axiom we consider a point $P \in \mathcal{Q}'$. By definition of $\mathcal{Q}'_P$ we have

$$\mathcal{Q}'_P = \{P\} \cup \{X \mid X \text{ is a point of } \mathbf{U} \text{ such that } XP \text{ is a tangent}\} = \mathcal{Q}_P \cap \mathbf{U}.$$

Since $\mathcal{Q}_P$ is a subspace of **P** of dimension at least $d-1$, the set $\mathcal{Q}_P \cap \mathbf{U}$ is a subspace of **U** whose dimension is at least $\dim(\mathbf{U}) - 1$. □

We say that $\mathcal{Q}$ **induces** a quadratic set in a subspace **U**.

Definition. Let $\mathcal{Q}$ be a quadratic set of **P**.

The **radical** of $\mathcal{Q}$ is the set $\text{rad}(\mathcal{Q})$ of all points $P \in \mathcal{Q}$ with the property that $\mathcal{Q}_P$ consists of all points of **P**.

We say that $\mathcal{Q}$ is **nondegenerate** if $\text{rad}(\mathcal{Q}) = \varnothing$, that is if for each point $P \in \mathcal{Q}$ its tangent space $\mathcal{Q}_P$ is a hyperplane of **P**.

Remark. In order to verify that a point is contained in the radical of a quadratic set, one has only to show that the tangents through that point span the whole space.

Example. A sphere in the Euclidean space is a nondegenerate quadratic set, while a cone is degenerate; its radical consists of exactly one point, its **vertex**.

4.1.2 Theorem. *Let* $\mathcal{Q}$ *be a quadratic set of* **P**.
(a) *The radical of* $\mathcal{Q}$ *is a linear subspace of* **P**.
(b) *Let* **U** *be a complement of* $\text{rad}(\mathcal{Q})$ *(that is a subspace* **U** *such that* $\mathbf{U} \cap \text{rad}(\mathcal{Q}) = \varnothing$ *and* $\langle \mathbf{U}, \text{rad}(\mathcal{Q})\rangle = \mathbf{P}$*). Then* $\mathcal{Q}' := \mathcal{Q} \cap \mathbf{U}$ *is a nondegenerate quadratic set of* **U**.
(c) $\mathcal{Q}$ *can be described as follows:* $\mathcal{Q}$ *consists of all points that lie on lines that join a point of* $\text{rad}(\mathcal{Q})$ *with a point of* $\mathcal{Q}' = \mathcal{Q} \cap \mathbf{U}$.

Proof. (a) Let $P, P' \in \text{rad}(\mathcal{Q})$. Let P'' be a third point on PP'. We have to show that P'' also lies in $\text{rad}(\mathcal{Q})$, that is that $\mathcal{Q}_{P''}$ contains all points of **P**.

First we observe that PP' is a tangent containing at least two points of $\mathcal{Q}$, therefore it is contained in $\mathcal{Q}$. Assume that $\mathcal{Q}_{P''}$ is only a hyperplane, which passes through PP'. Consider a line g through P'' not contained in $\mathcal{Q}_{P''}$. Then g is not a tangent and therefore contains another point R of $\mathcal{Q}$. Since the lines PR and P'R pass through P and P', respectively, they are tangents. Since they contain more than one point of $\mathcal{Q}$, they must be contained in $\mathcal{Q}$. Consider now a point T on PR with $T \neq P, R$ (see Figure 4.2).

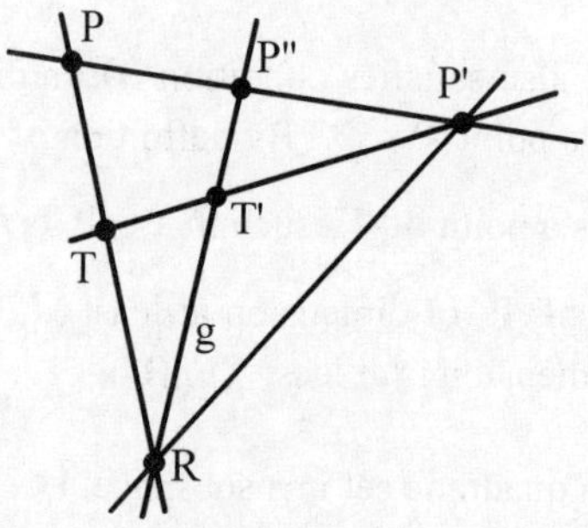

Figure 4.2

Then P'T is a tangent, and, since it contains two points of $\mathcal{Q}$, it is contained in $\mathcal{Q}$. Since P'T intersects the line P''R in some point T' with $T' \neq P'', R$, the line $g = P''R$ is incident with at least three points of $\mathcal{Q}$; therefore it is a tangent. This is a contradiction.

(b) Assume that $\mathcal{Q}'$ is degenerate. Then there is a point $P \in \mathcal{Q}'$ such that each line of **U** through P is a tangent. Since each line PR with $R \in \text{rad}(\mathcal{Q})$ is also a tangent, the tangents through P do not span only a hyperplane. Hence $\mathcal{Q}_P$ is the whole space, and P is contained in $\text{rad}(\mathcal{Q})$, contradicting the choice of **U**.

(c) follows directly from the definition of $\text{rad}(\mathcal{Q})$. □

Remark. In view of **4.1.2** we can restrict our attention to the study of *non*degenerate quadratic sets.

4.1.3 Lemma. *Let* $\mathcal{Q}$ *be a quadratic set of* **P**. *If* $\mathcal{Q}$ *is nondegenerate then for any two distinct points* $P, R \in \mathcal{Q}$ *we have* $\mathcal{Q}_P \neq \mathcal{Q}_R$.

In other words, the quadratic set that is induced by $\mathcal{Q}$ *in a tangent space* $\mathcal{Q}_P$ *has a radical that consists of just one point, namely* P.

For the *proof* we assume that for two distinct points P and Q we have $\mathcal{Q}_P = \mathcal{Q}_R$ =: **H**. Then the radical of the quadratic set $\mathcal{Q}'$ that is induced by $\mathcal{Q}$ in **H** contains at least the points P and R. Therefore, by **4.1.2**(a) each point of the line PR is contained in $\text{rad}(\mathcal{Q}')$.

Since $\mathfrak{Q}$ is nondegenerate each line through P that does not lie in **H** contains another point of $\mathfrak{Q}$. In particular there is a point S of $\mathfrak{Q}$ outside **H**. Consider the tangent space $\mathfrak{Q}_S$ at the point S; since $\mathfrak{Q}_S$ is a hyperplane, it intersects PR in some point T.

We shall show that $\mathfrak{Q}_T$ is the whole space: Since $T \in \mathrm{rad}(\mathfrak{Q}')$, $\mathfrak{Q}_T$ contains the hyperplane **H**. The line ST lies in $\mathfrak{Q}_S$, hence each point of ST is contained in $\mathfrak{Q}$. Thus ST also lies in $\mathfrak{Q}_T$. Therefore $\mathfrak{Q}_T$ contains the hyperplane **H** and the point S outside **H**, hence the whole point set. This means that $\mathfrak{Q}$ is degenerate, a contradiction. □

4.1.4 Lemma. *Let* $\mathfrak{Q}$ *be a nondegenerate quadratic set of* **P**.
(a) *If* $P \in \mathfrak{Q}$ *and* **W** *is a complement of* P *in* $\mathfrak{Q}_P$, *then* $\mathfrak{Q}' := \mathfrak{Q} \cap \mathbf{W}$ *is a nondegenerate quadratic set of* **W**.
(b) *If* **H** *is a hyperplane that is not a tangent hyperplane then* $\mathfrak{Q}' := \mathfrak{Q} \cap \mathbf{H}$ *is a nondegenerate quadratic set of* **H**.

Proof. (a) By **4.1.3**, $\mathrm{rad}(\mathfrak{Q} \cap \mathfrak{Q}_P) = \{P\}$. Thus, in view of **4.1.2**(b) $\mathfrak{Q}'$ is nondegenerate.
(b) Assume that there is a point $X \in \mathrm{rad}(\mathfrak{Q}')$. Then $\mathfrak{Q}'_X = \mathbf{H}$. Since **H** is not a tangent hyperplane, we have that $\mathfrak{Q}_X \neq \mathbf{H}$. This contradicts Lemma **4.1.1**, which says that $\mathfrak{Q}'_X = \mathfrak{Q}_X \cap \mathbf{H}$. □

4.2 The index of a quadratic set

A quadratic set $\mathfrak{Q}$ can be described very well by the dimension of its maximal $\mathfrak{Q}$-subspaces. This maximum dimension is connected to the 'index' of $\mathfrak{Q}$, which turns out to be the crucial parameter to describe $\mathfrak{Q}$.

Definition. Let $t-1$ be the maximum dimension of a $\mathfrak{Q}$-subspace of a quadratic set $\mathfrak{Q}$. Then the integer t is called the **index** of $\mathfrak{Q}$. The $\mathfrak{Q}$-subspaces of dimension $t-1$ are also called **maximal** $\mathfrak{Q}$-subspaces.

Examples. A cone and a hyperboloid in 3-dimensional real space have index 2, since they contain lines, but no planes. Any quadratic set that does not contain a line has index 1; for instance, a sphere has index 1.

The following lemma shows that the maximal $\mathfrak{Q}$-subspaces are 'uniformly distributed'.

4.2.1 Lemma. *Let $\mathfrak{Q}$ be a quadratic set of index t in* $\mathbf{P}$. *Then each point of $\mathfrak{Q}$ is on a maximal $\mathfrak{Q}$-subspace.*

More precisely: if P *is a point of $\mathfrak{Q}$ outside a $(t-1)$-dimensional $\mathfrak{Q}$-subspace* $\mathbf{U}$, *then there is a $(t-1)$-dimensional $\mathfrak{Q}$-subspace* $\mathbf{U}'$ *through* P *that intersects* $\mathbf{U}$ *in a $(t-2)$-dimensional subspace.*

Proof. The tangent hyperspace $\mathfrak{Q}_P$ at P intersects $\mathbf{U}$ in a subspace $\mathbf{V}$ of dimension $\geq t-2$. It follows that each line PX with $X \in \mathbf{V}$ is a tangent and therefore contained in $\mathfrak{Q}$. Thus $\mathbf{U}' := \langle P, \mathbf{V}\rangle$ is a $(t-1)$-dimensional $\mathfrak{Q}$-subspace. □

We formulate the assertion of **4.2.1** once more in the case $t = 2$: *through each point of $\mathfrak{Q}$ outside a $\mathfrak{Q}$-line* g *there is a $\mathfrak{Q}$-line that intersects* g.

We shall use the following technical lemma in the proof of the next theorem.

4.2.2 Lemma. *Let $\mathfrak{Q}$ be a quadratic set in* $\mathbf{P}$. *Let $\mathfrak{S}$ be a subset of $\mathfrak{Q}$ with the property that the line through any two points of $\mathfrak{S}$ is a $\mathfrak{Q}$-line. Then $\langle\mathfrak{S}\rangle$ is a $\mathfrak{Q}$-subspace.*

The *proof* consists of two tricks.
1st trick. W.l.o.g. $\mathfrak{S}$ is finite. For since any spanning set contains a basis there is a *finite* set $\mathfrak{S}_0 \subseteq \mathfrak{S}$ with $\langle\mathfrak{S}_0\rangle = \langle\mathfrak{S}\rangle$. It is therefore sufficient to show that $\langle\mathfrak{S}_0\rangle$ is a $\mathfrak{Q}$-subspace.
2nd trick. Induction by $|\mathfrak{S}|$. For $|\mathfrak{S}| = 0$, 1, or 2 the assertion follows trivially.

Suppose now that $|\mathfrak{S}| > 2$ and assume that the assertion is true for all sets of points with $|\mathfrak{S}|-1$ elements. We consider an arbitrary point $S \in \mathfrak{S}$. By induction, $\mathbf{V} := \langle\mathfrak{S}\setminus\{S\}\rangle$ is a $\mathfrak{Q}$-subspace. W.l.o.g. we have that $S \notin \mathbf{V}$. By hypothesis, for each point $R \in \mathfrak{S}\setminus\{S\}$ the line RS is a $\mathfrak{Q}$-line. Since these lines generate the subspace $\langle\mathfrak{S}\setminus\{S\}, S\rangle = \langle\mathfrak{S}\rangle$, the tangent space of $\mathfrak{Q}$ at the point S contains the subspace $\langle\mathfrak{S}\rangle = \langle\mathbf{V}, S\rangle$. Therefore, all lines XS with $X \in \mathbf{V}$ are contained in $\mathfrak{Q}$. It follows that $\langle\mathfrak{S}\rangle = \langle\mathbf{V}, S\rangle \subseteq \mathfrak{Q}$. □

4.2.3 Theorem. *Let $\mathfrak{Q}$ be a quadratic set in a d-dimensional projective space* $\mathbf{P}$, *and let* $\mathbf{U}$ *be a maximal $\mathfrak{Q}$-subspace. If $\mathfrak{Q}$ is nondegenerate, then there is a maximal $\mathfrak{Q}$-subspace that is skew to* $\mathbf{U}$.

Proof. Let t be the index of $\mathfrak{Q}$. We shall show more generally the following assertion: *if $j \in \{-1, \ldots, t-2\}$ then there is a maximal $\mathfrak{Q}$-subspace $\mathbf{U}_j$ such that* $\dim(\mathbf{U} \cap \mathbf{U}_j) = j$.

We proceed by induction on j.

If $j = t-2$ then the assertion follows from **4.2.1**.

Suppose now $0 \le j \le t-2$, and let $\mathbf{U}'$ be a maximal $\mathcal{Q}$-subspace with $\dim(\mathbf{U} \cap \mathbf{U}') = j$. We shall construct a maximal $\mathcal{Q}$-subspace $\mathbf{U}''$ with $\dim(\mathbf{U} \cap \mathbf{U}'') = j-1$.

First, we observe that there exists a point $P \in \mathcal{Q}$ such that $\langle \mathbf{U} \cap \mathbf{U}', P\rangle$ is not a $\mathcal{Q}$-subspace. Otherwise, any point of $\mathbf{U} \cap \mathbf{U}'$ would be in $\mathrm{rad}(\mathcal{Q})$; since $\dim(\mathbf{U} \cap \mathbf{U}') = j \ge 0$ this contradicts the fact that $\mathcal{Q}$ is nondegenerate.

By **4.2.1** there is a maximal $\mathcal{Q}$-subspace $\mathbf{W}$ of $\mathbf{P}$ through P intersecting $\mathbf{U}'$ in a subspace of dimension $t-2$. We claim that $\mathbf{W}$ satisfies our claim: Since $\mathbf{U} \cap \mathbf{U}' \not\subset \mathbf{W}$ we have that

$$\dim(\mathbf{W} \cap \mathbf{U} \cap \mathbf{U}') = j-1.$$

It is sufficient to show that

$$\mathbf{W} \cap \mathbf{U} = \mathbf{W} \cap \mathbf{U} \cap \mathbf{U}'.$$

Assume that there is a point $X \in \mathbf{W} \cap \mathbf{U}$ with $X \notin \mathbf{U}'$. Then the set

$$\mathcal{S} := (\mathbf{W} \cap \mathbf{U}') \cup (\mathbf{U}' \cap \mathbf{U}) \cup \{X\}$$

satisfies the hypothesis of Lemma **4.2.2**. Hence $\mathbf{M} := \langle \mathcal{S} \rangle$ is a $\mathcal{Q}$-subspace. This subspace contains the hyperplane $\mathbf{W} \cap \mathbf{U}'$ of $\mathbf{W}$ and the point $X \in \mathbf{W} \setminus \mathbf{U}'$, and hence the whole subspace $\mathbf{W}$. Thus $\mathbf{M} = \mathbf{W}$. So we would have $\mathbf{U} \cap \mathbf{U}' \subseteq \mathbf{M} = \mathbf{W}$, contradicting the choice of P. □

As a corollary we get the most important result of this section.

4.2.4 Theorem. *Let* $\mathcal{Q}$ *be a nondegenerate quadratic set of index* t *in a* d*-dimensional projective space* $\mathbf{P}$*. If* d *is even then*

$$t \le \frac{d}{2};$$

if d *is odd then*

$$t \le \frac{d+1}{2}.$$

Proof. By the preceding theorem there are two skew $(t-1)$-dimensional $\mathcal{Q}$-subspaces $\mathbf{U}$ and $\mathbf{U}'$. They satisfy $\dim(\mathbf{P}) \ge \dim(\mathbf{U}) + \dim(\mathbf{U}') - \dim(\mathbf{U} \cap \mathbf{U}')$, so $d \ge 2\cdot(t-1)+1$. □

4.3 Quadratic sets in spaces of small dimension

This section is devoted to a precise description of quadratic sets in projective planes and 3-dimensional projective spaces. In planes the following notion plays a central role.

Definition. A nonempty set $\mathcal{O}$ of points in a projective plane is called an **oval** if no three points of $\mathcal{O}$ are collinear and each point of $\mathcal{O}$ is on exactly one tangent.

4.3.1 Theorem. *Let $\mathcal{Q}$ be a quadratic set in a projective plane $\mathbf{P}$. Then $\mathcal{Q}$ is the empty set, just one point, one line, an oval, the set of points on two lines, or the whole set of points.*

Hence there is only one type of nonempty, nondegenerate quadratic sets in a projective plane, namely the ovals.

Proof. The assertion is true if $\mathcal{Q}$ contains at most one point.

If $\mathcal{Q}$ consists of more than one point but does not contain a line, then $\mathcal{Q}$ is an oval: Since $\mathcal{Q}$ has at least two points, each point P of $\mathcal{Q}$ is on at least one line that is *not* a tangent. Hence, $\mathcal{Q}_P$ is not the whole plane. Thus $\mathcal{Q}$ is nondegenerate. From this it follows that any point of $\mathcal{Q}$ is on exactly one tangent. Hence $\mathcal{Q}$ is an oval.

If $\mathcal{Q}$ has index at least 2 then the assertion is true, if there are at most two $\mathcal{Q}$-lines. If there are more than two $\mathcal{Q}$-lines, $\mathcal{Q}$ is the whole point set. For consider the point P of intersection of two $\mathcal{Q}$-lines. Then $\mathcal{Q}_P = \mathbf{P}$. If the third $\mathcal{Q}$-line passes through P then there is a line g not through P that intersects the three $\mathcal{Q}$-lines in distinct points. Hence g contains three points of $\mathcal{Q}$ and is therefore contained in $\mathcal{Q}$. Thus there is a $\mathcal{Q}$-line not through P and so $\mathcal{Q} = \mathcal{Q}_P = \mathbf{P}$. □

For the description of quadratic sets in 3-dimensional projective spaces we need three notions.

Definition. Let $\mathbf{P}$ be a d-dimensional projective spacc.

An **ovoid** is a nonempty set $\mathcal{O}$ of points of $\mathbf{P}$ satisfying the following properties:

- no three points of $\mathcal{O}$ are collinear;
- for each point $P \in \mathcal{O}$ the tangents through P cover exactly a hyperplane.

Now suppose $d = 3$.

A set $\mathcal{K}$ of points of $\mathbf{P}$ is called a **cone** if there are a plane π, an oval $\mathcal{O}$ in π, and a point $V \notin \pi$ such that $\mathcal{K}$ consists precisely of the points on the lines VX with $X \in \mathcal{O}$. We call V the **vertex** of the cone $\mathcal{K}$.

A **hyperboloid** is the set of points that are incident with the lines of a regulus (cf. Section **2.4**).

In other words, for a hyperboloid $\mathcal{H}$ there exist two sets $\mathcal{R}$, $\mathcal{R}'$ of mutually skew lines such that the following conditions are satisfied:

- each line of $\mathcal{R}$ intersects each line of $\mathcal{R}'$,
- through each point of a line of $\mathcal{R}$ there is a line of $\mathcal{R}'$ and conversely,
- $\mathcal{H}$ consists of all points on the lines of $\mathcal{R}$ (or $\mathcal{R}'$, respectively).

Remark. In **2.4.3** we have shown that in a 3-dimensional projective space a regulus (and therefore a hyperboloid) exists if and only if the underlying division ring is a field, hence commutative. In this case any hyperboloid can be described by a quadratic equation.

4.3.2 Theorem. *Let* $\mathcal{Q}$ *be a quadratic set in a 3-dimensional projective space* $\mathbf{P}$. *Then* $\mathcal{Q}$ *is a subspace, an ovoid, a cone, a hyperboloid, or the union of two hyperplanes.*

In particular, the nonempty, nondegenerate quadratic sets in a 3-dimensional projective space are precisely the ovoids and the hyperboloids.

Proof. Let us suppose that $\mathcal{Q}$ is not a subspace. Then $\mathcal{Q}$ contains at least two points.

If $\mathcal{Q}$ has index 1 then no three points of $\mathcal{Q}$ are collinear, in particular $\mathcal{Q}$ is nondegenerate. Therefore, $\mathcal{Q}$ is an ovoid.

Now we study the case where $\mathcal{Q}$ has index 2. This is the most difficult case, and we shall need several steps to solve it.

Step 1. We have that $\dim(\text{rad}(\mathcal{Q})) \leq 0$.

Assume that $\text{rad}(\mathcal{Q})$ contains a line g. Since $\mathcal{Q} \neq g$ there exists a point $P \in \mathcal{Q} \setminus g$. This implies $\langle P, g \rangle \subseteq \mathcal{Q}$, contradicting the fact that $\mathcal{Q}$ has only index 2.

Step 2. The number of $\mathcal{Q}$*-lines through a point of* $\mathcal{Q}$ *outside the radical is* 1 *or* 2.

This can be seen as follows. By **4.2.1** each point P of $\mathcal{Q}$ is on at least one $\mathcal{Q}$-line. Assume that P is on three $\mathcal{Q}$-lines. If these lines are not in a common plane then $\mathcal{Q}_P$ is the whole space, hence $P \in \text{rad}(\mathcal{Q})$, a contradiction. Hence the three lines lie in a common plane π. Since $\mathcal{Q} \cap \pi$ is a quadratic set it follows from **4.3.1** that $\mathcal{Q} \cap \pi$ is the whole plane π. Hence $\pi \subseteq \mathcal{Q}$, contradicting the fact that the index of $\mathcal{Q}$ equals 2.

Step 3. Each point of $\mathcal{Q}$ *that is not in* $\text{rad}(\mathcal{Q})$ *lies on the same number of* $\mathcal{Q}$*-lines.*

Assume that there are two points $P_1, P_2 \in \mathcal{Q}$ such that P_1 is on one $\mathcal{Q}$-line g_1 and P_2 is on two $\mathcal{Q}$-lines g_2, g_3.

Then the tangent plane $\mathcal{Q}_{P_1}$ passes through g_1, and the only points of $\mathcal{Q}$ in $\mathcal{Q}_{P_1}$ are those of g_1. The tangent plane $\mathcal{Q}_{P_2}$ is spanned by g_2 and g_3, and the points of $\mathcal{Q}$ in $\mathcal{Q}_{P_2}$ are those of g_2 and g_3. Therefore $\mathcal{Q}_{P_1}$ and $\mathcal{Q}_{P_2}$ are distinct planes, which meet in a line g.

First we look at g as a line in $\mathcal{Q}_{P_1}$. It either intersects g_1 in exactly one point or is equal to g_1. In the first case g is a line that contains precisely one point of $\mathcal{Q}$; if one considers g as line of $\mathcal{Q}_{P_2}$ then it becomes clear that this must be the point P_2. Thus we have $P_2 \in g_1$. In the second case it is $g = g_1$, hence in this case also P_2 lies on g_1. But for all points P on g_1 we have $\mathcal{Q}_P \supseteq \mathcal{Q}_{P_1}$, in particular $\mathcal{Q}_{P_2} \supseteq \mathcal{Q}_{P_1}$, a contradiction.

Now it is natural to distinguish two cases: either any nonradical point is on exactly one $\mathcal{Q}$-line or any such point is incident with exactly two $\mathcal{Q}$-lines.

Step 4. If each point of $\mathcal{Q}\setminus\text{rad}(\mathcal{Q})$ *is on just one* $\mathcal{Q}$*-line then* $\mathcal{Q}$ *is a cone.*

First we claim that in this case any two $\mathcal{Q}$-lines meet: Let g_1 and g_2 be two $\mathcal{Q}$-lines. Since the radical of $\mathcal{Q}$ consists of at most one point there is a point $P_1 \notin \text{rad}(\mathcal{Q})$ on g_1. By **4.2.1** the point P_1 must be joined with g_2 by a $\mathcal{Q}$-line. Since g_1 is the only $\mathcal{Q}$-line on P_1, g_1 must intersect the line g_2.

By the hypothesis of this step, the intersection of any two $\mathcal{Q}$-lines lies in $\text{rad}(\mathcal{Q})$. So the radical consists of exactly one point V, and all $\mathcal{Q}$-lines pass through V.

If π denotes a complement of V then, by **4.1.2**, $\mathcal{Q}$ induces a nondegenerate quadratic set $\mathcal{Q}'$ in π. Since $\mathcal{Q}$ has index 2 we have that $\mathcal{Q} \neq \mathcal{Q}'$. Moreover, $\mathcal{Q}'$ has index 1; so, by **4.3.1**, $\mathcal{Q} \cap \pi$ is an oval. Therefore $\mathcal{Q}$ is a cone.

Step 5. If any point of $\mathcal{Q}\setminus\text{rad}(\mathcal{Q})$ *is on exactly two* $\mathcal{Q}$*-lines then* $\mathcal{Q}$ *is a hyperboloid.*

Let g_1 be an arbitrary $\mathcal{Q}$-line. Since no three $\mathcal{Q}$-lines are in a common plane all $\mathcal{Q}$-lines that meet g_1 are skew; we denote the set of these lines by $\mathcal{R}_1$. Since any point of $\mathcal{Q}$ is connected by a $\mathcal{Q}$-line with g_1, the lines in $\mathcal{R}_1$ cover all points of $\mathcal{Q}$.

From this it follows that $\mathcal{Q}$ is nondegenerate. For each point of $\mathcal{Q}\setminus\text{rad}(\mathcal{Q})$ is on two $\mathcal{Q}$-lines, thus through each point of $\mathcal{Q}\setminus\text{rad}(\mathcal{Q})$ outside g_1 there is a $\mathcal{Q}$-line g_2 that does not intersect g_1. A hypothetical point X in $\text{rad}(\mathcal{Q})$ is w.l.o.g. not on g_1. Then all lines through X in $\langle X, g_1\rangle$ are tangents, hence $\mathcal{Q}$-lines. Therefore $\langle X, g_1\rangle \subseteq \mathcal{Q}$, a contradiction.

We are now ready to show that $\mathcal{R}_1$ is a regulus. For this it is sufficient to show that through any point P of a line $g \in \mathcal{R}_1$ there is a transversal to $\mathcal{R}_1$. Since $\mathcal{Q}$ is nondegenerate there is a uniquely determined $\mathcal{Q}$-line $h \neq g$ through

P. Since P is connected with any line of $\mathfrak{R}_1 \setminus \{g\}$ by a $\mathfrak{Q}$-line, the line h meets any line of $\mathfrak{R}_1$. Hence $\mathfrak{R}_1$ is a regulus.

This implies that $\mathfrak{Q}$ is a hyperboloid.

As an easy finish of this proof we consider the case that the index of $\mathfrak{Q}$ equals 3. (Note that the index of $\mathfrak{Q}$ is at most 3 since $\mathfrak{Q}$ is not the whole space.) Since $\mathfrak{Q}$ is not a subspace it must contain at least two planes. If $\mathfrak{Q}$ contains three planes then $\mathfrak{Q}$ is the whole space, a contradiction.

Thus we have proved Theorem **4.3.2** completely. □

4.4 Quadratic sets in finite projective spaces

Now we study quadratic sets in finite projective spaces. There will be a surprise: the upper bound for the index of a nondegenerate quadratic set (see **4.2.4**) is nearly the same as the lower bound. It turns out that there are only three types of nondegenerate quadratic sets in finite projective spaces.

In this section we suppose that $\mathbf{P} = PG(d, q)$ is a finite projective space of dimension d and order q. Let $\mathfrak{Q}$ be a quadratic set in $\mathbf{P}$.

4.4.1 Lemma. *For a point* $P \in \mathfrak{Q} \setminus \mathrm{rad}(\mathfrak{Q})$ *we denote by* $a\ (= a_P)$ *the number of* $\mathfrak{Q}$*-lines through* P. *Then the following assertions hold.*

(a) *If* $\mathfrak{Q}_P$ *is a hyperplane then* $\mathfrak{Q}_P$ *contains exactly* $aq + 1$ *points of* $\mathfrak{Q}$.

(b) *We have that* $|\mathfrak{Q}| = 1 + q^{d-1} + aq$; *in particular,* a *is independent of the choice of the point* $P \in \mathfrak{Q} \setminus \mathrm{rad}(\mathfrak{Q})$.

Proof. We observe that each line through P in $\mathfrak{Q}_P$ contains either no further point of $\mathfrak{Q}$ or exactly q further points of $\mathfrak{Q}$, while each line through P outside $\mathfrak{Q}_P$ has exactly one further point of $\mathfrak{Q}$.

On the a $\mathfrak{Q}$-lines through $P \in \mathfrak{Q} \setminus \mathrm{rad}(\mathfrak{Q})$ there are exactly $1 + aq$ points of $\mathfrak{Q}$. Since these are all the points of $\mathfrak{Q}$ in the tangent hyperplane $\mathfrak{Q}_P$, we have proved (a). All lines through P that are not contained in $\mathfrak{Q}_P$ intersect $\mathfrak{Q}$ in a second point. Since there are exactly q^{d-1} such lines, (b) also follows. □

Example. By **4.4.1** we can easily compute the numbers of points of quadratic sets in 2-dimensional and 3-dimensional projective spaces:

- An oval has exactly $q + 1$ points $(d = 2, a = 0)$.
- An ovoid has $q^2 + 1$ points $(d = 3, a = 0)$, a hyperboloid has $q^2 + 1 + 2q = (q + 1)^2$ points $(d = 3, a = 2)$, and a cone has $q^2 + 1 + q = q^2 + q + 1$ points $(d = 3, a = 1)$.

It is a remarkable fact that in finite projective spaces there exist only very few types of quadratic sets. This phenomenon can be seen for the first time in 4-dimensional projective spaces.

4.4.2 Theorem. *Any nonempty, nondegenerate quadratic set in* $\mathbf{P} = \mathrm{PG}(4, q)$ *has index* 2.

Proof. Let $\mathcal{Q}$ be a nonempty, nondegenerate quadratic set of $\mathbf{P}$. In view of **4.2.4** we have only to show that the index of $\mathcal{Q}$ is at least 2.

Assume that $\mathcal{Q}$ has index 1. Then $a = 0$, and, by **4.4.1**(b), $\mathcal{Q}$ has exactly $q^3 + 1$ points.
We claim: if $\mathbf{H}$ *is a hyperplane of* $\mathbf{P}$ *that contains at least two points of* $\mathcal{Q}$ *then the induced set* $\mathcal{Q}' = \mathcal{Q} \cap \mathbf{H}$ *is an ovoid.*

For $\mathcal{Q}'$ is a quadratic set of index 1 as well. Since $|\mathcal{Q}'| \geq 2$, $\mathcal{Q}'$ is nondegenerate. Therefore, the claim follows by **4.3.2**. In particular, $\mathcal{Q}'$ has exactly $q^2 + 1$ points.

We now consider the geometry consisting of the points of $\mathcal{Q}$ together with those hyperplanes that intersect $\mathcal{Q}$ in at least two points. We shall get a contradiction if we try to compute the number of these hyperplanes.

Through a fixed point P of $\mathcal{Q}$ there is the tangent hyperplane $\mathcal{Q}_P$; all other hyperplanes are of the type we are interested in. Hence through any point of $\mathcal{Q}$ there are exactly $q^3 + q^2 + q$ hyperplanes of interest.

Using this observation we are able to compute the number b of those hyperplanes:

$$b = \frac{|Q| \cdot (q^3 + q^2 + q)}{q^2 + 1},$$

since by our claim, each point of $\mathcal{Q}$ is counted $q^2 + 1$ times.

Since b is an integer, $q^2 + 1$ must divide the product $(q^3 + 1)\cdot(q^3 + q^2 + q)$. This is impossible, so we have a contradiction. □

The above theorem can be generalized to spaces of even dimension.

4.4.3 Theorem. *Any nonempty, nondegenerate quadratic set in* $\mathbf{P} = \mathrm{PG}(2t, q)$ *has index* t.

Proof. Let $\mathcal{Q}$ be a nonempty, nondegenerate quadratic set of $\mathbf{P}$. By **4.2.4** we have only to show that the index of $\mathcal{Q}$ is at least t. For this we proceed by induction on t.

The case $t = 1$ is treated in **4.3.1**, and the case $t = 2$ was dealt with in **4.4.2**. Therefore we suppose $t \geq 3$, and assume that the assertion is true for $t - 1$.

Our first *claim* is that the *index of* $\mathfrak{Q}$ *is at least* 2. For this we assume that the index of $\mathfrak{Q}$ is 1. Then each tangent hyperplane contains just one point of $\mathfrak{Q}$.

Consider two distinct points P, P' of $\mathfrak{Q}$, and let $\mathbf{U}$ be a $(2t-2)$-dimensional subspace of $\mathbf{P}$ through P and P'. By construction, the quadratic set $\mathfrak{Q}' = \mathfrak{Q} \cap \mathbf{U}$ is nonempty. For a point $R \in \mathfrak{Q}'$ we consider its tangent hyperplane $\mathfrak{Q}'_R$. By **4.1.1** we have that $\mathfrak{Q}'_R = \mathfrak{Q}_R \cap \mathbf{U}$. Since by assumption $\mathfrak{Q}_R$ contains only one point of $\mathfrak{Q}$ it follows that $\mathfrak{Q}'_R \neq \mathbf{U}$. Therefore, $\mathfrak{Q}'$ is a nonempty, nondegenerate quadratic set of index 1 in the $2(t-1)$-dimensional projective space $\mathbf{U}$ with $t \geq 2$, a contradiction.

Thus the index of $\mathfrak{Q}$ is at least 2. Now we consider a point $P \in \mathfrak{Q}$ and its tangent hyperplane $\mathbf{H} = \mathfrak{Q}_P$. By $\mathbf{W}$ we denote a complement of P in $\mathbf{H}$, that is a subspace of dimension $2t-2$ of $\mathbf{H}$ that does not contain P; let $\mathfrak{Q}' = \mathfrak{Q} \cap \mathbf{W}$ be the quadratic set induced by $\mathfrak{Q}$ in $\mathbf{W}$. Since the index of $\mathfrak{Q}$ is at least 2, there is at least one $\mathfrak{Q}$-line through P, in particular, $\mathfrak{Q}'$ is not empty. By **4.1.4**, $\mathfrak{Q}'$ is nondegenerate.

Therefore, by induction, the index of $\mathfrak{Q}'$ equals $t-1$. Thus $\mathfrak{Q}'$ contains a $\mathfrak{Q}$-subspace $\mathbf{U}$ of dimension $t-2$. It follows that the subspace $\langle P, \mathbf{U}\rangle$ has dimension $t-1$ and is contained in $\mathfrak{Q}$. Hence the index of $\mathfrak{Q}$ is at least t. □

As a corollary we get the following important theorem, which is due to Ernst Witt (1911–1991).

4.4.4 Theorem. *Let* $\mathfrak{Q}$ *be a nonempty, nondegenerate quadratic set of a finite projective space* $\mathbf{P}$ *of dimension* d. *Then there are only three possibilities for the index* s *of* $\mathfrak{Q}$: *if* d *is even then*

$$s = \frac{d}{2};$$

if d *is odd then*

$$s = \frac{d-1}{2} \quad or \quad s = \frac{d+1}{2}.$$

Proof. The case d even has been studied in **4.4.3**. Suppose therefore that $d = 2t+1$. By **4.2.4** the index of $\mathfrak{Q}$ is at most $t+1$.

It remains to show that $s \geq t$. In order to do this we consider a hyperplane $\mathbf{H}$ that contains at least one point of $\mathfrak{Q}$ and is not a tangent hyperplane. Such a hyperplane exists. Counting the incident point–hyperplane pairs $(P, \mathfrak{Q})$ with $P \in \mathfrak{Q}$ we get on the one hand $|\mathfrak{Q}|\cdot(q^{d-1} + \ldots + 1)$, since there are $q^{d-1} + \ldots + 1$ hyperplanes through any point of $\mathfrak{Q}$. On the other hand there are $|\mathfrak{Q}|$ tangent hyperplanes, each containing $1 + aq$ points of $\mathfrak{Q}$. Since $\mathfrak{Q}$ is nondegenerate not all points of a tangential hyperplane lie in $\mathfrak{Q}$, therefore $a < q^{d-2} + \ldots + 1$. Thus there must be a hyperplane containing a point of $\mathfrak{Q}$ that is not a tangent hyperplane.

Then the index s' of the quadratic set $\mathfrak{Q}' = \mathfrak{Q} \cap \mathbf{H}$ satisfies $s' \le s$. Moreover, by **4.1.4**, $\mathfrak{Q}'$ is nondegenerate and by construction, $\mathfrak{Q}'$ is not empty.

Now Theorem **4.4.3** yields $s' = t$, and therefore $s \ge s' = t$. □

4.5 Elliptic, parabolic, and hyperbolic quadratic sets

In the preceding section we proved that the index of any nonempty, nondegenerate quadratic set of a finite d-dimensional projective space is $\frac{d-1}{2}$, $\frac{d}{2}$, or $\frac{d+1}{2}$. The quadratic sets of such indices also play in general (not only in finite spaces) the main roles. We introduce the traditional names.

Definition. Let $\mathfrak{Q}$ be a nondegenerate quadratic set in a d-dimensional projective space $\mathbf{P}$. If d is even and the index of $\mathfrak{Q}$ is $\frac{d}{2}$ then $\mathfrak{Q}$ is called **parabolic**. If d is odd then $\mathfrak{Q}$ is called **elliptic** if the index of $\mathfrak{Q}$ is $\frac{d-1}{2}$, and **hyperbolic** if its index is $\frac{d+1}{2}$.

Examples. (a) We can reformulate Theorem **4.4.4** as follows. Any nonempty nondegenerate quadratic set of a finite projective space is elliptic, parabolic, or hyperbolic.

(b) The parabolic quadratic sets of a projective plane are precisely the ovals. In a 3-dimensional projective space, the elliptic quadratic sets are precisely the ovoids, the hyperboloids are precisely the hyperbolic quadratic sets.

Now we study the three most important types of nondegenerate quadratic sets in greater detail by looking at the structures they induce in hyperplanes. The results of this section show that one can ‘reduce’ the determination of nondegenerate quadratic sets in spaces of ‘big’ dimension to the investigation of quadratic sets in spaces of small dimension.

It is convenient to operate with the following general definition of a cone.

Definition. Let $\mathbf{H}$ be a hyperplane of a projective space $\mathbf{P}$, and denote by V a point outside $\mathbf{H}$. If $\mathfrak{Q}^*$ is a nondegenerate quadratic set of $\mathbf{H}$ then the quadratic set

$$\mathfrak{Q} := \bigcup_{X \in \mathfrak{Q}^*} (VX)$$

is called a **cone** with **vertex** V over $\mathfrak{Q}^*$.

We begin by studying the parabolic quadratic sets.

4.5.1 Theorem. *Let* $\mathfrak{Q}$ *be a parabolic quadratic set in a* $2t$*-dimensional projective space* $\mathbf{P}$ *with* $t \geq 2$.
(a) *Let* $\mathbf{H} = \mathfrak{Q}_P$ *be a tangent hyperplane. Then* $\mathfrak{Q}' := \mathfrak{Q} \cap \mathbf{H}$ *is a cone over a parabolic quadratic set.*
(b) *Let* $\mathbf{H}^*$ *be a hyperplane that is not a tangent hyperplane. Then* $\mathfrak{Q}^* := \mathfrak{Q} \cap \mathbf{H}^*$ *is an elliptic or a hyperbolic quadratic set.*

Proof. (a) Let $\mathbf{W}$ be a complement of P in $\mathbf{H}$, and define $\mathfrak{Q}'' := \mathfrak{Q} \cap \mathbf{W}$. By **4.1.4**, $\mathfrak{Q}''$ is a nondegenerate quadratic set. If $\mathbf{U}$ denotes a maximal $\mathfrak{Q}$-subspace through P then $\mathbf{U}'' := \mathbf{U} \cap \mathbf{W}$ has dimension $\dim(\mathbf{U}) - 1$. By **4.2.4**, $\mathbf{U}''$ is a maximal $\mathfrak{Q}''$-subspace, therefore $\mathfrak{Q}''$ is parabolic. Since by **4.1.3** the radical of $\mathfrak{Q}'$ consist of just one point, namely P, $\mathfrak{Q}'$ is a cone over $\mathfrak{Q}''$ with vertex P (cf. **4.1.2**).
(b) Since $\mathbf{H}^*$ is not a tangent hyperplane $\mathfrak{Q}^*$ is nondegenerate (cf. **4.1.4**). A maximal $\mathfrak{Q}$-subspace (which is a subspace of dimension $t-1$) intersects $\mathbf{H}^*$ in a subspace of dimension $t-1$ or $t-2$. This means that $\mathfrak{Q}^*$ is elliptic or hyperbolic. □

The above theorem seems to be very academic. But the following example shows its usefulness:

4.5.2 Corollary. *Let* $\mathfrak{Q}$ *be a nonempty, nondegenerate quadratic set in* $\mathbf{P} = PG(4, q)$. *Then* $\mathfrak{Q}$ *induces in any tangent hyperplane a cone, and in any other hyperplane an ovoid or a hyperboloid. Furthermore,* $\mathfrak{Q}$ *consist of exactly* $q^3 + q^2 + q + 1$ *points, the number of hyperplanes in which* $\mathfrak{Q}$ *induces an ovoid is* $q^2(q^2 - 1)/2$, *and the number of hyperplanes in which* $\mathfrak{Q}$ *induces a hyperboloid is* $q^2(q^2 + 1)/2$.

Proof. We know that $\mathfrak{Q}$ is parabolic. From this the first assertion follows in view of **4.5.1**. In particular we get that for each point $P \in \mathfrak{Q}$ the quadratic set induced in $\mathfrak{Q}_P$ is a cone with vertex P. Thus the number a of $\mathfrak{Q}$-lines through P equals $a = q + 1$. Hence, by **4.4.1** we have

$$|\mathfrak{Q}| = 1 + q^3 + a \cdot q = 1 + q^3 + (q+1)q.$$

Let $t = |\mathfrak{Q}| = q^3 + q^2 + q + 1$ be the number of tangent hyperplanes, and let h and e be the numbers of the 'hyperboloid hyperplanes' and the 'ovoid hyperplanes'. It follows that

$$t + h + e = q^4 + q^3 + q^2 + q + 1,$$

so

$$h + e = q^4.$$

Since a tangent hyperplane contains exactly $q^2 + q + 1$ points of $\mathcal{Q}$, while a hyperboloid hyperplane contains $(q + 1)^2$, and an ovoid hyperplane exactly $q^2 + 1$ points of $\mathcal{Q}$, and any point of $\mathcal{Q}$ is on exactly $q^3 + q^2 + q + 1$ hyperplanes, we see that

$$t \cdot (q^2 + q + 1) + h \cdot (q + 1)^2 + e \cdot (q^2 + 1) = |\mathcal{Q}| \cdot (q^3 + q^2 + q + 1).$$

This implies

$$\begin{aligned} h \cdot (q + 1)^2 + e \cdot (q^2 + 1) &= |\mathcal{Q}| \cdot (q^3 + q^2 + q + 1) - t \cdot (q^2 + q + 1) \\ &= q^3 \cdot (q^3 + q^2 + q + 1). \end{aligned}$$

Putting this together with the above equation we have

$$h \cdot (q^2 + 2q + 1) + (q^4 - h) \cdot (q^2 + 1) = q^3 \cdot (q^3 + q^2 + q + 1),$$

so

$$2q \cdot h = q^3(q^2 + 1).$$

From this it follows that

$$h = \frac{q^2(q^2 + 1)}{2} \quad \text{and} \quad e = q^4 - h = \frac{q^2(q^2 - 1)}{2}. \qquad \square$$

Now we study the particularly important hyperbolic case. For the elliptic case we refer to the exercises (see exercise 18).

4.5.3 Theorem. *Let $\mathcal{Q}$ be a hyperbolic quadratic set of a $(2t + 1)$-dimensional projective space $\mathbf{P}$ with $t \geq 2$. Then the following assertions are true.*
(a) *If $\mathbf{H} = \mathcal{Q}_P$ is a tangent hyperplane then $\mathcal{Q}' := \mathcal{Q} \cap \mathbf{H}$ is a cone over a hyperbolic quadratic set.*
(b) *Let $\mathbf{H}^*$ be a hyperplane that is not a tangent hyperplane. Then $\mathcal{Q}^* := \mathcal{Q} \cap \mathbf{H}$ is a parabolic quadratic set.*

Proof. (a) The proof is nearly literally the same as the proof of **4.5.1**(a):

Let $\mathbf{W}$ be a complement of P in $\mathbf{H}$, and define $\mathcal{Q}'' := \mathcal{Q} \cap \mathbf{W}$. By **4.1.4** , $\mathcal{Q}''$ is a nondegenerate quadratic set. If $\mathbf{U}$ denotes a maximal $\mathcal{Q}$-subspace through P (which means in particular that $\mathbf{U}$ has dimension t) then $\mathbf{U}'' := \mathbf{U} \cap \mathbf{W}$ has dimension $t - 1$. By **4.2.4**, $\mathbf{U}''$ is a maximal $\mathcal{Q}''$-subspace, therefore $\mathcal{Q}''$ is hyperbolic. Since by **4.1.3** the radical of $\mathcal{Q}'$ consists of just one point, namely P, $\mathcal{Q}'$ is a cone over $\mathcal{Q}''$ with vertex P (cf. **4.1.2**).

(b) The assertion follows from **4.1.4** and **4.4.3**. □

Also in this case we consider a particularly important special case in more detail:

4.5.4 Theorem. *Let* $\mathcal{Q}$ *be a hyperbolic quadratic set in* $\mathbf{P} = \mathrm{PG}(5, q)$. *Then* $a = (q+1)^2$ *and*

$$|\mathcal{Q}| = q^4 + q^3 + 2q^2 + q + 1 = (q^2 + q + 1)\cdot(q^2 + 1).$$

The *proof* follows from **4.5.3** together with **4.4.1**. □

A hyperbolic quadratic set has an extremely remarkable property: one can divide its maximal $\mathcal{Q}$-subspaces sets into two equivalence classes. The equivalence relation is given by the property that the corresponding subspaces intersect in a subspace of a certain parity.

In a 3-dimensional projective space we already know that the maximal $\mathcal{Q}$-subspaces of a hyperboloid are the lines of a regulus and its opposite regulus. If we define that two $\mathcal{Q}$-lines are equivalent if they are equal or skew we get two classes of $\mathcal{Q}$-lines.

Since the case $d = 5$ is particularly important (and will be studied in the next section in great detail) we shall first consider this case.

Definition. Let $\mathcal{Q}$ be a hyperbolic quadratic set of a 5-dimensional projective space $\mathbf{P}$. We say that two $\mathcal{Q}$-planes π_1, π_2 are **equivalent** (and we shall write $\pi_1 \sim \pi_2$) if π_1 and π_2 are equal or intersect each other in precisely one point.

Obviously, the relation $\sim$ is reflexive and symmetric. Surprisingly, it is also transitive:

4.5.5 Lemma. *Let* $\mathcal{Q}$ *be a hyperbolic quadratic set of a 5-dimensional projective space* $\mathbf{P}$. *Then the relation* $\sim$ *is an equivalence relation.*

Proof. We only have to show that $\sim$ is transitive. Let π_1, π_2, π_3 be three $\mathcal{Q}$-planes such that π_1 and π_2 intersect in precisely one point P, and π_2 and π_3 intersect in precisely one point R.

Since all lines of π_1 and π_2 through P are tangents it follows that $\pi_1, \pi_2 \subseteq \mathcal{Q}_P$, therefore $\mathcal{Q}_P = \langle \pi_1, \pi_2 \rangle$. Similarly it follows that $\mathcal{Q}_R = \langle \pi_2, \pi_3 \rangle$.

Case 1. P = R.

Consider a complement $\mathbf{W}$ of P in $\mathbf{H} = \mathcal{Q}_P$ and define $\mathcal{Q}' := \mathcal{Q} \cap \mathbf{W}$. Let $g_i = \pi_i \cap \mathbf{W}$ $(i = 1, 2, 3)$. By **4.5.3** we know that $\mathcal{Q}'$ is a hyperboloid. Since g_1 and g_2 are skew they belong to the same class of $\mathcal{Q}'$. Similarly, g_2 and g_3 are skew, so they belong to the same class of $\mathcal{Q}'$. From these together it follows that g_1 and g_3 also belong to the same class of $\mathcal{Q}'$. This means that $\pi_1 \cap \pi_3 = P$. Hence $\pi_1 \sim \pi_3$.

Case 2. $P \neq R$.

In this case the plane π_3 is not contained in $\mathcal{Q}_P$. (Otherwise $\langle P, \pi_3 \rangle$ would be a $\mathcal{Q}$-subspace.) Hence π_3 intersects the hyperplane $\mathcal{Q}_P$ in a line g_3.

Let $\mathbf{W}$ be a complement of P in $\mathcal{Q}_P$ containing g_3, and define $\mathcal{Q}' := \mathcal{Q} \cap \mathbf{W}$. Then $\mathcal{Q}'$ is a hyperboloid, and $g_1 := \pi_1 \cap \mathbf{W}$ and $g_2 := \pi_2 \cap \mathbf{W}$ are lines of the same class of $\mathcal{Q}'$. Since g_3 intersects the line g_2 in the point R, g_3 belongs to the other class. This implies that g_1 and g_3 also intersect each other in some point S. Thus, π_1 and π_3 intersect each other just in the point S. This implies $\pi_1 \sim \pi_3$. □

In general, the fewer equivalence classes an equivalence relation has, the 'stronger' it is. The strongest equivalence relations are those having just two equivalence classes.

4.5.6 Theorem. *Let $\mathcal{Q}$ be a hyperbolic quadratic set of a 5-dimensional projective space* **P**. *Then the set of all $\mathcal{Q}$-planes is partitioned into exactly two equivalence classes with respect to* $\sim$.

Proof. Let π_1 and π_2 be two $\mathcal{Q}$-planes that intersect in a line g. So they belong to different equivalence classes. We have to show that each $\mathcal{Q}$-plane π belongs to one of these equivalence classes.

The subspace $\mathbf{V} := \langle \pi_1, \pi_2 \rangle$ spanned by π_1 and π_2 has dimension 3, and the quadratic set induced by $\mathcal{Q}$ in $\mathbf{V}$ consists only of the points in π_1 and π_2.

W.l.o.g. we may assume that $\pi \neq \pi_1, \pi_2$. By the dimension formula, the plane π intersects the subspace $\mathbf{V}$, the intersection being contained in $\pi_1 \cup \pi_2$. Since π is not contained in $\mathbf{V}$, π cannot intersect both planes in different lines since otherwise $\pi \subseteq \mathbf{V}$. If π intersects both planes in one point not on g then the joining line is a $\mathcal{Q}$-line contained in $\mathbf{V}$, a contradiction.

Assume that π intersects $\mathbf{V}$ in the points of g. Consider a point P on g. By **4.5.3**, $\mathcal{Q}$ induces in any complement $\mathbf{W}$ of P in $\mathcal{Q}_P$ a hyperboloid. However, the point in which $\mathbf{V}$ and g intersect is on three distinct $\mathcal{Q}$-planes; thus, in $\mathbf{W}$, this point is on three distinct $\mathcal{Q}$-lines. This is a contradiction since $\mathcal{Q} \cap \mathbf{W}$ is a hyperboloid.

Hence π intersects either one of the planes π_1, π_2 in a line and the other in a point, or only one of the planes in exactly one point. In any case π belongs to the equivalence class of π_1 or to the equivalence class of π_2. □

With a little more technical effort one can generalize **4.5.5** to higher dimensions.

Definition. Let $\mathfrak{Q}$ by a hyperbolic quadratic set of a $(2t+1)$-dimensional projective space **P**. For any two maximal $\mathfrak{Q}$-subspaces $\mathbf{U}_1, \mathbf{U}_2$ we define:

$$\mathbf{U}_1 \sim \mathbf{U}_2 :\Leftrightarrow t + \dim(\mathbf{U}_1 \cap \mathbf{U}_2) \text{ is even.}$$

We shall deal with the case $t = 3$, and leave the general case as a challenging exercise for the reader (see exercise 19).

4.5.7 Theorem. *Let* $\mathfrak{Q}$ *be a hyperbolic quadratic set in a 7-dimensional projective space* **P**. *Then* $\sim$ *is an equivalence relation with exactly two equivalence classes.*

Proof. First we show that $\sim$ is an equivalence relation. In order to do this we have only to show that $\sim$ is transitive: Let $\mathbf{U}_1, \mathbf{U}_2, \mathbf{U}_3$ be 3-dimensional $\mathfrak{Q}$-subspaces such that $\mathbf{U}_1 \sim \mathbf{U}_2$ and $\mathbf{U}_2 \sim \mathbf{U}_3$. This means that $\dim(\mathbf{U}_1 \cap \mathbf{U}_2)$ and $\dim(\mathbf{U}_2 \cap \mathbf{U}_3)$ are odd. If $\mathbf{U}_1 \cap \mathbf{U}_3 = \varnothing$ then $\dim(\mathbf{U}_1 \cap \mathbf{U}_3) = -1$, hence $\mathbf{U}_1 \sim \mathbf{U}_3$.

Therefore we may assume that there is a point S in $\mathbf{U}_1 \cap \mathbf{U}_3$.

Case 1. $S \in \mathbf{U}_2$.

In this case we have that $\mathbf{U}_1, \mathbf{U}_2, \mathbf{U}_3 \subseteq \mathfrak{Q}_S$. Consider a complement **W** of S in $\mathfrak{Q}_S$. Then, by **4.5.3**, $\mathfrak{Q} \cap \mathbf{W}$ is a hyperbolic quadratic set. If we define $\mathbf{W}_i := \mathbf{U}_i \cap \mathbf{W}$ $(i = 1, 2, 3)$ we see that $\mathbf{W}_1 \sim \mathbf{W}_2$ and $\mathbf{W}_2 \sim \mathbf{W}_3$; thus, by **4.5.5** we have $\mathbf{W}_1 \sim \mathbf{W}_3$, as well. Therefore, $\mathbf{U}_1 \sim \mathbf{U}_3$.

Case 2. $S \notin \mathbf{U}_2$.

Then $\mathbf{U}_2 \not\subset \mathfrak{Q}_S$. Let $\mathbf{W}_2 := \mathbf{U}_2 \cap \mathfrak{Q}_S$, and consider a complement **W** of S in $\mathfrak{Q}_S$ that passes through $\mathbf{W}_2$; finally let $\mathbf{W}_i := \mathbf{U}_i \cap \mathbf{W}$ $(i = 1, 3)$. Since

$$\mathbf{W}_2 \cap \mathbf{W}_i = \mathbf{W}_2 \cap \mathbf{U}_i = \mathbf{U}_2 \cap \mathbf{U}_i$$

$\dim(\mathbf{W}_2 \cap \mathbf{W}_i)$ is odd; therefore $\mathbf{W}_2$ is not equivalent to $\mathbf{W}_i$ $(i = 1, 3)$. Thus $\mathbf{W}_1$ and $\mathbf{W}_2$ as well as $\mathbf{W}_2$ and $\mathbf{W}_3$ are in distinct equivalence classes. Since by **4.5.6** there are only two equivalence classes we have that $\mathbf{W}_1 \sim \mathbf{W}_3$ and therefore $\mathbf{U}_1 \sim \mathbf{U}_3$.

It remains to show that there are exactly two equivalence classes. Let $\mathbf{U}_1, \mathbf{U}_2$ be two 3-dimensional $\mathfrak{Q}$-subspaces that intersect in a plane π. We have to show that each 3-dimensional $\mathfrak{Q}$-subspace $\mathbf{U}_0$ is equivalent to $\mathbf{U}_1$ or to $\mathbf{U}_2$.

The subspace $\mathbf{V} := \langle \mathbf{U}_1, \mathbf{U}_2 \rangle$ has dimension 4, and we have $\mathfrak{Q} \cap \mathbf{V} = \mathbf{U}_1 \cup \mathbf{U}_2$. Moreover, $\mathbf{U}_0$ intersects the subspace **V** in at least one point of $\mathbf{U}_1 \cup \mathbf{U}_2$.

Assume that $\mathbf{U}_0 \cap \mathbf{V} \subseteq \pi$. Let P be a point of $\mathbf{U}_0 \cap \mathbf{U}_1 \cap \mathbf{U}_2$. Then we have that $\mathbf{U}_0, \mathbf{U}_1, \mathbf{U}_2 \subseteq \mathcal{Q}_P$. Hence in a complement $\mathbf{W}$ of P in $\mathcal{Q}_P$, the subspaces $\mathbf{W}_i := \mathbf{U}_i \cap \mathbf{W}$ $(i = 0, 1, 2)$ are three maximal subspaces, where $\mathbf{W}_1$ is not equivalent to $\mathbf{W}_2$. By assumption, $\mathbf{U}_0$ intersects $\mathbf{U}_1$ and $\mathbf{U}_2$ in a subspace contained in $\mathbf{U}_1 \cap \mathbf{U}_2$. Thus, $\mathbf{W}_0$ also intersects $\mathbf{W}_1$ and $\mathbf{W}_2$ in a subspace contained in $\mathbf{W}_1 \cap \mathbf{W}_2$. In particular $\mathbf{W}_0 \cap \mathbf{W}_1$ and $\mathbf{W}_0 \cap \mathbf{W}_2$ have the same dimension, which contradicts **4.5.6**.

In view of $\mathcal{Q} \cap \mathbf{V} = \mathbf{U}_1 \cup \mathbf{U}_2$ the subspace $\mathbf{U}_0$ cannot contain a point of $\mathbf{U}_1 \setminus \pi$ as well as a point of $\mathbf{U}_2 \setminus \pi$. Thus, w.l.o.g. we have

$$\mathbf{U}_0 \cap \mathbf{V} \subseteq \mathbf{U}_1.$$

Now there are three possibilities: either $\mathbf{U}_0$ intersects the plane π (and therefore $\mathbf{U}_2$) in one line and we have that $\mathbf{U}_0 \sim \mathbf{U}_2$, or $\mathbf{U}_0$ intersects π in just one point (and therefore $\mathbf{U}_1$ in a line) and we have that $\mathbf{U}_0 \sim \mathbf{U}_1$, or $\mathbf{U}_0 \cap \pi = \varnothing$; then also $\mathbf{U}_0 \cap \mathbf{U}_2 = \varnothing$, so $\mathbf{U}_0 \sim \mathbf{U}_2$. In any case $\mathbf{U}_0$ belongs to the equivalence class of $\mathbf{U}_1$ or that of $\mathbf{U}_2$. □

To conclude this section we shall present a diagram of a hyperbolic quadratic set. (We have introduced diagrams in Section **1.7**.)

Let us consider a hyperbolic quadratic set $\mathcal{Q}$ of a 5-dimensional projective space **P**. Then the points, lines, and planes of $\mathcal{Q}$ form a geometry of rank 3. Moreover, the residue of each plane – the points and lines incident with that plane – is a projective plane, and the residue of each point P – the $\mathcal{Q}$-lines and $\mathcal{Q}$-planes incident with P – is a hyperbolic quadratic set of a 3-dimensional projective space. Since all $\mathcal{Q}$-lines through P lie in $\mathcal{Q}_P$ and since, by **4.5.4**, $\mathcal{Q} \cap \mathcal{Q}_P$ is a cone over a hyperbolic quadratic set of a 3-dimensional projective space, the assertion follows.

So, from an incidence-geometric point of view, the residue of a point of $\mathcal{Q}$ consists of a regulus and its opposite regulus. In order to present a diagram for $\mathcal{Q}$, we therefore need a symbol for those geometries. Such an incidence structure is a special case of a 'generalized quadrangle', for generalized quadrangles the following symbol has been introduced:

Definition. A **generalized quadrangle** is a rank 2 geometry consisting of points and lines such that the following properties are satisfied:

– Any two distinct points are on at most one common line.

– All lines are incident with the same number of points; all points are incident with the same number of lines.
– If P is a point outside a line g then there is precisely one line through P intersecting g.

The last property is the most important and is the justification of the name generalized 'quadrangle'. Examples of generalized quadrangles are the quadrangle and a hyperbolic quadratic set of a 3-dimensional projective space (see Figure 4.3).

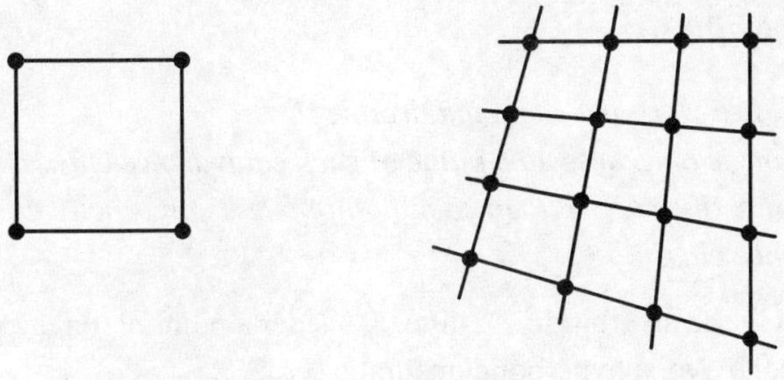

Figure 4.3 Examples of generalized quadrangles

There are much more interesting examples of generalized quadrangles (see exercise 22). The reader who wants to get an impression of this very active branch of research should consult [PaTh85].

Now we may formulate

4.5.8 Theorem. *The geometry consisting of the points, lines, and planes of a hyperbolic quadratic set of a 5-dimensional projective space has the following diagram.*

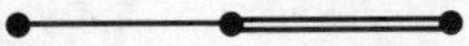

□

4.6 The Klein quadratic set

Definition. A hyperbolic quadratic set of a 5-dimensional projective space is also called a **Klein quadratic set** (Christian Felix Klein (1849–1925)).

The Klein quadratic set is particularly important since a 3-dimensional projective space is disguised in it. We shall now reveal this secret.

Let $\mathcal{C}_1$ and $\mathcal{C}_2$ be the two equivalence classes of planes of a Klein quadratic set $\mathcal{Q}$ of a 5-dimensional projective space **P**. We define a geometry **S** as follows.

– The *points* of **S** are the planes of $\mathcal{C}_1$;
– the *lines* of **S** are the points of $\mathcal{Q}$;
– the *planes* of **S** are the planes of $\mathcal{C}_2$.
– The *incidence* between a line of **S** and a point or a plane of **S** is induced by the incidence of **P**; a point π_1 of **S** and a plane π_2 of **S** are incident if the planes π_1 and π_2 of **P** are not disjoint. (Then, by **4.5.5**, they intersect each other in a line of **P**.)

4.6.1 Lemma. *Let $\mathcal{Q}$ be the Klein quadratic set.*
(a) *Each $\mathcal{Q}$-line is on exactly one plane of each equivalence class.*
(b) *If **P** is finite of order q then each point of $\mathcal{Q}$ is on exactly $q+1$ planes of each equivalence class.*

Proof. (a) Let g be an arbitrary $\mathcal{Q}$-line. Consider a point P on g. We know that $\mathcal{Q}_P \cap \mathcal{Q}$ is a cone over a hyperbolic quadratic set $\mathcal{Q}''$.

The line g meets $\mathcal{Q}''$ in some point, which is incident with one line h, h' of each of the two classes of $\mathcal{Q}''$. It follows that $\langle P, h\rangle$ and $\langle P, h'\rangle$ are the planes of the two equivalence classes through g.
(b) follows similarly. □

The above mentioned highly remarkable connection is expressed in the following theorem.

4.6.2 Theorem. *The geometry **S** is a 3-dimensional projective space; more precisely, **S** is isomorphic to a 3-dimensional subspace of **P**.*

Proof. We proceed in several steps.
*Claim 1: Any two distinct points of **S** are incident with exactly one line of **S**.*

This directly follows from **4.5.5**, which says that any two planes of $\mathcal{C}_1$ intersect in precisely one point of $\mathcal{Q}$.
*Claim 2: If two points π, π' of **S** are incident with a plane π_2 of **S** then any point of the line incident with π and π' is also incident with π_2. (Hence π_2 is a linear set.) Moreover, the structure of points and lines of **S** incident with π_2 is a projective plane.*

The fact that π and π' are incident with π_2 means that π and π' intersect the plane π_2 of **P** in lines g and g'. The intersection P of g and g' is the common point of π and π'. This point P of **P** is the line in **S** through π and

π'. Since $\pi_2 \in \mathcal{C}_2$, any further plane from $\mathcal{C}_1$ through P intersects π_2 in a line, thus it is also incident with π_2. This means that each point of **S** that is incident with the line of **S** through π and π' is also incident with π_2.

By **4.6.1**, each line of π_2 is on exactly one plane of $\mathcal{C}_1$. Thus the structure of points and lines of **S** that are incident with π_2 is exactly the structure of lines and points of **P** on π_2. This is the dual plane of π_2. Since π_2 is Desarguesian it follows by **3.4.2** and **2.3.5** that $(\pi_2)^\Delta$ is isomorphic to π_2.

Claim 3: Any three points of **S** *that are not on a common line of* **S** *are incident with precisely one plane of* **S**.

Let π_1, π_2, π_3 be three points of **S** that are not incident with a common line of **S**. By definition, these are three planes from $\mathcal{C}_1$ not through a common point of $\mathcal{Q}$. Thus the points

$$P_1 := \pi_2 \cap \pi_3,\ P_2 := \pi_1 \cap \pi_3,\ P_3 := \pi_1 \cap \pi_2$$

are three distinct points of $\mathcal{Q}$. (They are even noncollinear, since otherwise π_1, π_2, π_3 would be incident with the line through P_1, P_2, P_3.)

We shall show: *the plane* π *from* $\mathcal{C}_2$ *through* P_1 *and* P_2 *also passes through* P_3: Since π is from $\mathcal{C}_2$, π_1 and π_2 are from $\mathcal{C}_1$ and since $\pi \cap \pi_1 \neq \varnothing$ and $\pi \cap \pi_2 \neq \varnothing$, the plane π intersects the planes π_1 and π_2 of P in lines g_1 and g_2. These lines of π intersect in some point X. This point satisfies

$$X = g_1 \cap g_2 = (\pi \cap \pi_1) \cap (\pi \cap \pi_2) \subseteq \pi_1 \cap \pi_2 = P_3.$$

Therefore $X = P_3$, and hence the point P_3 lies in π.

This implies that the plane π of **S** is incident with the points π_1, π_2, π_3 of **S**. There is no further plane with this property, since any such plane would have a line in common with π_2 and a line in common with π_3; hence it would pass through P_1. Similarly, it follows that this plane is through P_2 and P_3; hence it would be equal to π.

(1), (2), and (3) imply that **S** is a projective space (see also exercise 24). We have to convince ourselves what the dimension of **S** is.

Claim 4: The projective space **S** *has dimension* 3.

For this we show that each line of **S** and each plane of **S** are incident with a common point of **S**. Then it follows that each plane is a hyperplane, so **S** has dimension 3.

Therefore, let P be a line of **S** and π_2 a plane of **S**, which w.l.o.g. are not incident. This means that P is a point of $\mathcal{Q}$ and π_2 a plane in $\mathcal{C}_2$ such that P is outside π_2.

The tangent hyperplane $\mathcal{Q}_P$ intersects π_2 in a line g. Then $\pi_1 := \langle P, g\rangle$ is a $\mathcal{Q}$-plane, which is in $\mathcal{C}_1$, since π_1 and π_2 have a line in common. Hence π_1 is a point of **S** that is incident with the line P and the plane π_1 of **S**. □

This view of Klein quadratic sets enables us to describe hyperbolic quadratic sets extremely significantly by another diagram.

4.6.3 Theorem. *Let $\mathcal{Q}$ be a hyperbolic quadratic set of a 7-dimensional projective space. Then the points, lines, and maximal subspaces of the two equivalence classes form a Buekenhout–Tits geometry of rank* 4 *having the following diagram:*

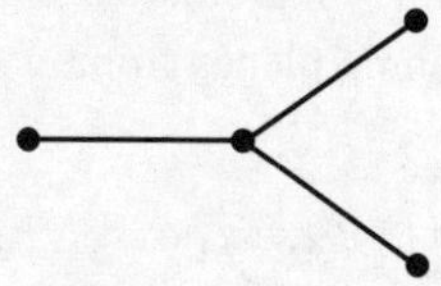

Here, two 3*-dimensional $\mathcal{Q}$-subspaces are incident if they intersect in a plane.*

Proof. The diagram says that the residual of any point and any maximal subspace must be a 3-dimensional projective space.

It is clear that the residual of any maximal $\mathcal{Q}$-subspace is a 3-dimensional projective space.

Therefore we only have to show that the residual of a point P is a 3-dimensional projective space. We shall show that this residual is isomorphic to the projective space **S**. In essence, this follows from **4.6.2**. Since $\mathcal{Q}$ induces in a complement **W** of P in $\mathcal{Q}_P$ a Klein quadratic set $\mathcal{Q}'$ the following assertions are true.

- The $\mathcal{Q}$-lines through P correspond to the points of $\mathcal{Q}'$, hence to the lines of **S**;
- the 3-dimensional $\mathcal{Q}$-subspaces through P correspond to the planes of the two equivalence classes $\mathcal{Q}'$, hence to the points and planes of **S**.

Thus the assertion follows in fact from **4.6.2**. □

Remark. Here comes a magic word, *triality*. Very roughly, this means that one can rotate the above diagram by 120°, and it remains the diagram of a hyperbolic quadratic set in the 7-dimensional space. More precisely, given the hyperbolic quadratic set $\mathcal{Q}$ in a 7-dimensional projective space with its two equivalence classes $\mathcal{M}_1$ and $\mathcal{M}_2$ of 3-dimensional subspaces contained in Q, we define a new geometry consisting of *points*, *lines*, *planes* and *solids* in the following way:

- *points* are the elements of $\mathcal{M}_1$,
- *lines* are the lines contained in $\mathcal{Q}$,
- *planes* are the the incident point–plane pairs (P, π), where $\pi \in \mathcal{M}_2$,

– *solids* are the points of $\mathfrak{Q}$ along with the elements of $\mathfrak{M}_2$.
– Incidence is defined as in the quadratic set with the following exceptions: a *point* and a *solid* are incident if the corresponding 3-dimensional subspaces intersect in a plane, a *plane* (P, π) is incident with all the elements that are incident with P and π.

Then one can show that the new geometry is again a hyperbolic quadratic set in the 7-dimensional projective space. For details see e.g. [Cam92].

Remark. If one looks at the geometries coming from quadratic sets from an abstract point of view, one gets the so-called **polar spaces**. The fundamental paper for this beautiful theory is [BuSh74].

The game is not yet over. Now we turn the tables. We start from a 3-dimensional projective space $\mathbf{P}(V)$ and construct in an analytic way a Klein quadratic set, which will turn out to be an essentially uniquely determined quadric.

Before doing this we have to introduce the analytic notion of a 'quadric'.

4.7 Quadrics

Quadrics are the prototypes of quadratic sets. We have already considered quadrics in special cases (see **2.4.4**); now we shall study quadrics in general.

Definition. Let V be a vector space over a (commutative) field F. A map $q: V \to F$ is called a **quadratic form** of V if the following are true:
(i) $q(a \cdot v) = a^2 \cdot q(v)$ for all $v \in V$ and all $a \in F$,
(ii) the map $B: V \times V \to F$ defined by

$$B(v, w) := q(v + w) - q(v) - q(w)$$

is a symmetric bilinear form.

4.7.1 Lemma. *Let* $\{v_1, \ldots, v_n\}$ *be a basis of the vector space* V.
(a) *If* $a_{ij} \in F$ *then a quadratic form is defined by the following rule:*

$$q(\sum_{i=1}^{n} b_i v_i) := \sum_{i,j=1}^{n} a_{ij} \cdot b_i b_j.$$

(b) *Conversely: for any quadratic form* q *there are elements* $a_{ij} \in F$ *such that for all* $v = \sum_{i=1,\ldots,n} b_i v_i \in V$ *we have*

$$q(\sum_{i=1}^{n} b_i v_i) := \sum_{i,j=1}^{n} a_{ij} \cdot b_j.$$

Proof. (a) For any $b \in F$ we have

$$q(b\cdot v) = q(\sum_{i=1}^{n}(bb_i)\cdot v_i) := \sum_{i,j=1}^{n} a_{ij}\cdot(bb_i)\cdot(bb_j) = b^2\cdot \sum_{i,j=1}^{n} a_{ij}\cdot b_ib_j = b^2\cdot q(v).$$

Moreover,

$$\begin{aligned} B(v, w) &= q(v + w) - q(v) - q(w) \\ &= \sum_{i,j=1}^{n} a_{ij}\cdot(b_i + c_i)\cdot(b_j + c_j) - \sum_{i,j=1}^{n} a_{ij}\cdot b_ib_j - \sum_{i,j=1}^{n} a_{ij}\cdot c_ic_j \\ &= \sum_{i,j=1}^{n} a_{ij}\cdot(b_ic_j + b_jc_i). \end{aligned}$$

Now it is easy to verify that B is a symmetric bilinear form.
(b) will be handled in exercise 27. □

Remark. The definition of a quadratic form is so complicated only because one wants to include the case of characteristic 2. If the characteristic of F is different from 2, then one can reconstruct the bilinear form from the quadratic form (see exercise 26).

Definition. We call the quadratic form q of V **nondegenerate** if the following is true: if $q(v) = 0$ and $B(x, v) = 0$ for all $x \in V$ then $v = o$.

If q is a quadratic form of V then $q(v) = 0$ implies $q(a\cdot v) = 0$ for all $a \in F$. This is the basis of the following definition.

Definition. Let q be a quadratic form of the vector space V. The **quadric** of the projective space **P**(V) corresponding to q is the set of all points $\langle v\rangle$ of **P**(V) with $q(v) = 0$.

In view of the above lemma one can now construct as many quadrics as one likes: Given a d-dimensional projective space over the field F one takes a homogeneous polynomial f over F in $d+1$ variables. Then the 'corresponding' quadric consists precisely of the points with homogenous coordinates $(a_0\colon a_1\colon \ldots\colon a_d)$ with $f(a_0, a_1, \ldots, a_d) = 0$.

For instance, the polynomial $x_0^2 - x_1^2 - x_2^2$ yields a quadric in the plane, $x_0^2 - x_1^2 - x_2^2 - x_3^2$ and $x_0^2 + x_1^2 - x_2^2 - x_3^2$ provide quadrics in 3-dimensional space. In exercise 28 you are invited to recognize these quadrics in the case $F = \mathbf{R}$.

Our aim is to show that all quadrics are quadratic sets. First we prove the if-three-then-all axiom.

4.7.2 Lemma. *Let* q *be a quadratic form of a vector space* V, *and let* $\mathfrak{Q}$ *be the corresponding quadric in* **P**(V). *Then, if a line* g *contains three points of* $\mathfrak{Q}$, *each point of* g *lies in* $\mathfrak{Q}$.

Proof. Let $P = \langle v \rangle$, $R = \langle w \rangle$, and S be three different points on g in $\mathfrak{Q}$. W.l.o.g. we may assume that $S = \langle v + a \cdot w \rangle$ with $a \neq 0$. Then we have

$$\begin{aligned} 0 = q(v + a\cdot w) &= B(v, a\cdot w) + q(v) + q(a\cdot w) \\ &= a\cdot B(v, w) + q(v) + a^2\cdot q(w) \\ &= a\cdot B(v, w). \end{aligned}$$

Since $a \neq 0$ we have $B(v, w) = 0$. Now for each point $X = \langle v + b\cdot w \rangle \neq R$ on g we see

$$\begin{aligned} q(v + b\cdot w) &= B(v, b\cdot w) + q(v) + q(b\cdot w) \\ &= b\cdot B(v, w) + q(v) + b^2\cdot q(w) \\ &= b\cdot B(v, w) = 0. \end{aligned}$$

Hence each point of g is in $\mathfrak{Q}$. □

Definition. Let q be a quadratic form of the vector space V. For a vector $v \in V\setminus\{o\}$ we define

$$\langle v \rangle^{\perp} := \{x \in V \mid B(x, v) = 0\}.$$

4.7.3 Lemma. *Let* q *be a quadratic form of the vector space* V, *and let* $\mathfrak{Q}$ *be the corresponding quadric of* **P**(V). *Then for each vector* $v \in V\setminus\{o\}$ *we have:*

(a) $\langle v \rangle^{\perp}$ *is a subspace of* V, *hence also a subspace of* **P**(V).

(b) *Either* $\langle v \rangle^{\perp}$ *is a hyperplane, or we have* $\langle v \rangle^{\perp} = V$.

(c) *Suppose that* $\langle v \rangle \in \mathfrak{Q}$. *Then each line in* $\langle v \rangle^{\perp}$ *through* $\langle v \rangle$ *that contains another point* $\langle w \rangle \in \mathfrak{Q}\setminus\{\langle v \rangle\}$ *contains only points of* $\mathfrak{Q}$. *This means that each line in* $\langle v \rangle^{\perp}$ *through* $\langle v \rangle$ *is a tangent of* $\mathfrak{Q}$.

(d) *Suppose that* $\langle v \rangle \in \mathfrak{Q}$. *Then each line through* $\langle v \rangle$ *that is not contained in* $\langle v \rangle^{\perp}$ *intersects* $\mathfrak{Q}$ *in precisely one further point.*

Proof. (a) follows directly from the fact that B is a bilinear form.

(b) It is sufficient to show that any line $g = \langle u, w \rangle$ of **P**(V) intersects the subspace $\langle v \rangle^{\perp}$ in at least one vector different from o: We may assume that

$\langle w\rangle \notin \langle v\rangle^\perp$. Thus, $B(w, v) \neq 0$. Let $X = \langle u + a\cdot w\rangle$ be an arbitrary point on g different from $\langle w\rangle$. Then

$$\begin{aligned} & X \in \langle v\rangle^\perp \\ \Leftrightarrow\ & B(u + a\cdot w, v) = 0 \\ \Leftrightarrow\ & B(u, v) + a\cdot B(w, v) = 0 \\ \Leftrightarrow\ & a = -\frac{B(u, v)}{B(w, v)}. \end{aligned}$$

(c) Let $\langle v + a\cdot w\rangle$ be an arbitrary point $\neq \langle w\rangle$ on the line $g = \langle v, w\rangle$. We have that

$$\begin{aligned} q(v + a\cdot w) &= B(v, a\cdot w) + q(v) + q(a\cdot w) \\ &= a\cdot B(v, w) + q(v) + a^2\cdot q(w) = 0. \end{aligned}$$

(d) Suppose that $\langle w\rangle \in \mathbf{P}\setminus\langle v\rangle^\perp$, and let $\langle w + a\cdot v\rangle$ be an arbitrary point on the line $\langle v, w\rangle$ that is different from $\langle v\rangle$. Then it follows that

$$\begin{aligned} q(w + a\cdot v) &= B(w, a\cdot v) + q(w) + q(a\cdot v) \\ &= a\cdot B(w, v) + q(w). \end{aligned}$$

From $q(w + a\cdot v) = 0$ it follows that $a = -q(w)/B(w, v)$. Hence the line $\langle v, w\rangle$ intersects the quadric $\mathcal{Q}$ in the points $\langle w - q(w)/B(w, v)\cdot v\rangle$ and $\langle v\rangle$. □

4.7.4 Theorem. *Each quadric is a quadratic set.*

Proof. By **4.7.2**, the if-three-then-all axiom holds. In view of Lemma **4.7.3** for any point $\langle v\rangle$ of a quadric, the subspace $\langle v\rangle^\perp$ is the corresponding tangent hyperplane. □

Remark. A fundamental theorem of F. Buekenhout says that any nondegenerate quadratic set is a quadric or an ovoid (see [Buek69a]). We have already proved this for hyperbolic quadratic sets in 3-dimensional projective spaces. In Section **4.8** we shall prove the analogous statement for Klein quadratic sets.

In the cases $d = 2$ and $d = 3$ the corresponding theorems were first proved by B. Segre [Seg54] and A. Barlotti and Panella. The theorem of B. Segre is particularly remarkable:

4.7.5 Theorem. *Any oval in a finite Desarguesian projective plane of odd order is a conic, that is a nonempty, nondegenerate quadric in a projective plane.*

4.8 Plücker coordinates

Let **P**(V) be a 3-dimensional projective space over a field F such that the points and planes are represented by homogeneous coordinates. Our aim is to represent the lines by homogeneous coordinates as well. The method we shall present in the following is due to Julius Plücker (1801–1868).

Definition. Let g be a line of **P**(V) that passes through the points $(x_0\colon x_1\colon x_2\colon x_3)$ and $(y_0\colon y_1\colon y_2\colon y_3)$. We define the following elements:

$$p_{01} = \begin{vmatrix} x_0 & x_1 \\ y_0 & y_1 \end{vmatrix},\ p_{02} = \begin{vmatrix} x_0 & x_2 \\ y_0 & y_2 \end{vmatrix},\ p_{03} = \begin{vmatrix} x_0 & x_3 \\ y_0 & y_3 \end{vmatrix},$$

$$p_{23} = \begin{vmatrix} x_2 & x_3 \\ y_2 & y_3 \end{vmatrix},\ p_{31} = \begin{vmatrix} x_3 & x_1 \\ y_3 & y_1 \end{vmatrix},\ p_{12} = \begin{vmatrix} x_1 & x_2 \\ y_1 & y_2 \end{vmatrix}.$$

Then the 6-tuple $(p_1\colon p_2\colon p_3\colon p_4\colon p_5\colon p_6) := (p_{01}\colon p_{02}\colon p_{03}\colon p_{23}\colon p_{31}\colon p_{12})$ is called the **Plücker coordinates** of the line g.

In the following lemma we shall collect the basic properties of Plücker coordinates.

4.8.1 Lemma. (a) *Up to a scalar multiple, the Plücker coordinates of a line* g *are independent of the choice of the two points on* g; *in other words, the Plücker coordinates of a line are welldefined.*
(b) *The Plücker coordinates of a line are determined only up to a scalar multiple; therefore they are homogeneous coordinates.*
(c) *The Plücker coordinates* $P = (p_1\colon \ldots\colon p_6)$ *of a line satisfy the following quadratic equation:*

$$p_1 p_4 + p_2 p_5 + p_3 p_6 = 0. \qquad (*)$$

(*d*) *Suppose that* $p_1, \ldots, p_6$ *are elements of* F *such that at least one* $p_i \neq 0$. *If the* $p_1, \ldots, p_6$ *satisfy the equation* (*) *then there is a line of* **S** *having Plücker coordinates* $(p_1, \ldots, p_6)$.

Proof. (a) Let $Z = (z_0\colon z_1\colon z_2\colon z_3) = a\cdot(x_0\colon x_1\colon x_2\colon x_3) + (y_0\colon y_1\colon y_2\colon y_3)$ be an arbitrary point on g = XY different from X. Then the Plücker coordinates $(q_1\colon \ldots\colon q_6)$ of the line through X and Z are obtained as follows:

$$q_1 = \begin{vmatrix} x_0 & x_1 \\ a\cdot x_0 + y_0 & a\cdot x_1 + y_1 \end{vmatrix} = \begin{vmatrix} x_0 & x_1 \\ y_0 & y_1 \end{vmatrix} = p_1,$$

etc. Hence the Plücker coordinates with respect to X and Z are the same as those with respect to X and Y. Thus the Plücker coordinates of g are independent of the choice of the two points.
(b) If one replaces $(x_0: x_1: x_2: x_3)$ by $a\cdot(x_0: x_1: x_2: x_3)$ then the Plücker coodinates of the line through $a\cdot(x_0: x_1: x_2: x_3)$ and $(y_0: y_1: y_2: y_3)$ are $a\cdot(p_1: \ldots: p_6)$.
(c) We define

$$F(P) := p_1 p_4 + p_2 p_5 + p_3 p_6 = p_{01} p_{23} + p_{02} p_{31} + p_{03} p_{12}.$$

By patient computations one gets after a while

$$\begin{aligned} F(P) &= \begin{vmatrix} x_0 & x_1 \\ y_0 & y_1 \end{vmatrix} \cdot \begin{vmatrix} x_2 & x_3 \\ y_2 & y_3 \end{vmatrix} + \begin{vmatrix} x_0 & x_2 \\ y_0 & y_2 \end{vmatrix} \cdot \begin{vmatrix} x_3 & x_1 \\ y_3 & y_1 \end{vmatrix} + \begin{vmatrix} x_0 & x_3 \\ y_0 & y_3 \end{vmatrix} \cdot \begin{vmatrix} x_1 & x_2 \\ y_1 & y_2 \end{vmatrix} \\ &= (x_0 y_1 - y_0 x_1)\cdot(x_2 y_3 - y_2 x_3) \\ &\quad + (x_0 y_2 - y_0 x_2)\cdot(x_3 y_1 - y_3 x_1) \\ &\quad + (x_0 y_3 - y_0 x_3)\cdot(x_1 y_2 - y_1 x_2) \\ &= x_0 x_2 y_1 y_3 - x_0 x_3 y_1 y_2 - x_1 x_2 y_0 y_3 + x_1 x_3 y_0 y_2 \\ &\quad + x_0 x_3 y_1 y_2 - x_0 x_1 y_2 y_3 - x_2 x_3 y_0 y_1 + x_1 x_2 y_0 y_3 \\ &\quad + x_0 x_1 y_2 y_3 - x_0 x_2 y_1 y_3 - x_1 x_3 y_0 y_2 + x_2 x_3 y_0 y_1 \\ &= 0. \end{aligned}$$

(*d*) W.l.o.g. $p_{01} \neq 0$. We define the points X' and Y' of **S** as follows:

$$X' := (0: p_{01}: p_{02}: p_{03}),\quad Y' := (-p_{01}: 0: p_{12}: -p_{31}).$$

Then X' and Y' are two distinct points of **S**. The Plücker coordinates of the line X'Y' are computed as follows:

$$p_{01}' = \begin{vmatrix} 0 & p_{01} \\ -p_{01} & 0 \end{vmatrix} = p_{01}\cdot p_{01},$$

$$p_{02}' = \begin{vmatrix} 0 & p_{02} \\ -p_{01} & p_{12} \end{vmatrix} = p_{01}\cdot p_{02},$$

$$p_{03}' = \begin{vmatrix} 0 & p_{03} \\ -p_{01} & p_{31} \end{vmatrix} = p_{01}\cdot p_{03},$$

$$p_{23}' = \begin{vmatrix} p_{02} & p_{03} \\ p_{12} & -p_{31} \end{vmatrix} = -p_{02}\cdot p_{31} - p_{12}\cdot p_{03} = p_{01}\cdot p_{23}$$

in view of (*),

$$p_{31}' = \begin{vmatrix} p_{03} & p_{01} \\ -p_{31} & 0 \end{vmatrix} = p_{01} \cdot p_{31},$$

$$p_{12}' = \begin{vmatrix} p_{01} & p_{02} \\ 0 & p_{12} \end{vmatrix} = p_{01} \cdot p_{12}.$$

Putting these together we get

$$(p_{01}' : \ldots : p_{12}') = p_{01} \cdot (p_{01} : \ldots : p_{12}).$$

Hence the line through X' and Y' has the Plücker coordinates $(p_{01} : \ldots : p_{12})$. □

Our next aim is to show that looking at the Plücker coordinates of two lines one can see whether these lines intersect each other or not.

4.8.2 Lemma. *Let* g *and* h *be two lines of* **S** *with Plücker coordinates* P = $(p_1 : \ldots : p_6)$ *and* Q = $(q_1 : \ldots : q_6)$. *Then* g *and* h *are skew if and only if the expression*

$$F(P, Q) = p_1 q_4 + p_2 q_5 + p_3 q_6 + p_4 q_1 + p_5 q_2 + p_6 q_3$$

is different from zero.

Proof. Let X, Y, S, T be points such that g = XY and h = ST,

$$X = (x_0 : x_1 : x_2 : x_3), \quad Y = (y_0 : y_1 : y_2 : y_3),$$

$$S = (s_0 : s_1 : s_2 : s_3), \quad T = (t_0 : t_1 : t_2 : t_3).$$

By **2.3.1** we know that

$$g = h \text{ or } g \cap h \neq \emptyset$$

$$\Leftrightarrow \{X, Y, S, T\} \text{ is not a basis of } \mathbf{S}$$

$$\Leftrightarrow \begin{vmatrix} x_0 & x_1 & x_2 & x_3 \\ y_0 & y_1 & y_2 & y_3 \\ s_0 & s_1 & s_2 & s_3 \\ t_0 & t_1 & t_2 & t_3 \end{vmatrix} = 0.$$

We develop this determinant by the first row and get

$$\begin{vmatrix} x_0 & x_1 & x_2 & x_3 \\ y_0 & y_1 & y_2 & y_3 \\ s_0 & s_1 & s_2 & s_3 \\ t_0 & t_1 & t_2 & t_3 \end{vmatrix}$$

$$= x_0 \begin{vmatrix} y_1 & y_2 & y_3 \\ s_1 & s_2 & s_3 \\ t_1 & t_2 & t_3 \end{vmatrix} - x_1 \begin{vmatrix} y_0 & y_2 & y_3 \\ s_0 & s_2 & s_3 \\ t_0 & t_2 & t_3 \end{vmatrix} + x_2 \begin{vmatrix} y_0 & y_1 & y_3 \\ s_0 & s_1 & s_3 \\ t_0 & t_1 & t_3 \end{vmatrix} - x_3 \begin{vmatrix} y_0 & y_1 & y_2 \\ s_0 & s_1 & s_2 \\ t_0 & t_1 & t_2 \end{vmatrix}$$

$$= x_0 y_1 \begin{vmatrix} s_2 & s_3 \\ t_2 & t_3 \end{vmatrix} - x_0 y_2 \begin{vmatrix} s_1 & s_3 \\ t_1 & t_3 \end{vmatrix} + x_0 y_3 \begin{vmatrix} s_1 & s_2 \\ t_1 & t_2 \end{vmatrix}$$

$$- x_1 y_0 \begin{vmatrix} s_2 & s_3 \\ t_2 & t_3 \end{vmatrix} + x_1 y_2 \begin{vmatrix} s_0 & s_3 \\ t_0 & t_3 \end{vmatrix} - x_1 y_3 \begin{vmatrix} s_0 & s_2 \\ t_0 & t_2 \end{vmatrix}$$

$$+ x_2 y_0 \begin{vmatrix} s_1 & s_3 \\ t_1 & t_3 \end{vmatrix} - x_2 y_1 \begin{vmatrix} s_0 & s_3 \\ t_0 & t_3 \end{vmatrix} + x_2 y_3 \begin{vmatrix} s_0 & s_1 \\ t_0 & t_1 \end{vmatrix}$$

$$- x_3 y_0 \begin{vmatrix} s_1 & s_2 \\ t_1 & t_2 \end{vmatrix} + x_3 y_1 \begin{vmatrix} s_0 & s_2 \\ t_0 & t_2 \end{vmatrix} - x_3 y_2 \begin{vmatrix} s_0 & s_1 \\ t_0 & t_1 \end{vmatrix}$$

$$= (x_0 y_1 - x_1 y_0) \begin{vmatrix} s_2 & s_3 \\ t_2 & t_3 \end{vmatrix} + (x_2 y_0 - x_0 y_2) \begin{vmatrix} s_1 & s_3 \\ t_1 & t_3 \end{vmatrix}$$

$$+ (x_0 y_3 - x_3 y_0) \begin{vmatrix} s_1 & s_2 \\ t_1 & t_2 \end{vmatrix} + (x_1 y_2 - x_2 y_1) \begin{vmatrix} s_0 & s_3 \\ t_0 & t_3 \end{vmatrix}$$

$$+ (x_3 y_1 - x_1 y_3) \begin{vmatrix} s_0 & s_2 \\ t_0 & t_2 \end{vmatrix} + (x_2 y_3 - x_3 y_2) \begin{vmatrix} s_0 & s_1 \\ t_0 & t_1 \end{vmatrix}$$

$$= p_{01} \cdot q_{23} + p_{02} \cdot q_{31} + p_{03} \cdot q_{12} + p_{12} \cdot q_{03} + p_{31} \cdot q_{02} + p_{23} \cdot q_{01}$$
$$= p_1 \cdot q_4 + p_2 \cdot q_5 + p_3 \cdot q_6 + p_4 \cdot q_1 + p_5 \cdot q_2 + p_6 \cdot q_3$$
$$= F(P, Q).$$

This is the assertion. □

We now study the function F considered in the proof of **4.8.1**,

$$F(P) = p_1 p_4 + p_2 p_5 + p_3 p_6,$$

where $P = (p_1 : \ldots : p_6)$.

4.8.3 Lemma. *Let* $\mathrm{P} = (p_1 : \ldots : p_6)$ *and* $\mathrm{Q} = (q_1 : \ldots : q_6)$ *be Plücker coordinates of two lines of* **S**. *Then for all* $a, b \in \mathrm{F}$ *we have*

$$\mathrm{F}(a\cdot\mathrm{P} + b\cdot\mathrm{Q}) = a^2\cdot\mathrm{F}(\mathrm{P}) + b^2\cdot\mathrm{F}(\mathrm{Q}) + ab\cdot\mathrm{F}(\mathrm{P}, \mathrm{Q}).$$

Proof. We simply expand the left-hand side:

$$\begin{aligned}
&\mathrm{F}(a\cdot\mathrm{P} + b\cdot\mathrm{Q})\\
&= \mathrm{F}(ap_1 + bq_1, \ldots, ap_6 + bq_6)\\
&= (ap_1 + bq_1)\cdot(ap_4 + bq_4) + (ap_2 + bq_2)\cdot(ap_5 + bq_5)\\
&\quad + (ap_3 + bq_3)\cdot(ap_6 + bq_6)\\
&= a^2\cdot(p_1 p_4 + p_2 p_5 + p_3 p_6) + b^2\cdot(q_1 q_4 + q_2 q_5 + q_3 q_6)\\
&\quad + ab\cdot(p_1 q_4 + q_1 p_4 + p_2 q_5 + q_2 p_5 + p_3 q_6 + q_3 p_6)\\
&= a^2\cdot\mathrm{F}(\mathrm{P}) + b^2\cdot\mathrm{F}(\mathrm{Q}) + ab\cdot\mathrm{F}(\mathrm{P}, \mathrm{Q}).
\end{aligned}$$

□

Definition. Let **P** be a 5-dimensional projective space represented by homogeneous coordinates. The set $\mathcal{K}$ of points $\mathrm{P} = (p_1 : \ldots : p_6)$ satisfying $\mathrm{F}(\mathrm{P}) = 0$ is called the **Klein quadric** (sometimes also **Plücker quadric**).

We shall show that the Klein quadric justly bears its name. More precisely, we shall prove that $\mathcal{K}$ is a quadric and forms a Klein quadratic set. We shall also show that any Klein quadratic set is a quadric.

4.8.4 Theorem. *Let* **P** *be a 5-dimensional projective space. Then* $\mathcal{K}$ *is a hyperbolic quadric, hence a Klein quadratic set.*

Proof. By **4.7.1**, F is a quadratic form, so $\mathcal{K}$ is a quadric. Next we show that $\mathcal{K}$ is nondegenerate. In order to do this we have to show that for each point $\mathrm{P} \in \mathcal{K}$ its tangent space is only a hyperplane. The tangent space consists of the points

$$\langle \mathrm{P} \rangle^{\perp} := \{\mathrm{X} \in \mathbf{P} \mid \mathrm{B}(\mathrm{X}, \mathrm{P}) = 0\},$$

where $\mathrm{B}(\mathrm{X}, \mathrm{P}) = \mathrm{F}(\mathrm{X} + \mathrm{P}) - \mathrm{F}(\mathrm{X}) - \mathrm{F}(\mathrm{P})$. Using the function F defined in Lemma **4.8.2** it follows in view of **4.8.3** that the tangent space consists precisely of those points $\mathrm{X} \in \mathbf{P}$ that fulfil

$$\mathrm{B}(\mathrm{X}, \mathrm{P}) = \mathrm{F}(\mathrm{X}, \mathrm{P}) = 0.$$

Since $\mathrm{P} \neq (0{:}\ 0{:}\ 0{:}\ 0{:}\ 0{:}\ 0)$, this is the equation of a hyperplane.

Hence $\mathcal{K}$ is a nondegenerate quadric, and in particular a nondegenerate quadratic set in **P**. We now show that $\mathcal{K}$ has index 3. For this it is sufficient to show that $\mathcal{K}$ contains a plane.

We consider the point $P = (1: 0: 0: 0: 0: 0)$. The tangent hyperplane $\mathcal{K}_P$ contains all points $X = (0: x_2: x_3: 0: 0: 0)$ of the line g passing through (0: 1: 0: 0: 0: 0) and (0: 0: 1: 0: 0: 0). Since any point of g lies in $\mathcal{K}$, $\langle P, g\rangle$ is a $\mathcal{K}$-plane.

So $\mathcal{K}$ is a hyperbolic quadratic set. □

Our next aim is to prove that conversely each Klein quadratic set is a quadric. For this we consider in a 5-dimensional projective space the set $\mathfrak{M}$ of quadruples $(\alpha, \alpha_1, \alpha_2, \alpha_3)$ of planes with the following properties: there is a plane β such that

(i) α is skew to β,

(ii) each plane α_i intersects β in a line g_i $(i = 1, 2, 3)$,

(iii) the lines g_1, g_2, g_3 are distinct and and do not pass through a common point,

(iv) the subspaces $\langle\alpha_1, \alpha_2\rangle$, $\langle\alpha_2, \alpha_3\rangle$, and $\langle\alpha_3, \alpha_1\rangle$ are distinct 4-dimensional subspaces of P.

First we show that each hyperbolic quadratic set contains such a quadruple.

4.8.5 Lemma. *Let* **P** *be a 5-dimensional projective space and* $\mathcal{Q}$ *be a Klein quadratic set in* **P**. *Then* $\mathcal{Q}$ *contains a quadruple* $(\alpha, \alpha_1, \alpha_2, \alpha_3)$ *of* $\mathcal{Q}$*-planes of the above defined set* $\mathfrak{M}$.

Proof. By Theorem **4.6.2** any Klein quadratic set is a 3-dimensional projective space **S**. In this space, there exist three noncollinear points P_1, P_2, P_3 outside a plane π. The noncollinear points of **S** are planes of $\mathcal{Q}$ mutually intersecting in a point, but not having a common point. We denote these planes by $\alpha_1, \alpha_2, \alpha_3$ and the plane spanned by their points of intersection by β. Therefore, each plane α_i intersects β in a line g_i $(i = 1, 2, 3)$ and these lines are distinct and form a triangle. The plane π of **S** corresponds to a plane α of $\mathcal{Q}$ which is disjoint from α_i $(i = 1, 2, 3)$ because the points P_i are outside π.

It remains to show that condition (iv) holds. For this we assume that $\langle\alpha_1, \alpha_2\rangle = \langle\alpha_2, \alpha_3\rangle$. Then the tangent hyperplane **H** with respect to $\alpha_1 \cap \alpha_2$ contains the three planes $\alpha_1, \alpha_2, \alpha_3$. But this is impossible, as **H** is a cone over a hyperbolic quadratic set (see **4.5.3**). □

4.8.6 Lemma. *Let* **P** *be a 5-dimensional projective space, and denote by* G *its group of collineations. Then* G *acts transitively on* $\mathfrak{M}$.

Proof. Let $(\alpha', \alpha_1', \alpha_2', \alpha_3')$ and $(\alpha, \alpha_1, \alpha_2, \alpha_3)$ be two elements of $\mathfrak{M}$.

Since G acts transitively on pairs of skew planes (see exercise 32), there is an element of G mapping α' onto α and β' onto β. So we can identify α' with α, and β' with β.

Next we show that the subgroup of G fixing α and β induces all central collineations of β: Each central collineation of β may be extended to a central collineation of **P** whose axis contains α. Since the group generated by the central collineations acts transitively on the triangles, there is a collineation mapping g_1' onto g_1, g_2' onto g_2, and g_3' onto g_3.

Now we have the following situation. Let α_i, α_i' be planes through g_i such that $(\alpha, \alpha_1', \alpha_2', \alpha_3')$ and $(\alpha, \alpha_1, \alpha_2, \alpha_3)$ satisfy the hypothesis of the lemma. We have to show that there is an element of G fixing α and β and mapping α_i' onto α_i $(i = 1, 2, 3)$.

First, we map α_1' onto α_1 in such a way that α and g_i are fixed. Observe that the 3-dimensional subspace $\langle\alpha_1, \alpha_1'\rangle$ intersects the plane α in a point P. This point has the property that there is a line through it intersecting α_1 and α_1' in different points. Furthermore, let **H** be a hyperplane through β that contains neither α_1' nor α_1. Then there is a central collineation γ with centre P and axis **H** mapping α_1' onto α_1. Because the centre P of this central collineation is a point of α the plane α remains fixed as a whole. Furthermore, the points of **H**, in particular the points of g_i, are fixed.

Next, we map α_2' onto α_2. We consider a hyperplane **H** through $\langle\beta, \alpha_1\rangle$ that contains neither α_2' nor α_2, and the point P on α that is also contained in $\langle\alpha_2, \alpha_2'\rangle$. As above, there is a central collineation γ with axis **H** and centre P mapping α_2' onto α_2, fixing α as a whole and all points of α_1 and β.

Finally, we have to map α_3' onto α_3. For this, let **H** be the hyperplane through α_1 and α_2. By definition of the set $\mathfrak{M}$, neither α_3' nor α_3 is contained in **H**, but $g_3 = \alpha_1 \cap \alpha_2$ lies in **H**. Let P be the intersection of $\langle\alpha_3, \alpha_3'\rangle$ and α. Then there is a central collineation γ with axis **H** and centre P mapping α_3' onto α_3, fixing α as a whole and the points of α_1 and α_2.

This finishes the proof. □

4.8.7 Lemma. *Let* **P** *be a 5-dimensional projective space and* $\mathfrak{Q}_1$ *and* $\mathfrak{Q}_2$ *hyperbolic quadratic sets. If* $\mathfrak{Q}_1$ *and* $\mathfrak{Q}_2$ *coincide in a set* $\{\alpha, \alpha', \alpha''\}$ *of planes, where* α' *and* α'' *intersect in exactly one point* P *and* α *is disjoint from* α' *and* α'', *then* $\mathfrak{Q}_1$ *and* $\mathfrak{Q}_2$ *coincide in their points in the tangential hyperplane* $\langle\alpha', \alpha''\rangle$ *of* P.

Proof. Let **H** be the tangential hyperplane of $\mathfrak{Q}_1$ in P. Then $\mathbf{H} = \langle\alpha', \alpha''\rangle$, because α_1 and α_2 are contained in **H**. The hyperplane **H** intersects α in a line g. Let **W** be a complement of P in **H** containing g. By **4.5.3** we know that $\mathfrak{Q}' = \mathfrak{Q}_1 \cap \mathbf{W}$ is a regulus. The regulus $\mathfrak{Q}'$ contains the three lines $\alpha \cap \mathbf{H} = g$ and $\alpha_i \cap \mathbf{H}$ $(i = 1, 2)$, therefore, by **2.4.3**, it is uniquely determined by these lines.

Therefore, $\mathcal{Q}' = \mathcal{Q}_1 \cap \mathbf{W} = \mathcal{Q}_2 \cap \mathbf{W}$. Because $\mathcal{Q}_1 \cap \mathbf{H}$ and $\mathcal{Q}_2 \cap \mathbf{H}$ are cones over this regulus, $\mathcal{Q}_1$ and $\mathcal{Q}_2$ coincide in their points in $\mathbf{H}$. □

4.8.8 Theorem. *Each Klein quadratic set is a quadric.*

Proof. Let $\mathcal{Q}_1$ be a Klein quadratic set and $\mathcal{Q}_2$ a Klein quadric. By Theorem **4.8.4** it follows that $\mathcal{Q}_2$ is also a Klein quadratic set. We show that there is a collineation of $\mathbf{P}$ mapping $\mathcal{Q}_1$ onto $\mathcal{Q}_2$. This proves that $\mathcal{Q}_1$ is a quadric.

By Lemma **4.8.5** and Lemma **4.8.6** we know that $\mathcal{Q}_1$ and $\mathcal{Q}_2$ each contain an element of $\mathcal{M}$ and that there is a collineation mapping one element of $\mathcal{M}$ onto the other. Therefore we assume that $\mathcal{Q}_1$ and $\mathcal{Q}_2$ contain the same element $(\alpha, \alpha_1, \alpha_2, \alpha_3)$ of $\mathcal{M}$. In the following we shall show that any two hyperbolic quadratic sets having such a quadruple in common are equal. This proves the theorem.

Let $P_1 = \alpha_2 \cap \alpha_3$, $P_2 = \alpha_1 \cap \alpha_3$, $P_3 = \alpha_1 \cap \alpha_2$, and $g_1 = P_2P_3$, $g_2 = P_1P_3$, $g_3 = P_1P_2$.

By Lemma **4.8.7** both quadratic sets coincide in their points in the tangential hyperplanes at P_i $(i = 1, 2, 3)$. In the tangential hyperplane $\mathbf{H} = \langle \alpha_1, \alpha_2 \rangle$ at P_3 the quadratic sets are cones over a regulus. One equivalence class of this cone contains the planes α_1 and α_2. Since each point X of the line g_3 lies on one plane of this equivalence class, there are two planes through X that have $\mathcal{Q}_1$ and $\mathcal{Q}_2$ in common intersecting only in X, namely the plane of the equivalence class through X and α_3. Exactly one of the planes of this equivalence class intersects α. We denote the point of intersection of this plane with g_3 by X_3. Thus, for all points $\neq X_3$ of g_3 the conditions of Lemma **4.8.7** are satisfied and so both quadratic sets coincide in their points in the tangential hyperplanes of these points as well.

In the same way we define X_1 and X_2. Using the same arguments as above, we see that both quadratic sets coincide in their points in the tangential hyperplanes in all points $\neq X_i$ of g_i $(i = 1, 2, 3)$.

Let R be an arbitrary point of $\mathcal{Q}_1$, and denote by $\mathbf{H}$ the tangential hyperplane of R. Then $\mathbf{H}$ intersects each line g_i, $i = 1, 2, 3$. If $\mathbf{H}$ intersects the line g_i in the point X_i, then $\langle X_1, X_2, X_3 \rangle = \langle P_1, P_2, P_3 \rangle$ is contained in $\mathbf{H}$. Therefore, in all cases, there is a point $X \neq X_i$ on some line g_i $(i = 1, 2, 3)$ contained in $\mathbf{H}$. Thus, the line RX is a $\mathcal{Q}_1$-line, and therefore contained in the tangential hyperspace of X. Moreover, the point R lies in the tangential hyperspace of X and thus also in $\mathcal{Q}_2$. □

4.9 Application: storage reduction for cryptographic keys

Let us consider a network in which any two participants can communicate with each other. This communication should be secret: for any two participants it should be possible to exchange messages in encrypted form.

In order to do this one applies a so-called symmetric enciphering algorithm. Such an algorithm assigns to any **data** d a **ciphertext** c in such a way that a secret key K is involved. More formally we can describe this procedure as follows.

Let $\mathcal{D}$, $\mathcal{M}$, and $\mathcal{K}$ be sets; the elements of $\mathcal{D}$ are called **plaintexts** (or **data**), the elements of $\mathcal{M}$ are the **ciphertexts** (or **messages**), and the elements of $\mathcal{K}$ are called **keys**. An **enciphering algorithm** is a map $f\colon \mathcal{D} \times \mathcal{K} \to \mathcal{M}$ with the property that for any $k \in \mathcal{K}$ the map $f_k\colon \mathcal{D} \to \mathcal{M}$ that is defined by

$$f_k(d) := f(d, k)$$

is invertible; we denote the inverse map of f_k by f_k^{-1}.

We consider the following model (see Figure 4.4): Sender and recipient agree upon a key $k \in \mathcal{K}$ and keep it secret. The sender **enciphers** the message m using f_k and sends the ciphertext $c = f_k(m)$ to the recipient. Using f_k^{-1}, the recipient is able to **decipher** the ciphertext c:

$$f_k^{-1}(c) = f_k^{-1}f_k(m) = m.$$

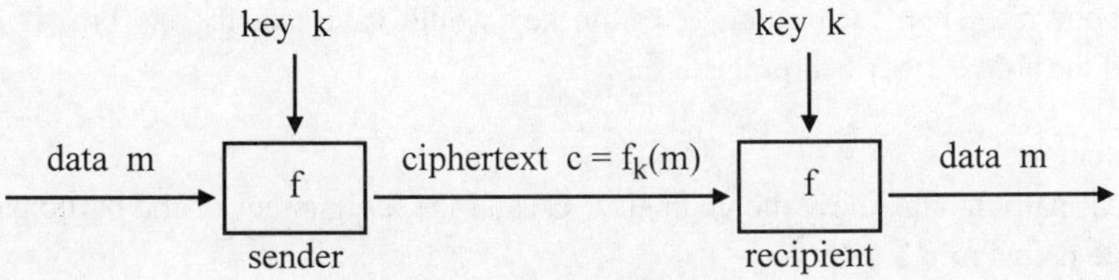

Figure 4.4 Enciphering and deciphering a message

Roughly speaking the security of an enciphering algorithm is based on the fact that an attacker does not know the actual key and, therefore, cannot decipher the ciphertext.

Clearly, there are enciphering algorithms of quite different levels of security. One requirement is that even if an attacker knows a large amount of ciphertexts it should not be possible to compute the key. There exist many investigations concerning this problem, which we will omit here.

We shall study a problem that is completely independent of the algorithms used: in order that any two of v participants can secretly communicate with each other they have to have a secret key that must not be known to any third person (otherwise this person could decipher the ciphertext).

Thus every participant has to store $v-1$ keys. This provides a serious problem if one takes into consideration that in practice values such as, for instance, $v = 10000$ often occur. Note that these keys have to be stored in such a way that they cannot be read in an unauthorized way.

The question is whether the participants can store fewer secret data without destroying the overall security. The idea being that from the specific secret data of two participants their communication key will be computed.

We shall consider the following model, which is based on geometrical notions. Let $\mathbf{G} = (\mathcal{P}, \mathcal{B})$ be a rank 2 geometry with points and blocks. We identify the participants with the points of $\mathbf{G}$. A key distribution centre assigns to each block B secret information k_B, which will be called a **prekey**. A participant obtains such a prekey k_B if and only if 'his' point is incident with the block B. If $\mathbf{G}$ has the property that each point is on exactly r blocks then each participant obtains exactly r prekeys.

If two participants P and Q want to secretly communicate then they compute their communication key k_{PQ} from the secrets k_B of those blocks B that are incident with P and Q. The computation of k_{PQ} is performed using a public procedure. For instance, if the prekeys are binary sequences, one can concatenate the prekeys in a predefined order. A particularly simple case is when any two points of $\mathbf{G}$ are on a constant number λ of blocks and each prekey consists of just one bit. Then each communication key would have exactly λ bits, if one used the above described procedure.

To sum up:

– All participants know the geometry $\mathbf{G}$ and the assignment of the participants to the points of $\mathbf{G}$.

– Each participant P knows the prekeys belonging to the blocks B through P.

– In order to compute the communication key between P and Q one applies a public procedure on the prekeys k_B, where B runs through all blocks through P and Q. Note that this is possible for P as well as for Q.

How can one 'measure' the security of such a procedure? Does it depend on the geometry $\mathbf{G}$? As measure for the security of this procedure we choose the number of fraudulent participants that have to pool their prekeys in order to compute the communication key between P and Q.

More precisely we define the number $k(\mathbf{G})$ as the largest number k having the following property: for no pair {P, Q} of participants is there a set of k participants that does not contain P or Q such that using their prekeys these participants can compute the communication key k_{PQ} of P and Q. In this case one says that the system is **resistant** against a collusion of $k(\mathbf{G})$ participants.

It turns out that a particularly good system is the geometry constructed from an ovoid.

4.9.1 Theorem. *Let* $\mathcal{O}$ *be an ovoid of a finite projective space* $\mathbf{P} = \mathrm{PG}(3, q)$, *and let* $\mathbf{G}$ *be the geometry that consists of the points of* $\mathcal{O}$ *and of the planes that intersect* $\mathcal{O}$ *in an oval.*
(a) *Through any point of* $\mathcal{O}$ *there are exactly* $r = q^2 + q$ *'oval planes'.*
(b) $k(\mathbf{G}) = q$.

Proof. (a) Any point of $\mathcal{O}$ is on exactly $q^2 + q + 1$ planes, exactly one of which is a tangent plane.
(b) Since no point $\neq$ P, Q is incident with two planes through P and Q, one needs one point on each of these $q + 1$ planes. Thus we have $k(\mathbf{G}) = q$. □

4.9.2 Corollary. *Let* $\mathbf{G}$ *be the geometry from* **4.9.1**. *Suppose that each prekey consists of exactly one bit. Then each participant has to store exactly* $q^2 + q$ *secret bits. In the naive model each participant had to store* $q^2 \cdot (q + 1)$ *bits. Thus, one can reduce the amount of memory by a factor* q.

Proof. Since each point of $\mathbf{G}$ is on exactly $q^2 + q$ blocks the first assertion follows. Moreover, the length of a communication key is equal to the number of blocks through two points, hence $q + 1$.

In the naive model, in order to store communication keys of length $q + 1$ any participant has to store $(v - 1) \cdot (q + 1) = q^2 \cdot (q + 1)$ bits. □

An example will illustrate the above assertion. Let $q = 127$ (this will result in a common key length.) Then, in a network with $127^2 + 1 = 16130$ participants, each participant has to store just 16256 bits, fewer than 1 % of the 2064512 bits that each participant would have to store in the naive model. The system tolerates a collusion of up to 127 participants without weakening the security.

Exercises

1 Let g_1 and g_2 be two lines of a projective plane of order ≥ 3, let P be the intersection point, and let P_i be a point $\neq$ P on g_i $(i = 1, 2)$. Show that the set of points on g_1 and g_2 that are different from P, P_1, P_2 is a set of points that fulfils the tangent-space axiom, but not the if-three-then-all axiom.

2 Give an example of a set of points in a projective space that satisfies the if-three-then-all axiom, but not the tangent-space axiom.

3 Let $\mathcal{Q}$ be a degenerate quadratic set that is not a subspace. Show that there exist two distinct points $P, R \in \mathcal{Q}\setminus\mathrm{rad}(\mathcal{Q})$ such that $\mathcal{Q}_R = \mathcal{Q}_P$.

4 Let $\mathcal{Q}$ be a degenerate quadratic set that is not a subspace. Show that for any two distinct points $P, R \in \mathcal{Q}\setminus\mathrm{rad}(\mathcal{Q})$ with $\mathcal{Q}_R = \mathcal{Q}_P$ it follows that the line PR intersects $\mathrm{rad}(\mathcal{Q})$.

5 Let $\mathbf{P}$ be a finite projective plane of order n. A ***k*-arc** of $\mathbf{P}$ is a set of k points of $\mathbf{P}$ no three of which are collinear.
Show that $k \leq n+2$.

6 Show that if a finite projective plane of order n contains an $(n+2)$-arc then n is even.

7 Let $\mathbf{P}$ be a projective plane of order n. Show that a k-arc is an oval if and only if $k = n+1$.

8 Show that in $PG(3, 2)$ each set of five points in general position is an ovoid.

9 Let $\mathcal{O}$ be an ovoid of a finite projective space $\mathbf{P} = PG(3, q)$.
(a) Compute the number of tangent planes and the number of planes that intersect $\mathcal{O}$ in an oval.
(b) Is there a plane that contains no point of $\mathcal{O}$?

10 Show that in the real affine plane the set of points (x, y) with $x^4 + y^4 = r$ is an oval, but not a quadric.

11 Is there an ovoid in the real 3-dimensional space that is not a quadric?

12 Show that in any real projective space there are ovoids that are not quadrics.

13 Show that in each infinite projective space of finite dimension there is an ovoid that is not a quadric. [This is a rather difficult exercise, use transfinite induction.]

14 (a) Show that each sphere in $\mathbf{R}^n$ is an ovoid.
(b) In d-dimensional real projective space, are there nondegenerate quadrics that are neither elliptic nor hyperbolic?

15 Show that each nonempty, nondegenerate quadric of a projective space $\mathbf{P}$ spans the whole space $\mathbf{P}$.

16 Is the vertex of a cone over a quadratic set uniquely determined?

17 Let $\mathfrak{Q}$ be a cone with vertex V over the quadratic set $\mathfrak{Q}^*$. Then show that $\mathfrak{Q}$ induces in all hyperplanes not through V isomorphic quadratic sets.

18 Let $\mathfrak{Q}$ be an elliptic quadratic set. Characterize the quadratic set $\mathfrak{Q}' := \mathfrak{Q} \cap \mathbf{H}$, where (a) **H** is a tangent hyperplane and (b) **H** is a nontangent hyperplane.

19 Show the analogue of **4.5.7** for a hyperbolic quadratic set of a $(2t+1)$-dimensional projective space ($t \geq 4$).

20 Analogously to **4.5.8** provide a diagram of a hyperbolic quadratic set of a $(2t+1)$-dimensional projective space.

21 Construct explicitly the geometry consisting of the points and $\mathfrak{Q}$-lines of a parabolic quadratic set $\mathfrak{Q}$ in PG(4, 2).

22 Show that the points and lines of a nondegenerate quadratic set of 4-dimensional projective space are a generalized quadrangle.

23 Show that each quadratic set has the **one-or-all property**: if a point P lies outside a line g, then P is joined to exactly one point of g or to all points of g.

24 Look for the exact place where in the proof of **4.6.2** the Veblen–Young axiom has been verified.

25 Analogously to **4.6.3** provide a diagram of a hyperbolic quadratic set of a $(2t+1)$-dimensional projective space ($t \geq 4$).

26 Let V be a vector space over a field F with char(F) $\neq$ 2. Show: if B is a bilinear form of V, then by $q(\mathrm{v}) := \frac{1}{2} \cdot \mathrm{B}(\mathrm{v}, \mathrm{v})$ a quadratic form is defined.

27 Prove **4.7.1**(b):
Let $\{\mathrm{v}_1, \ldots, \mathrm{v}_n\}$ be a basis of the vector space V. Then for any quadratic form q there are elements $a_{ij} \in \mathrm{F}$ such that for all $\mathrm{v} = \sum_{i=1,\ldots,n} a_i \mathrm{v}_i \in \mathrm{V}$ one has

$$q\left(\sum_{i=1}^{n} a_i \mathrm{v}_i\right) = \sum_{i,j=1}^{n} a_{ij} \cdot a_i a_j .$$

28 Do you know the quadric in the real projective plane that is given by the polynomial $x_0{}^2 - x_1{}^2 - x_2{}^2$? Which are the quadrics defined by $x_0{}^2 - x_1{}^2 - x_2{}^2 - x_3{}^2$, $x_0{}^2 + x_1{}^2 - x_2{}^2 - x_3{}^2$, and $x_0{}^2 + x_1{}^2 + x_2{}^2 - x_3{}^2$ in a real 3-dimensional space?

29 Determine in each of the above cases the equation of the tangent through a point of the quadric.

30 Show that if a finite d-dimensional projective space contains an ovoid then $d \leq 3$.

31 Let F be a division ring, let $\mathbf{P} = \mathbf{P}(V)$ be the 3-dimensional projective space over F, and denote by g the line with equations $x_0 = 0$ and $x_1 = 0$. Furthermore, let $\mathbf{A} = \mathbf{A}(V)$ be the 4-dimensional affine space over F.
For any point $P = (a, b, c, d)$ of $\mathbf{A}$ we define

$$\varphi(P) := \langle (1, 1, a, b), (1, 0, c, d) \rangle.$$

(a) Show that φ is a bijective map of the points of $\mathbf{A}$ onto the lines of $\mathbf{P}$ that are skew to g.
(b) How do the lines of $\mathbf{A}$ look in this representation in $\mathbf{P}$?

32 Let $\mathbf{P}$ be a projective space, and denote by G its group of collineations. Show that G acts transitively on pairs of skew planes.

True or false?

☐ The empty set is a nondegenerate quadratic set.

☐ Any union of two hyperplanes is a degenerate quadratic set.

☐ Any union of two subspaces is a degenerate quadratic set.

☐ There is a quadratic set having exactly two points.

☐ There is a quadratic set having exactly three points.

☐ Each elliptic quadratic set is nondegenerate.

☐ Any set of points is contained in a quadratic set.

☐ Any set of points is contained in a nondegenerate quadratic set.

☐ Any plane not through the vertex of a cone in a 3-dimensional projective space hits the cone in an oval.

Let $\mathcal{Q}$ be a nonempty, nondegenerate quadratic set of a projective space.

☐ The number of tangent hyperplanes equals the number of points of $\mathcal{Q}$ if and only if the index of $\mathcal{Q}$ is 1.

☐ PG(4, q) contains an ovoid.

☐ $\mathfrak{Q}$ has index $t \Leftrightarrow d = 2t$.

☐ There is always a hyperplane **H** such that $\mathfrak{Q} \cap \mathbf{H}$ is degenerate.

☐ If **H** is a tangent hyperplane then $\dim(\mathrm{rad}(\mathfrak{Q} \cap \mathbf{H})) = 0$.

☐ If **P** is finite then, if $\mathfrak{Q}$ has index 1001, then $2001 \le d \le 2003$.

☐ In a real d-dimensional projective space each quadric has index 1.

You should know the following notions

Tangent, quadratic set, radical, nondegenerate, index, oval, ovoid, hyperboloid, cone, cone over a quadratic set, parabolic, elliptic, hyperbolic, generalized quadrangle, Klein quadratic set, quadratic form, bilinear form, quadric, Plücker coordinates.

5 Applications of geometry to coding theory

When data are stored or transmitted random errors will often occur. It is the aim of coding theory to detect or even correct those errors. In the last fifty years coding theory has proved to be of both practical and theoretical importance. There exists an elaborate theory however, we shall restrict ourselves to the basic facts. The interested reader is referred to the standard literature, for instance the ‘bible’ of coding theory [MWSl83]; as an introduction we recommend [Hill86]. Surprisingly enough, when codes are constructed one often uses structures of a geometrical or combinatorial nature. In this chapter we will present the foundations of coding theory and some of its links to geometry. For the connections to design theory see [AsKe92].

5.1 Basic notions of coding theory

All considerations in coding theory are based on the following communication model. A sender wants to transmit data to a recipient. These data are transmitted via a channel that, despite any amount of care, might not transmit the data unaltered – there might be random noise (cf. Figure 5.1), usually due to circumstances beyond the sender’s control. Probably everybody has experienced the irritation caused by poor reception due to ‘atmospheric noise’ during a favourite TV programme.

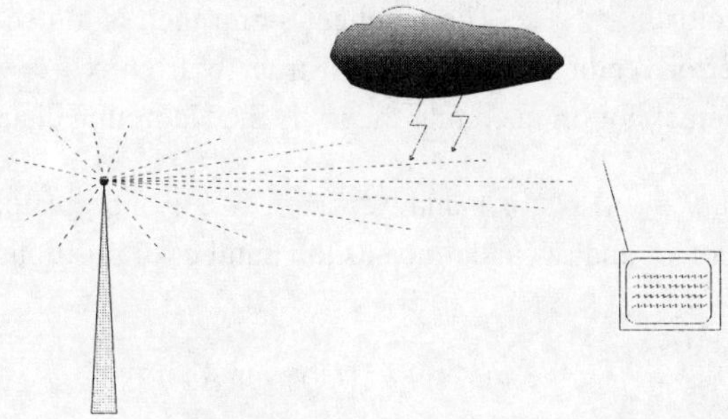

Figure 5.1 Atmospheric noise

Our communication model looks like Figure 5.2. The sender **encodes** data d into a **message** c (also called a **codeword**); this codeword will be transmitted over the channel. The recipient **decodes** the message and tries to detect whether errors have occurred or not. If one uses 'error correcting codes' then the original data can again be reconstructed. (This means that in Figure 5.2 we have d = d'.)

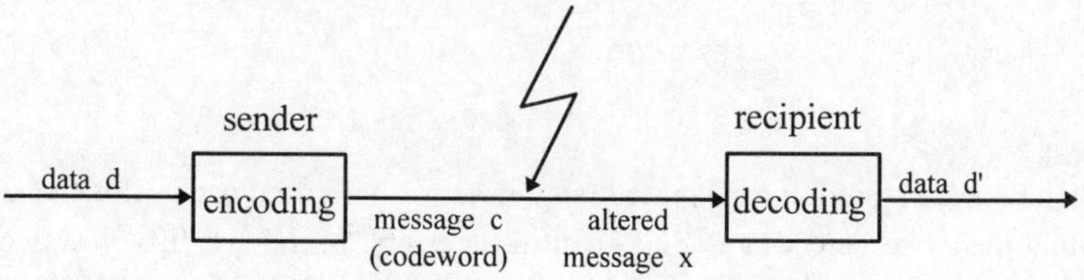

Figure 5.2 Encoding and decoding of data

Remark. Terminology is not uniform. In order to be consistent with the next chapter we shall restrict the term 'message' to the data that were originally transmitted.

The errors that are dealt with in coding theory are *random* errors, which excludes alterations perpetrated by our next-door neighbours. The errors we investigate are alterations of the symbols. Therefore we do not consider 'errors' such as the deletion or addition of symbols. (There are also methods to master those errors; see for instance [Bla83].)

Our first aim is to make precise what we mean for a code to be able to correct errors.

In this chapter a **message** is always a binary string of length n, hence an element of the vector space $\mathbf{V} := \{0, 1\}^n$. (In this chapter $\mathbf{V}$ will always denote this vector space.)

The problem we want to study can be described as follows. The channel adds to the transmitted vector c (the 'message') an **error vector** e, so the recipient receives the vector x = c + e. The recipient's aim then is to decode x, that is, to determine the error vector in order to reconstruct c from x.

The fundamental notion in coding theory is the Hamming distance.

Definition. Let $v = (v_1, \ldots, v_n)$ and $w = (w_1, \ldots, w_n)$ be vectors of $\mathbf{V}$. The **distance** $d(v, w)$ of v and w is defined as the number of positions in which v and w differ:

$$d(v, w) = |\{i \mid v_i \neq w_i\}|.$$

The distance d is often called the **Hamming distance**, in honour of Richard W. Hamming, one of the fathers of coding theory.

First we convince ourselves that d bears the name 'metric' not unjustly.

5.1.1 Lemma. *The function* d *is a metric on* $\mathbf{V}$.

Proof. (1) Since $d(\mathrm{v}, \mathrm{w})$ enumerates positions, we have $d(\mathrm{v}, \mathrm{w}) \geq 0$; moreover $d(\mathrm{v}, \mathrm{w}) = 0$ if and only if v and w differ in no position, that is if they are equal.
(2) Obviously we have $d(\mathrm{v}, \mathrm{w}) = d(\mathrm{w}, \mathrm{v})$.
(3) The proof of the *triangle inequality* is a little bit more tricky. We must show that for all $\mathrm{u}, \mathrm{v}, \mathrm{w} \in \mathbf{V}$ we have

$$d(\mathrm{u}, \mathrm{w}) \leq d(\mathrm{u}, \mathrm{v}) + d(\mathrm{v}, \mathrm{w}).$$

We may assume that u and w differ precisely at the first $a = d(\mathrm{u}, \mathrm{w})$ positions. Among those a positions, let there be b positions in which v and w differ; furthermore, let there be c positions outside the first a positions in which v and w differ (see Figure 5.3). Of course we have $d(\mathrm{v}, \mathrm{w}) = b + c$.

$$\begin{array}{ll}
\mathrm{u} & \overbrace{\mathrm{x\ x\ x\ x\ x}}^{a}\ \mathrm{x\ x\ x\ x\ x\ x\ x} \\
\mathrm{w} & \mathrm{o\ o\ o\ o\ o\ x\ x\ x\ x\ x\ x\ x} \\
\mathrm{v} & \underbrace{\mathrm{x\ x\ x}}_{b}\ \mathrm{o\ o}\ \underbrace{\mathrm{o\ o\ o}}_{c}\ \mathrm{x\ x\ x\ x}
\end{array}$$

Figure 5.3

From Figure 5.3 we see $d(\mathrm{u}, \mathrm{v}) = a - b + c$. It follows that

$$d(\mathrm{u}, \mathrm{v}) + d(\mathrm{v}, \mathrm{w}) = a - b + c + b + c = a + 2c \geq a = d(\mathrm{u}, \mathrm{w}). \qquad \square$$

Whenever a mathematician has a metric, he defines 'spheres' with respect to this metric. We shall later see that those spheres are very useful for describing codes.

Definition. Let $\mathrm{v} \in \mathbf{V}$, and let r be a nonnegative integer. Then

$$\mathrm{S}_r(\mathrm{v}) := \{\mathrm{x} \in \mathbf{V} \mid d(\mathrm{x}, \mathrm{v}) \leq r\}$$

is called the **sphere** (sometimes **Hamming sphere**) of **radius** r with **centre** v.

Now we are able to define what an error correcting code is.

Definition. Let t be a positive integer. A subset $\mathbf{C}$ of $\mathbf{V} = \{0, 1\}^n$ is called a ***t*-error correcting code**, if any two distinct elements $\mathrm{v}, \mathrm{w} \in \mathbf{C}$ satisfy

$$d(\mathrm{v}, \mathrm{w}) \geq 2t + 1.$$

In other words, a subset $\mathbf{C} \subseteq \mathbf{V}$ is a t-error correcting code, if the **minimum distance**

$$d(\mathbf{C}) := \min \{d(\mathrm{c}, \mathrm{c}') \mid \mathrm{c}, \mathrm{c}' \in \mathbf{C}, \mathrm{c} \neq \mathrm{c}'\}$$

of $\mathbf{C}$ is at least $2t + 1$. We call the elements of a code its **codewords**.

In order to explain the idea behind this definition, we need a little lemma.

5.1.2 Lemma. *Let* $\mathbf{C}$ *be a* t*-error correcting code. Then*
(a) *for each vector* $\mathrm{v} \in \mathbf{V}$ *there is at most one codeword* $\mathrm{c} \in \mathbf{C}$ *such that* $d(\mathrm{v}, \mathrm{c}) \leq t$;
(b) *the spheres* $\mathrm{S}_t(\mathrm{c})$ *with* $\mathrm{c} \in \mathbf{C}$ *are mutually disjoint.*

Proof. (a) Assume that there are two distinct $\mathrm{c}, \mathrm{c}' \in \mathbf{C}$ and a vector $\mathrm{v} \in \mathbf{V}$ with $d(\mathrm{v}, \mathrm{c}) \leq t$ and $d(\mathrm{v}, \mathrm{c}') \leq t$. Using the triangle inequality we get

$$d(\mathrm{c}, \mathrm{c}') \leq d(\mathrm{c}, \mathrm{v}) + d(\mathrm{v}, \mathrm{c}') \leq 2t,$$

contradicting $d(\mathbf{C}) \geq 2t + 1$.
(b) follows directly from (a). □

Now we can justify the name of t-error correcting code: For messages that are transmitted we use only codewords. If during the transmission of a codeword c there occur at most t errors, then the received vector x has a distance of at most t from c. In view of **5.1.2**(a) there is just one codeword having distance at most t from x. The recipient **decodes** x to c.

Here the idea of spheres is particularly helpful: the fact that during the transmission at most t errors occur means that the received vector still lies inside the sphere $\mathrm{S}_t(\mathrm{c})$. Since by the above lemma any two spheres whose centres are distinct codewords are disjoint, the received vector x can be uniquely decoded to the centre of the sphere containing x.

Remark. If there are more than t errors per codeword, then the received vector usually will not be decoded correctly. In practice one proceeds as follows. First one estimates how many errors would occur in the channel, then chooses t accordingly, and finally constructs a t-error correcting code.

Now we are able to say exactly what is the aim of coding theory: to construct codes that
– have a big minimum distance (and therefore good error correcting properties) and
– allow an efficient decoding algorithm.

We conclude this section with an example of a code. Although this code is still rather small, it serves to show that the tools developed so far are not sufficient to investigate codes. This code will serve as the running example in this chapter.

5.1.3 Theorem. *The following* 16 *vectors from* $\mathbf{V} = \{0, 1\}^7$ *form a* 1*-error correcting code:*

0000000	1111111
1110000	0001111
1001100	0110011
1000011	0111100
0101010	1010101
0100101	1011010
0011001	1100110
0010110	1101001

□

In exercise 2 you are invited to investigate this code.

5.2 Linear codes

Our approach so far is fairly naive and not very practical. The impracticability consists in listing the code (one has to store each codeword), determining the minimal distance (a problem of order $|\mathbf{C}|^2$), and – last but not least – decoding (for each received vector one has to compute its distances from all codewords).

As a start toward an applicable theory, we say the magic words, 'linear codes'.

Definition. A code $\mathbf{C} \subseteq \mathbf{V}$ is called **linear**, if $\mathbf{C}$ is a *subspace* of the vector space $\mathbf{V}$ (and not only a subset of the set $\mathbf{V}$). In this case $\mathbf{C}$ has a dimension, which will often be denoted by the letter k; we call $\mathbf{C}$ a **linear** $[n, k]$**-code**.

One advantage of linear codes is immediate. To work with $\mathbf{C}$ one has to know only a *basis* of $\mathbf{C}$.

Definition. Let $c_1, \ldots, c_k$ be a basis of the linear $[n, k]$-code $\mathbf{C}$. Then the $k \times n$ matrix G whose ith row is the basis vector c_i $(i = 1, \ldots, n)$ is called a **generator matrix** of $\mathbf{C}$.

To store a generator matrix one needs k vectors, a gigantic saving over the 2^k vectors that constitute $\mathbf{C}$.

Example. A generator matrix of the code in Example **5.1.3** is

$$\mathrm{G} = \begin{pmatrix} 1 & 1 & 1 & 0 & 0 & 0 & 0 \\ 1 & 0 & 0 & 1 & 1 & 0 & 0 \\ 1 & 0 & 0 & 0 & 0 & 1 & 1 \\ 0 & 1 & 0 & 1 & 0 & 1 & 0 \end{pmatrix}.$$

Also, determining the minimal distance of a linear code is easy compared to the general case.

Definition. The **weight** $w(\mathrm{x})$ of a vector $\mathrm{x} \in \mathbf{V}$ is the number of nonzero positions of x. In other words

$$w(\mathrm{x}) = d(\mathrm{x}, \mathrm{o}).$$

The **minimum weight** $w(\mathbf{C})$ of a code $\mathbf{C}$ is defined as

$$w(\mathbf{C}) := \min \{w(\mathrm{c}) \mid \mathrm{c} \in \mathbf{C}, \mathrm{c} \neq \mathrm{o}\}.$$

5.2.1 Lemma. *Let* $\mathbf{C}$ *be a linear code. Then*

$$d(\mathbf{C}) = w(\mathbf{C}).$$

Proof. For any code we have that

$$\begin{aligned} d(\mathbf{C}) &= \min \{d(\mathrm{c}, \mathrm{c}') \mid \mathrm{c}, \mathrm{c}' \in \mathbf{C}, \mathrm{c} \neq \mathrm{c}'\} \\ &\leq \min \{d(\mathrm{c}, \mathrm{o}) \mid \mathrm{c} \in \mathbf{C}, \mathrm{c} \neq \mathrm{o}\} = w(\mathbf{C}). \end{aligned}$$

For the proof of the other inequality it is sufficient to show that there is a codeword c_0 of weight $d(\mathbf{C})$. Let $\mathrm{c}, \mathrm{c}' \in \mathbf{C}$ with $d(\mathrm{c}, \mathrm{c}') = d(\mathbf{C})$. Then we have

$$w(\mathrm{c} - \mathrm{c}') = d(\mathrm{c} - \mathrm{c}', \mathrm{o}) = d(\mathrm{c} - \mathrm{c}', \mathrm{c}' - \mathrm{c}') = d(\mathrm{c}, \mathrm{c}') = d(\mathbf{C}).$$

Since $\mathbf{C}$ is linear, we have that $\mathrm{c}_0 := \mathrm{c} - \mathrm{c}' \in \mathbf{C}$. This proves the assertion. □

The lemma implies that in order to determine the minimum distance (and therefore the error correcting quality) of a linear code **C**, one has only to calculate the minimum weight of **C**; for this one needs at most $|\mathbf{C}|$ steps.

To explain how to decode a linear code, we have to introduce a few notions.

Definition. Let $\mathbf{C} \subseteq \mathbf{V}$ be a code. The **dual code** $\mathbf{C}^{\perp}$ is defined as follows:

$$\mathbf{C}^{\perp} := \{\mathrm{v} \in \mathbf{V} \mid \mathrm{v} \cdot \mathrm{c} = 0 \text{ for all } \mathrm{c} \in \mathbf{C}\};$$

where $\mathrm{v} \cdot \mathrm{c}$ is the 'inner product' of the vectors $\mathrm{v} = (v_1, \ldots, v_n)$ and $\mathrm{c} = (c_1, \ldots, c_n)$, that is

$$\mathrm{v} \cdot \mathrm{c} = v_1 c_1 + v_2 c_2 + \ldots + v_n c_n.$$

If $\mathrm{v} \cdot \mathrm{c} = 0$ then we shall also say that v and c are **orthogonal**.

Remark. Sorry, the terminology is well established, so we are stuck with calling $\mathbf{C}^{\perp}$ the 'dual' code of **C**, although 'orthogonal code' would be preferable – and $\mathbf{C}^{\perp}$ has little meaningful to do with the 'dual vector space' of **C**. One wonders if 'duel code' had been intended, conjuring up an image of crossed swords.

5.2.2 Lemma. *If* **C** *is a linear* $[n, k]$*-code then* $\mathbf{C}^{\perp}$ *is a subspace of* **V** *of dimension* $n - k$.

Proof. Independently of whether or not **C** is linear, $\mathbf{C}^{\perp}$ is always a subspace of **V**. We have to show that the dimension of $\mathbf{C}^{\perp}$ is $n - k$.

For this we consider a generator matrix G of **C** with rows $\mathrm{c}_1, \ldots, \mathrm{c}_k$. Then we have

$$\mathbf{C}^{\perp} = \{\mathrm{v} \in \mathbf{V} \mid \mathrm{v} \cdot \mathrm{c}_i = 0,\ i = 1, \ldots, k\}.$$

Looking naively at the last line one sees that we seek those $\mathrm{v} = (v_1, \ldots, v_n) \in \mathbf{V}$ that are solutions of the homogeneous system of linear equations with coefficient matrix G. As everybody knows from linear algebra, the dimension of the solution space equals $n - \mathit{rank}(\mathrm{G})$. Since the rows of G form a basis of **C**, necessarily G has rank k. Hence we have $\dim(\mathbf{C}^{\perp}) = n - k$. □

Of course one can iterate the process of constructing the dual code – constructing the dual of the dual code. The following lemma shows that this yields nothing new, and hence there is no point in iterating this process.

5.2.3 Lemma. *Let* **C** *be a linear code. Then*

$$\mathbf{C}^{\perp\perp} = \mathbf{C}.$$

Proof. First we show that $\mathbf{C} \subseteq \mathbf{C}^{\perp\perp}$: The set $\mathbf{C}^{\perp\perp}$ consists of all those vectors that are orthogonal to all vectors of $\mathbf{C}^{\perp}$; these include the vectors of $\mathbf{C}$, since $\mathbf{C}^{\perp}$ is the set of all vectors that are orthogonal to each vector of $\mathbf{C}$. (Hint: Read the last sentence once more, but very slowly. Then you will understand it.)

Applying **5.2.2** to $\mathbf{C}^{\perp}$ gives

$$\dim(\mathbf{C}^{\perp\perp}) = n - \dim(\mathbf{C}^{\perp}) = n - (n-k) = k = \dim(\mathbf{C}).$$

From these results together it follows that $\mathbf{C}^{\perp\perp} = \mathbf{C}$. □

Definition. Let $\mathbf{C} \subseteq \mathbf{V}$ be a linear $[n, k]$-code. A matrix H whose rows form a basis of the dual code $\mathbf{C}^{\perp}$ is called a **parity check matrix** of $\mathbf{C}$.

Since $\mathbf{C}^{\perp}$ has dimension $n-k$, any parity check matrix of $\mathbf{C}$ has exactly $n-k$ rows and n columns.

In order to convince ourselves that decoding linear codes is significantly simpler than decoding a code in general, we need a further notion, the syndrome of a code.

Definition. Let H be a parity check matrix of the linear code $\mathbf{C} \subseteq \mathbf{V}$. For any vector $\mathrm{v} \in \mathbf{V}$ we define its **syndrome** $\mathrm{s(v)}$ as

$$\mathrm{s(v)} := \mathrm{v} \cdot \mathrm{H}^{\mathrm{T}},$$

where H^{T} is the transpose of H. (Hence, a syndrome is a binary vector of length $n-k$.)

Using syndromes one can easily describe a linear code.

5.2.4 Lemma. *If* $\mathbf{C}$ *is a linear code with parity check matrix* H *then*

$$\mathbf{C} = \{\mathrm{v} \in \mathbf{V} \mid \mathrm{s(v)} = \mathrm{o}\}.$$

Proof. Let $\mathrm{v} \in \mathbf{V}$ be an arbitrary vector. Then

$$\begin{aligned} \mathrm{s(v)} = \mathrm{o} &\Leftrightarrow \mathrm{v} \cdot \mathrm{H}^{\mathrm{T}} = \mathrm{o} \\ &\Leftrightarrow \mathrm{v} \text{ is orthogonal to all vectors of a basis of } \mathbf{C}^{\perp} \\ &\Leftrightarrow \mathrm{v} \in \mathbf{C}^{\perp\perp} \\ &\Leftrightarrow \mathrm{v} \in \mathbf{C} \quad \text{in view of } \mathbf{5.2.3}. \end{aligned}$$

□

The following crucial observation says that a syndrome $\mathrm{s(v)}$ depends only on the coset containing v.

5.2.5 Lemma. *Let* H *be a parity check matrix of a linear code* $\mathbf{C} \subseteq \mathbf{V}$. *Then for all vectors* $v, w \in V$ *we have*

$$s(v) = s(w) \Leftrightarrow v + \mathbf{C} = w + \mathbf{C}.$$

Proof.
$$\begin{aligned} s(v) = s(w) &\Leftrightarrow v \cdot H^T = w \cdot H^T \\ &\Leftrightarrow v \cdot H^T - w \cdot H^T = o \\ &\Leftrightarrow (v - w) \cdot H^T = o \\ &\Leftrightarrow v - w \in \mathbf{C} \quad \text{by } \mathbf{5.2.4} \\ &\Leftrightarrow v + \mathbf{C} = w + \mathbf{C}. \end{aligned}$$ □

Now we are able to describe how to decode using a linear code. First one has to represent the cosets of $\mathbf{C}$ by suitable vectors.

Definition. Let $\mathbf{C} \subseteq \mathbf{V}$ be a linear code. A vector is called a **leader** of a coset of $\mathbf{C}$ if it has minimum weight among all vectors in this coset.

In general, coset leaders are not unique. However, we have the following assertion which shows the importance of this concept.

5.2.6 Lemma. *Let* $\mathbf{C} \subseteq \mathbf{V}$ *be a linear* t*-error correcting code. Then*
(a) *each vector of* $\mathbf{V}$ *whose weight is at most* t *is leader of some coset;*
(b) *the leader of a coset that contains a vector of weight at most* t *is uniquely determined.*

Proof. We prove (a) and (b) simultaneously. Let v be a vector for which $w(v) \le t$. Consider an arbitrary vector $v' \in v + \mathbf{C}$ with $v' \neq v$. We have to show that v' has weight at least $t + 1$.

Since v and v' are in the same coset of $\mathbf{C}$, we have $v - v' \in \mathbf{C}$. Since $v \neq v'$, $v - v' \neq o$, hence $w(v - v') \ge 2t + 1$ by definition of a t-error correcting code. It follows that

$$\begin{aligned} 2t + 1 &\le w(v - v') = d(v - v', o) = d(v, v') \\ &\le d(v, o) + d(o, v') = w(v) + w(v') \\ &\le t + w(v'), \end{aligned}$$

and therefore $w(v') \ge t + 1$. □

Lemma **5.2.6** yields a decoding algorithm: One first determines the coset of $\mathbf{C}$ containing the received vector x. The error vector of x will be the leader of that coset. One then adds the leader to x and gets the codeword back.

This procedure can be organized a little more efficiently.

Syndrome-Decoding: *Let* $\mathbf{C} \subseteq \mathbf{V}$ *be a linear* $[n, k]$*-code that is* t*-error correcting. One computes a list of all coset leaders and the corresponding syndromes. When a vector* x *is received one*
- *computes its syndrome* s(x),
- *looks up this syndrome in the list of all syndromes,*
- *finds the corresponding coset leader* e *and*
- *decodes* x *to* c := x + e.

Lemma **5.2.6** guarantees that this decoding procedure works correctly if at most t errors occur.

In order to illustrate syndrome-decoding we look back to Example **5.1.3**. First we have to determine a parity check matrix. In the next section we shall prove that

$$\mathrm{H} = \begin{pmatrix} 1 & 0 & 0 & 1 & 1 & 0 & 1 \\ 0 & 1 & 0 & 1 & 0 & 1 & 1 \\ 0 & 0 & 1 & 0 & 1 & 1 & 1 \end{pmatrix}$$

is a parity check matrix of **C**. (But you should convince yourself now.)

The coset leaders are the vectors 0000000, 0000001, 0000010, ... Therefore the list of coset leaders and syndromes is the following:

coset leader	syndrome
0000000	000
0000001	111
0000010	011
0000100	101
0001000	110
0010000	001
0100000	010
1000000	100

If, for instance, the vector x = 0010001 has been received we compute its syndrome s(x) = 110. Then, from the list we find the error vector e = 0001000, and we get as codeword

$$\mathrm{c} = \mathrm{x} + \mathrm{e} = 0010001 + 0001000 = 0011001.$$

5.3 Hamming codes

We shall now study the Hamming codes. Here, for the first time in this chapter we shall use projective geometry. Hamming codes can best be defined via a parity check matrix.

Definition. Let r be a positive integer and $n := 2^r - 1$. Consider a binary $r \times (2^r - 1)$ matrix H whose columns are the binary r-tuples different from o. We define the code Ham(r) by

$$\text{Ham}(r) := \{\mathrm{c} = (c_1, \ldots, c_n) \in \{0, 1\}^n \mid \mathrm{c} \cdot \mathrm{H}^\mathrm{T} = \mathrm{o}\}.$$

In other words, the codewords of Ham(r) are exactly those vectors c for which $\mathrm{c} \cdot \mathrm{H}^\mathrm{T}$ is the zero-vector of length r.

Ham(r) is called the (**binary**) **Hamming code** of length $n = 2^r - 1$.

The reader may verify that the code studied in Example **5.1.3** is the Hamming code Ham(3).

Since the matrix H in the definition of a Hamming code has rank r (see exercise 6), it follows from **5.2.2** that Ham(r) has dimension $2^r - 1 - r$, so that Ham(r) is a linear $[2^r - 1, 2^r - 1 - r]$-code.

After the construction of a code, the first question one has to answer is how many errors it can correct.

5.3.1 Theorem. *The Hamming codes are* 1*-error correcting codes.*

Proof. In view of **5.2.1** we have to show that the minimum weight of Ham(r) is at least 3.

First we assume that Ham(r) contains a vector c of weight 1. Let c have a 1 at the ith position and 0 elsewhere. By definition of Ham(r) it then follows that

$$\mathrm{c} \cdot \mathrm{H}^\mathrm{T} = \mathrm{o}.$$

This means that the ith column of H must be o, a contradiction. Hence Ham(r) does not have minimum weight 1.

Now let us assume that Ham(r) contains a vector that has 1 in the ith and jth positions and 0 elsewhere. Then the sum of the ith and jth columns of H must be zero. This is a contradiction since the columns of H are distinct. Therefore Ham(r) does not have minimum weight 2. □

Hamming codes are not just arbitrary 1-error correcting codes but, in a certain sense, the best possible such codes, namely those that are most densely packed.

Definition. A t-error correcting $\mathbf{C} \subseteq \mathbf{V}$ is called **perfect** if any vector of $\mathbf{V}$ has distance t or less from (exactly) one codeword.

One can express this as follows. A code $\mathbf{C}$ is perfect if it satisfies

$$\bigcup_{c \in \mathbf{C}} S_t(c) = \mathbf{V},$$

that is, if the spheres of radius t centred around the codewords fill the whole vector space $\mathbf{V}$.

At first glance it is hard to believe that perfect codes exist, and, in fact, they are rather scarce. However, we shall show that Hamming codes are perfect. For this, the following lemma will be useful.

5.3.2 Lemma. *Let* $\mathbf{V} = \{0, 1\}^n$, *and let* $\mathbf{C} \subseteq \mathbf{V}$ *be a* 1*-error correcting code. Then we have*

$$|\mathbf{C}| \leq \frac{2^n}{n+1}$$

with equality if and only if $\mathbf{C}$ *is perfect.*

Proof. First we calculate the number of vectors in a sphere $S_1(c)$ around a codeword c: The sphere consists of c itself and all vectors of distance 1 from c – all those vectors that differ from c at exactly one position. Since c has n positions, there are precisely n vectors of distance 1 from c. Hence

$$|S_1(c)| = 1 + n.$$

Since $\mathbf{C}$ is a 1-error correcting code, the spheres $S_1(c)$ around the codewords c are mutually disjoint. Therefore the spheres of radius 1 centred around the codewords cover exactly

$$|\mathbf{C}| \cdot (n + 1)$$

vectors of $\mathbf{V}$. Since $\mathbf{V}$ consists of exactly 2^n vectors we get

$$|\mathbf{C}| \cdot (n + 1) \leq 2^n.$$

Equality holds if and only if any vector of $\mathbf{V}$ lies in a sphere of radius 1 around a codeword, that is if and only if $\mathbf{C}$ is perfect. □

As a corollary we get that any perfect 1-error correcting code $\mathbf{C} \subseteq \{0, 1\}^n$ has a length n of the form

$$n = 2^r - 1,$$

because $|\mathbf{C}| \cdot (n + 1) = 2^n$ implies that $n + 1$ is a divisor of 2^n.

5.3.3 Theorem. *The Hamming codes are perfect* 1*-error correcting codes.*

Proof. Since $\dim(\mathrm{Ham}(r)) = 2^r - 1 - r$ it follows that

$$|\mathrm{Ham}(r)| = 2^{2^r - 1 - r}.$$

In view of $n = 2^r - 1$ this implies

$$|\mathrm{Ham}(r)| \cdot (n+1) = 2^{2^r - 1 - r} \cdot 2^r = 2^{2^r - 1} = 2^n.$$

Thus, by **5.3.2**, $\mathrm{Ham}(r)$ is perfect. □

In Hamming codes, syndrome-decoding can be arranged in such a way that one does not need additional storage for the coset leaders. For this we recall the definition of the matrix H: For H we could choose *any* $r \times (2^r - 1)$ matrix whose columns are the nonzero binary r-tuples; the order in which the columns were listed would not yet be important. Now we interpret the columns of H as binary representations of the integers $1, \ldots, 2^r - 1$ and order the columns in such a way that the ith column s_i is the representation of the integer i. So, for instance, the last column consists only of 1s.

5.3.4 Theorem. *Let* $\mathrm{Ham}(r)$ *be the code that is constructed using the matrix* H *whose* i*th column* s_i *is the binary representation of the integer* i*. Then for each vector* $v \in \mathbf{V} \backslash \mathbf{C}$ *its syndrome* $s(v)$ *is the value* s_i *for which* $v - e_i \in \mathbf{C}$*. (Here,* e_i *is the vector of weight* 1 *that has a* 1 *at the* i*th position.)*

In other words, the syndrome of an erroneous vector shows the position where the error occurred.

Proof. Let e_i be the vector of weight 1 that has a 1 at the ith position. Since the code $\mathrm{Ham}(r)$ is perfect, any vector $v \in \mathbf{V} \backslash \mathbf{C}$ is of the form $v = c + e_i$ for a suitable codeword c and some integer i. This implies

$$s(v) = v \cdot H^T = (c + e_i) \cdot H^T = c \cdot H^T + e_i \cdot H^T = e_i \cdot H^T = s_i.$$

Thus $s(v)$ is the ith column of H; since this represents the number i, one can localise the error using $s(v)$. □

Therefore, the decoding algorithm is extremely simple:

When a vector x has been received, one simply computes its syndrome $s(x)$, reads this r-tuple as an integer i, then adds the error vector e_i to x to get the corresponding codeword.

A simpler algorithm is hardly conceivable!

At the beginning of this section we promised that Hamming codes are connected to geometry. An example should make this clear. We consider the Hamming code Ham(3); a complete list of its codewords can be found in **5.1.3**.

On the other hand we consider the projective plane **P** of order 2; we order its points as shown in Figure 5.4.

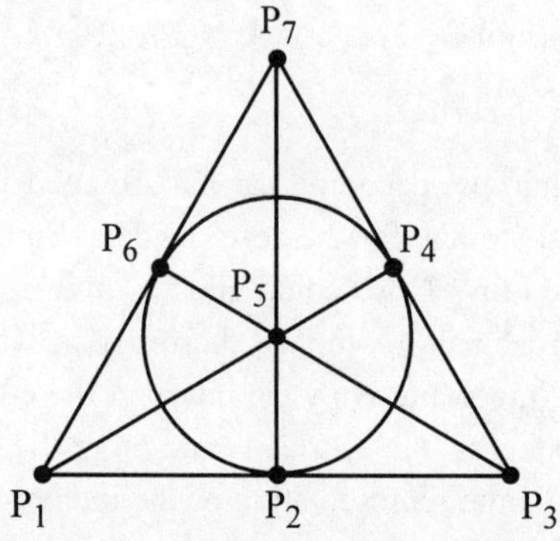

Figure 5.4 The projective plane of order 2

Since any codeword of Ham(3) has length 7 we can assign to any codeword c a set $\mathbb{S}$(c) of points of **P**, where a point P_i is in $\mathbb{S}$(c) if and only if c has a 1 in its *i*th position. (In other words, the vector c is the characteristic vector of $\mathbb{S}$(c).) For instance, the codeword c = 1110000 is mapped onto the set $\mathbb{S}$(c) = $\{P_1, P_2, P_3\}$.

What are the point sets of **P** that correspond to codewords of Ham(3)? One easily checks the amazing fact that these sets are – apart from the empty set and the whole point set – precisely the *lines* of **P** and the *complements of the lines* of **P**!

Considering this representation one sees geometrically – and therefore convincingly – that Ham(3) is a perfect 1-error correcting code: for this, we have only to show that any set of points that does not represent a codeword can be made into a codeword by adding or subtracting just one point.

Let's try. Well, vectors of weight 1, 2, or 6 are easy to decode. A vector of weight 3 that is not a codeword corresponds to a triangle of **P**; by adding one point this can made into a quadrangle, hence to the complement of some line (see exercise 4). A vector of weight 4 that is not a codeword corresponds to a line together with a point outside that line; by subtracting this point, the vector can be decoded to a line. Finally, let c be a vector of weight 5. Then the point set $\mathbb{S}$(c) must contain a line, and one easily verifies that $\mathbb{S}(c)$ is the union of two lines. By removing the intersection of these lines one decodes $\mathbb{S}(c)$ to a quadrangle.

In the last part of this section we deal with the extended Hamming codes.

Let Ham(r) be the binary Hamming code of length $2^r - 1$. From this, we get the **extended Hamming code** Ham(r)* by extending each codeword of Ham(r) by one position; the entry at this position is made 0 or 1 in order to achieve an even total number of 1s in each codeword.

For instance, the code Ham(3)* consists of the following codewords:

00000000	11111111
11100001	00011110
10011001	01100110
10000111	01111000
01010101	10101010
01001011	10110100
00110011	11001100
00101101	11010010

5.3.5 Theorem. Ham(r)* *is a linear* $[2^r, 2^r - 1 - r]$*-code with minimum distance* 4.

Proof. First we show that Ham(r)* is a subspace of $\mathbf{V}^* = GF(2)^{2^r}$. Let c_1^* and c_2^* be two arbitrary elements of Ham(r)*, and let c_1 and c_2 be the corresponding codewords of Ham(r). Since $c_1^* + c_2^*$ coincides in the first $2^r - 1$ positions with $c_1 + c_2$, we have only to show the following: the last position of $c_1^* + c_2^*$ is 1 if $w(c_1 + c_2)$ is odd and 0 otherwise.

If $c_1 + c_2$ has odd weight then w.l.o.g. c_1 has odd weight and c_2 has even weight. Thus the last entry of c_1^* is 1 and the last entry of c_2^* is 0. Therefore the last entry of $c_1^* + c_2^*$ is 1. If $c_1 + c_2$ has even weight then c_1 and c_2 have both even or both odd weight. In either case c_1^* and c_2^* have the same last entry and thus the last entry of $c_1^* + c_2^*$ is 0. Therefore in both cases $c_1^* + c_2^*$ is a codeword, and so Ham(r)* is a vector space.

Clearly, Ham(r)* has the same dimension as Ham(r) since the two vector spaces have the same number of elements.

Finally, we show that Ham(r)* has minimum weight 4. Since $w(\text{Ham}(r)) = 3$, we trivially have $w(\text{Ham}(r)^*) \geq 3$. Assume that $w(\text{Ham}(r)^*) = 3$. Then there would exist a vector of Ham(r)* of weight 3; this is impossible since any vector of Ham(r)* has even weight. □

5.3.6 Theorem. *Suppose that* H *is a parity check matrix of* Ham(r). *Then, by the following operations, one can obtain a parity check matrix* H* *of* Ham(r)*:
– *adjoin to each row of* H *a position whose entry is* 0,
– *adjoin an additional row that consists entirely of* 1*s*.

Example. Ham(3)* has the following parity check matrix:

$$\mathrm{H}^* = \begin{pmatrix} 1 & 0 & 0 & 1 & 1 & 0 & 1 & 0 \\ 0 & 1 & 0 & 1 & 0 & 1 & 1 & 0 \\ 0 & 0 & 1 & 0 & 1 & 1 & 1 & 0 \\ 1 & 1 & 1 & 1 & 1 & 1 & 1 & 1 \end{pmatrix}.$$

Proof. In order to construct a parity check matrix of Ham(r)*, we have to consider the dual code. By **5.3.5** the dual code of Ham(r)* is a linear $[2^r, r+1]$-code. Therefore, any parity check matrix of Ham(r)* is an $(r+1)\times 2^r$ matrix. It is therefore sufficient to show that the rows of H* are linearly independent and are codewords of the dual code of Ham(r)*.

Since H is a parity check matrix, its rows are linearly independent. Consequently the first r rows of H* are also linearly independent. The fact that the last entry in each of the first r rows is 0, while the last entry in the last row is 1, implies that all rows of H* are linearly independent. Consequently, H* has rank $r+1$.

Since the rows of H are codewords of the dual code of Ham(r), by construction the first r rows of Ham(r)* are codewords of the dual code of Ham(r)*. By definition, each codeword of Ham(r)* has an even number of 1s. Therefore the product of a codeword of Ham(r)* with the last row of H* is also 0. □

5.4 MDS codes

One of the central questions in coding theory concerns codes with a big minimum distance. In the framework of linear codes, this question can be made precise as follows.

Let n and r be positive integers. What is the maximum possible minimum distance of a linear $[n, n-r]$-code?

This question has a surprisingly simple answer.

5.4.1 Theorem ('Singleton bound'). *Let* d *be the minimum distance of a linear* $[n, n-r]$*-code. Then*

$$d \leq r+1.$$

Proof. Let $\mathbf{C}$ be a linear $[n, n-r]$-code. Since $\mathbf{C}$ is linear we have only to show that $\mathbf{C}$ satisfies $w(\mathbf{C}) \leq r+1$.

For this we consider a generator matrix G of $\mathbf{C}$. By elementary operations we can transform G into a matrix G' in standard form, which is a generator matrix of $\mathbf{C}$ as well:

$$\mathrm{G'} = \left(\begin{array}{ccccc|c} 1 & & & & & \\ & 1 & & 0 & & \\ & & \ddots & & & \mathrm{G}^* \\ & 0 & & \ddots & & \\ & & & & 1 & \end{array}\right).$$
$$\underbrace{\qquad\qquad}_{n-r} \quad \underbrace{\qquad}_{r}$$

Since any row vector of G^* has at most r nonzero elements, any row of G' has weight at most $1+r$. Thus we have

$$w(\mathbf{C}) \leq r+1. \qquad \square$$

One can also read this theorem in another way. If the length n and the minimum distance d are given, a linear code can have a dimension of at most $n-d+1$. Thus, the number of codewords is limited.

The case of equality in the above theorem is of particular interest.

Definition. A linear $[n, n-r]$-code $\mathbf{C}$ is called an **MDS code** (**maximum distance separable**) if it satisfies

$$w(\mathbf{C}) = r+1.$$

Most astonishingly the question at the beginning of this section and, in particular, questions concerning the existence of MDS codes can be translated into a very interesting geometrical problem, which was studied long before its connection to coding theory had been noticed. For this translation we need the generalization of Theorem **5.3.1**.

5.4.2 Lemma. *Let* $\mathbf{C}$ *be a linear code of length* n *with parity check matrix* H. *Then*

$$d(\mathbf{C}) \geq d \;\Leftrightarrow\; \textit{each set of } d-1 \textit{ columns of } \mathrm{H} \textit{ is linearly independent.}$$

Proof. Since $\mathbf{C}$ is linear we have $d(\mathbf{C}) = w(\mathbf{C}) =: w$.

'$\Leftarrow$': Let $c = (c_1, \ldots, c_n)$ be a codeword of minimum weight w. Then

$$\mathrm{c}\cdot \mathrm{H}^{\mathrm{T}} = \mathrm{o},$$

and so

$$c_1\mathrm{h}_1 + c_2\mathrm{h}_2 + \ldots + c_n\mathrm{h}_n = \mathrm{o},$$

where $\mathrm{h}_1, \mathrm{h}_2, \ldots, \mathrm{h}_n$ are the columns of H. Therefore there are w linearly dependent columns in H. Therefore it follows that $w > d - 1$.
'$\Rightarrow$': Assume that there is a set $\{\mathrm{h}_{i_1}, \mathrm{h}_{i_2}, \ldots, \mathrm{h}_{i_s}\}$ of $s \le w - 1$ linearly dependent columns of H. Consider the vector x that has 1 in positions i_j $(j = 1, \ldots, s)$ and 0 elsewhere. Then

$$\mathrm{x}\cdot \mathrm{H}^{\mathrm{T}} = \mathrm{h}_{i_1} + \mathrm{h}_{i_2} + \ldots + \mathrm{h}_{i_s} = \mathrm{o},$$

hence $\mathrm{x} \in \mathbf{C}$. But this contradicts $w(\mathrm{x}) = s \le w - 1$. Therefore any $w - 1$ columns of H are linearly independent. □

Now we introduce the geometrical counterpart of a linear code.

Definition. A set of n points in a projective space **P** is called an (n, s)**-set** if s is the largest integer for which every subset of s points is independent.

Examples. An $(n, 3)$-set is a set of n points, no three of which are collinear, but at least four of them lie in a common plane. An $(n, d + 1)$-set of a projective space of dimension d is a set of n points in general position.

5.4.3 Theorem. *Let n and r be positive integers. Then a linear $[n, n - r]$-code with minimum distance d exists if and only if there exists an $(n, d - 1)$-set in* $\mathbf{P} = \mathrm{PG}(r - 1, 2)$.

Proof. Let **C** be a linear $[n, n - r]$-code with minimum distance d, and let H be a parity check matrix of **C**.

By the preceding lemma, the columns of H are n binary r-tuples, any $d - 1$ of which are linearly independent. We consider these n vectors as homogeneous coordinates of points in $\mathrm{PG}(r - 1, 2)$. This set $\mathfrak{M}$ of n points has the property that any $d - 1$ are independent. Should d points of $\mathfrak{M}$ be independent, then, again by Lemma **5.4.2**, the code **C** would have minimum distance at least $d + 1$. From these results together it follows that $\mathfrak{M}$ is an $(n, d - 1)$-set.

Conversely we suppose that there is an $(n, d - 1)$-set $\mathfrak{M}$ in $\mathbf{P} = \mathrm{PG}(r - 1, 2)$. We consider the homogeneous coordinates of the n points of $\mathfrak{M}$ as column vectors of an $n \times r$ matrix H. Then H has the property that any $d - 1$ of its columns are linearly independent. By Lemma **5.4.2**, the code **C** that is defined by

$$\mathbf{C} := \{x \in \{0, 1\}^n \mid x \cdot H^T = o\}$$

has minimum weight at least d. Since there exist d linearly dependent columns of H we have $d(\mathbf{C}) \le d$. From these results together it follows that $d(\mathbf{C}) = d$. □

Example. Using **5.3.5** we see that the extended Hamming code $\mathrm{Ham}(r)^*$ corresponds to a set of 2^r points in $\mathrm{PG}(r, 2)$ no three of which are collinear. One example for such a set is the set of points outside some hyperplane. (In $\mathrm{PG}(d, q)$ the points outside some hyperplane have the property that no $q + 1$ of them are collinear; since in our case we have $q = 2$ the assertion follows.)

In **5.4.5**(a) we shall show that the points outside some hyperplane are the only examples of such point sets.

Definition. (a) A set of points in a projective space is called a **cap** if no three of its points are collinear.

(b) A set of points in a d-dimensional projective space $\mathbf{P}$ is called an **arc** if any $d + 1$ of its points form a basis of $\mathbf{P}$. An arc having k points is also called a k-**arc**.

So, in a projective plane, any arc is a cap and conversely. Moreover, each cap is an $(n, 3)$-set; an arc in a d-dimensional projective space is an $(n, d + 1)$-set.

In Section **4.3** we defined ovals in a projective plane; these are examples of arcs and caps. Other examples of caps are ovoids in 3-dimensional projective spaces. In **5.4.5** and **5.4.7** we shall show that in many cases these examples are the caps with the maximum possible number of points. Examples of arcs are the normal rational curves which we have already met in Section **2.5**. In a projective space of order q, these curves have exactly $q + 1$ points.

Now we are able to express the existence of MDS codes in a geometric language (cf. **5.4.3**).

5.4.4 Corollary. *Let n and r be positive integers. Then a linear MDS $[n, n - r]$-code with minimum distance* 4 *exists if and only if in* $\mathrm{PG}(r - 1, 2)$ *there exists a cap with precisely n points.* □

We may therefore rephrase the problem of constructing good codes with a prescribed large minimum distance as follows:

Let d and r be positive integers. Determine the largest number n such that there is an $(n, d - 1)$-set in $\mathrm{PG}(r - 1, q)$. We denote this maximum n by $\max_{d-1}(r, q)$.

For instance, by **5.4.4**, the number $\max_r(r, 2)$ is the largest length n of a linear MDS $[n, n-r]$-code.

The determination of the numbers $\max_r(r, q)$ is an active (though difficult) area of research inside finite geometry. We present some of the fundamental results. Further results can be found in Hirschfeld [Hir79].

Let's start with something easy. What is $\max_2(r, q)$? Well, by definition $\max_2(r, q)$ is the largest possible number of points in $PG(r-1, q)$ such that any two of them are independent – but this is the number of *all* points of $PG(r-1, q)$. Hence it follows that

$$\max_2(r, q) = q^{r-1} + \ldots + q + 1.$$

In particular we have

$$\max_2(r, 2) = 2^{r-1} + \ldots + 2 + 1 = 2^r - 1.$$

The corresponding code is the Hamming $[n, n-r]$-code.

While the determination of $\max_s(r, q)$ is trivial in the case $s = 2$, it is not completely solved for $s = 3$ (that is $d = 4$). First we look at two easy cases.

5.4.5 Theorem. (a) $\max_3(r, 2) = 2^{r-1}$.

(b) $\max_3(3, q) = \begin{cases} q+1, & \textit{when } q \textit{ is odd,} \\ q+2, & \textit{when } q \textit{ is even.} \end{cases}$

Proof. (a) Let $\mathfrak{M}$ be an $(n, 3)$-set in $\mathbf{P} = PG(r-1, 2)$. Consider a point $P \in \mathfrak{M}$. The number of lines through P is

$$2^{r-2} + \ldots + 2 + 1 = 2^{r-1} - 1.$$

Since on each of these lines there is at most another point of $\mathfrak{M}$ we have

$$\max_3(r, 2) \le 2^{r-1}.$$

Since the points outside some hyperplane of $\mathbf{P}$ form an $(n, 3)$-set with $n = 2^{r-1}$ points, it follows that

$$\max_3(r, 2) = 2^{r-1}.$$

(b) Let $\mathbf{P} = PG(2, q)$ be the Desarguesian projective plane of order q, and let $\mathfrak{M}$ be an $(n, 3)$-set in $\mathbf{P}$.

Since each of the $q+1$ lines through a point of $\mathfrak{M}$ contains at most another point of $\mathfrak{M}$ it follows that

$$\max_3(3, q) \le q + 1 + 1 = q + 2.$$

We show now that $\max_3(3, q) = q + 2$ implies that q is even.

Suppose therefore that $\max_3(3, q) = q + 2$, and consider a $(q + 2, 3)$-set of $\mathbf{P}$. Then any line through some point of $\mathfrak{M}$ contains another point of $\mathfrak{M}$. This means that any line contains either 0 or exactly 2 points of $\mathfrak{M}$. Consider now a point Q $\notin \mathfrak{M}$. Since any line through Q has 0 or 2 points in common with $\mathfrak{M}$, $|\mathfrak{M}|$ must be even. Hence also $q = |\mathfrak{M}| - 2$ is even.

Since any conic is a $(q + 1)$-arc, it follows that $\max_3(3, q) \ge q + 1$. It remains to show that $\max_3(3, q) = q + 2$ for q even. This will be proved in the next theorem. □

5.4.6 Theorem (Qvist, [Qvi52]). *Let $\mathfrak{A}$ be a $(q + 1)$-arc in a finite projective plane of order q. If q is even then there exists a point X (the* ***nucleus*** *of $\mathfrak{A}$) such that all tangents of $\mathfrak{A}$ pass through X. This means that $\mathfrak{A}$ can be extended to a $(q + 2)$-arc $\mathfrak{A} \cup \{X\}$.*

In particular, any projective plane PG(2, q) *with q even has a $(q + 2)$-arc.*

Proof. Let P be a point of $\mathfrak{A}$. Then, by the definition of an arc, any of the $q + 1$ lines through P contains at most one further point of $\mathfrak{A}$. Since $|\mathfrak{A}| = q + 1$ there are precisely q lines through P that are incident with two points of $\mathfrak{A}$; therefore, there is exactly one tangent through P. Thus, $\mathfrak{A}$ has $q + 1$ tangents.

Let P, Q be two points of $\mathfrak{A}$ and let R be a point of PQ not on $\mathfrak{A}$. Then through R there is at least one tangent. (Any line connecting R to a point of $\mathfrak{A}$ contains one or two points of $\mathfrak{A}$; because $|\mathfrak{A}|$ is odd, there must be a tangent.) Since each of the $q + 1$ points of PQ contains at least one of the $q + 1$ tangents of $\mathfrak{A}$, every point of the secant PQ has exactly one tangent.

Consequently, two tangents of $\mathfrak{A}$ must meet in a point X that is not a secant. In other words, every line joining X to a point of $\mathfrak{A}$ must be a tangent so that all $q + 1$ tangents pass through X. □

One can rephrase the above theorems as follows. Let $\mathbf{P}$ be a finite Desarguesian projective plane of order q. If q is odd then the arcs with a maximum number of points are precisely the ovals (by the theorem of Segre it also follows that these are precisely the conics). If q is even, then each oval can be uniquely extended to a $(q + 2)$-arc; these arcs are called **hyperovals**.

In 3-dimensional projective spaces the situation becomes more difficult (and more interesting).

5.4.7 Theorem. *If* $q > 2$, *then we have*

$$\max_3(4, q) = q^2 + 1.$$

Proof. An ovoid is a set of $q^2 + 1$ points no three of which are collinear. Therefore we have

$$\max_3(4, q) \geq q^2 + 1$$

The reverse inequality is much more difficult. Let $\mathcal{M}$ be a set of points no three of which are collinear. We suppose that $|\mathcal{M}| \geq q^2 + 1$ and show that we have $|\mathcal{M}| = q^2 + 1$. Furthermore, we may assume that $\mathcal{M}$ is maximal – there is no set $\mathcal{M}'$ of points no three of which are collinear such that $\mathcal{M}$ is properly contained in $\mathcal{M}'$.

The first case is simple.

Step 1. If q *is odd then* $\mathcal{M}$ *is an ovoid.*

Let P and Q be two arbitrary points of $\mathcal{M}$. Since, by **5.4.5**, any plane through P and Q contains at most $q + 1$ points of $\mathcal{M}$ we see that

$$|\mathcal{M}| \leq 2 + (q + 1)\cdot(q + 1 - 2) = q^2 + 1.$$

It follows that $|\mathcal{M}| = q^2 + 1$.

From now on we suppose that q is even. This case is much more intricate.

Step 2. Through each point of $\mathcal{M}$ *there is at least one tangent line.*

Assume that each line through a point P of $\mathcal{M}$ contains two points of $\mathcal{M}$. Then it follows that $|\mathcal{M}| = q^2 + q + 2$ and no point of $\mathcal{M}$ would be on a tangent. So any plane that contains at least one point of $\mathcal{M}$ would intersect $\mathcal{M}$ in exactly $q + 2$ points. We consider a line g that contains no point of $\mathcal{M}$. (Through any point outside $\mathcal{M}$ there is at least one such line, since otherwise we have $|\mathcal{M}| = 2(q^2 + q + 1)$.) If n denotes the number of planes through g intersecting $\mathcal{M}$ in $q + 2$ points then

$$n(q + 2) = |\mathcal{M}| = q^2 + q + 2.$$

Thus $q + 2$ would divide $(q + 2)\cdot(q - 1) + 4$ $(= q^2 + q + 2)$, hence also 4. This is a contradiction to our hypothesis $q > 2$.

Step 3. Let t *be a tangent that touches* $\mathcal{M}$ *in* P. *Then any plane through* t *contains at most* $q + 1$ *points of* $\mathcal{M}$; *moreover, there is at least one such 'oval plane' (a plane having* $q + 1$ *points of* $\mathcal{M}$) *through* t.

For, since a plane containing $q + 2$ points of $\mathcal{M}$ has no tangent, no plane through t has $q + 2$ points of $\mathcal{M}$. If every plane through t were to contain q or fewer points of $\mathcal{M}$, we would get the following contradiction:

$$|\mathcal{M}| \le 1 + (q+1)\cdot(q-1) = q^2.$$

Step 4. Let π *be an oval plane, let* $\mathcal{M}_\pi$ *be the set of points of* $\mathcal{M}$ *in* π, *and let* X *be the nucleus of* $\mathcal{M}_\pi$. *Then there is a* **secant** *of* $\mathcal{M}$ *(this is a line intersecting* $\mathcal{M}$ *in two points) through* X.

Otherwise $\mathcal{M} \cup \{X\}$ would be a set of points no three of which are collinear, contradicting the maximality of $\mathcal{M}$.

Step 5. Let π *and* X *be as in the above step, and denote by* s *a secant through* X. *Then every plane through* s *is an oval plane; in particular we have* $|\mathcal{M}| = q^2+1$.

For any plane π' through s intersects π in a line through X, hence in a tangent (of $\mathcal{M}_\pi$). Therefore π' contains at most $q+1$ points of $\mathcal{M}$. If one of the planes through s were to contain fewer than $q+1$ points of $\mathcal{M}$, then it would follow that

$$|\mathcal{M}| \le 2 + q - 2 + (q+1-1)\cdot(q+1-2) = q^2.$$

Putting these results together, the theorem is proved completely. □

Remark. One can show that $\mathcal{M}$ is in fact an ovoid if $|\mathcal{M}| = q^2+1$ (see for instance [Beu83], [HaHe76], §12).

5.5 Reed–Muller codes

A class of codes that is most important in theory as well as in practice are the so-called Reed–Muller codes. These Reed–Muller codes are particularly easy to describe using affine geometry, as we now see.

Let $\mathbf{A} = AG(d, 2)$ be the affine space of dimension d and order 2, whose points we shall label arbitrarily $P_1, P_2, \ldots, P_{2^d}$. Using this labelling we can associate with any set $\mathcal{M}$ of points of **A** its **characteristic vector** $\chi(\mathcal{M})$ as follows:

$$\chi(\mathcal{M}) = (a_1, \ldots, a_{2^d}) \quad \text{where } a_i = \begin{cases} 1, & \text{if } P_i \in \mathcal{M}, \\ 0 & \text{otherwise.} \end{cases}$$

In this way we may identify subsets of the point set of A with binary vectors of length 2^d. We shall not bother to distinguish between $\mathcal{M}$ and $\chi(\mathcal{M})$; in other words, any subset of points of **A** is a vector of $\{0, 1\}^{2^d}$.

Definition. The ***r*th order Reed–Muller code (of A)** is the code $\mathbf{C} \subseteq \{0, 1\}^{2^d}$ that is generated by all $(d-r)$-dimensional subspaces of **A**.

Examples. Consider the case $d = 3$. We picture AG(3, 2) as shown in Figure 5.5.

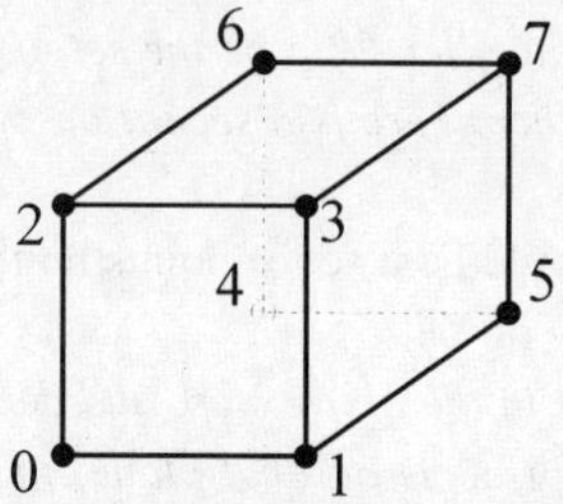

Figure 5.5 The affine space AG(3, 2)

(a) $r = 1$: The characteristic vectors of the 14 planes are the following:

11110000	00001111
11001100	00110011
11000011	00111100
10101010	01010101
10100101	01011010
10011001	01100110
10010110	01101001

If we consider all spans we find the following vectors:

00000000	11111111
11110000	00001111
11001100	00110011
11000011	00111100
10101010	01010101
10100101	01011010
10011001	01100110
10010110	01101001

(b) $r = 2$: The characteristic vectors of the lines are precisely the subsets having two elements.

Now we describe the Reed–Muller codes in another way. The set of all subsets of the point set of **A** forms a GF(2)-vector space **W** with the following vector addition:

$$X + Y := (X \cup Y) \setminus (X \cap Y).$$

One calls $X + Y$ the **symmetric difference** of the sets X and Y (cf. exercise 13).

The rth order Reed–Muller code is then simply the subspace of $\mathbf{W}$ spanned by those vectors that correspond to $(d-r)$-dimensional subspaces of $\mathbf{A}$. In other words, one obtains the elements of the Reed–Muller code if one starts with suitable $(d-r)$-dimensional subspaces and forms certain symmetric differences.

In the above examples, the respective minimum weights are 4 and 2, and the vectors of minimum weight correspond respectively to the planes and lines of $\mathbf{A}$. In other words, the minimum-weight vectors correspond to the generating subspaces of the Reed–Muller code. In the following central theorem we shall generalize this fact.

5.5.1 Theorem. *Let* $\mathbf{C}$ *be the rth order Reed–Muller code. Then* $\mathbf{C}$ *satisfies*

$$d(\mathbf{C}) = 2^{d-r}.$$

A codeword of $\mathbf{C}$ *has weight* 2^{d-r} *if and only if it is the characteristic vector of some* $(d-r)$*-dimensional subspace of* $\mathbf{A}$.

Proof. We prove the theorem by induction on $d-r$.

First we assume $d-r = 1$. Here we have to show that no single point can be represented as a symmetric difference of lines of $\mathbf{A} = AG(d, 2)$. For this, let $\mathcal{M}$ be an arbitrary set of lines of $\mathbf{A}$. We shall use the fact that a point lies in the span of $\mathcal{M}$ (with respect to the symmetric difference) if and only if it is on an odd number of lines of $\mathcal{M}$ (see exercise 17).

Now we count the lines of $\mathcal{M}$: For a point P denote by r_P the number of lines of $\mathcal{M}$ through P (the **degree** of P); we see that

$$\sum_{P \in \mathbf{A}} r_P = 2 \cdot |\mathcal{M}|.$$

Therefore $\sum_{P \in \mathbf{A}} r_P$ is an even number. Since the sum of the even degrees is also even, we conclude that

$$\sum_{\substack{P \in \mathbf{A} \\ r_P \text{ odd}}} r_P$$

is also an even number. Since a sum of odd integers is even if and only if the number of summands is even, we see that the number of points with odd degree is even. In particular there could never be a unique point of odd degree.

Therefore the minimum weight is larger than 1. Since any pair of points form a line of $AG(d, 2)$ we have proved the case $d-r=1$.

Suppose now $d-r>1$, and assume that the assertion is true for all Reed–Muller codes with $d'-r'<d-r$.

Let $\mathcal{M}$ be an arbitrary set of $(d-r)$-dimensional subspaces of $\mathbf{A} = AG(d, 2)$. Let $\mathcal{H}$ be the set of points that is generated (with respect to the symmetric difference) by the elements of $\mathcal{M}$. We have to show that $\mathcal{H}$ contains at least 2^{d-r} points with equality if and only if these points form a $(d-r)$-dimensional subspace.

In the following we suppose that $|\mathcal{H}| \le 2^{d-r}$; we have to show that $\mathcal{H}$ is the point set of a subspace of dimension $d-r$.

Step 1. Let $\mathbf{U}$ *be an s-dimensional subspace of* $\mathbf{A}$, *where* $r+1 \le s \le d-1$. *Then* $\mathcal{H} \cap \mathbf{U}$ *is a codeword of the rth order Reed–Muller code of* $\mathbf{U}$. *By induction, this means that* $\mathcal{H} \cap \mathbf{U}$ *is either empty or contains at least* 2^{s-r} *points, where equality holds if and only if* $\mathcal{H} \cap \mathbf{U}$ *consists of the points of an* $(s-r)$*-dimensional subspace of* $\mathbf{U}$.

This can be seen as follows. Let

$$\mathcal{M}' = \mathcal{M}(\mathbf{U}) = \{\mathbf{M} \cap \mathbf{U} \mid \mathbf{M} \in \mathcal{M}, \mathbf{M} \cap \mathbf{U} \neq \varnothing\}$$

be the set of all nonempty subspaces of $\mathbf{U}$ that are the intersection of $\mathbf{U}$ with some element of $\mathcal{M}$. First we observe that the symmetric difference of the subspaces in $\mathcal{M}'$ equals $\mathcal{H} \cap \mathbf{U}$ since the subspaces of $\mathcal{M}$ that are disjoint from $\mathbf{U}$ do not contribute any points to $\mathcal{H} \cap \mathbf{U}$.

Moreover, $\mathcal{M}'$ consists of subspaces of $\mathbf{U}$ whose dimension is at least $s-r$. For by the dimension formula **1.3.11**, from the projective point of view any subspace $\mathbf{M} \in \mathcal{M}$ intersects the subspace $\mathbf{U}$ in a projective subspace of dimension at least $s-r$. If $\mathbf{M}$ and $\mathbf{U}$ are not parallel, the intersection is not contained in the hyperplane at infinity. Therefore $\mathbf{M}$ and $\mathbf{U}$ intersect each other also in $\mathbf{A}$ in an (affine) subspace of dimension at least $s-r$.

Since any subspace whose dimension exceeds $s-r$ is the disjoint union of (parallel) subspaces of dimension $s-r$, the symmetric difference of subspaces of $\mathcal{M}'$ is a codeword of the Reed–Muller code of order r of $\mathbf{U}$.

Step 2. Let P *be a point of* $\mathcal{H}$. *Then, for any* $i \in \{0, 1, \ldots, r\}$ *there is an i-dimensional subspace* $\mathbf{U}_i$ *through* P *intersecting* $\mathcal{H}$ *just in* P.

Trivially, the assertion is true for $i=0$. Suppose that $i \ge 1$, and assume that the assertion is true for $i-1$. Consider the subspace $\mathbf{U} = \mathbf{U}_{i-1}$. Assume that any i-dimensional subspace through $\mathbf{U}$ would intersect $\mathcal{H}$ in P and at least one further point. Since there are exactly $2^{d-i} + \ldots + 2 + 1 = 2^{d-i+1} - 1$ subspaces of dimension i through $\mathbf{U}$, we would have $|\mathcal{H}| \ge 2^{d-i+1} > 2^{d-r}$. This contradiction shows that there is some i-dimensional subspace $\mathbf{U}_i$ through $\mathbf{U}$ intersecting $\mathcal{H}$ just in P.

Step 3. $|\mathcal{H}| = 2^{d-r}$.

We consider the subspace $\mathbf{U} := \mathbf{U}_r$. By step 1, $\mathcal{H}$ induces in any $(r+1)$-dimensional subspace through $\mathbf{U}_r$ a Reed–Muller code of order r. So, by induction, any such subspace contains at least $2^{r+1-r} = 2$ points of $\mathcal{H}$, and hence at least one further point of $\mathcal{H}$. Since there are precisely $2^{d-r-1} + \ldots + 1 = 2^{d-r} - 1$ subspaces of dimension $r+1$ through $\mathbf{U}$, it follows that

$$|\mathcal{H}| \geq 1 + 2^{d-r} - 1 = 2^{d-r}.$$

We have reached our initial goal. From now on we shall suppose that $|\mathcal{H}| = 2^{d-r}$.

Step 4. The points of $\mathcal{H}$ form a $(d-r)$-dimensional affine subspace.

We temporarily adopt some terminology: we call an r-dimensional subspace **good**, if it contains precisely one point of $\mathcal{H}$.

By induction on $s \in \{r, \ldots, d\}$ we shall show the following assertion: *if* $\mathbf{W}_s$ *is an s-dimensional subspace that contains at least one good subspace, then* $\mathbf{W}_s \cap \mathcal{H}$ *is an $(s-r)$-dimensional subspace.*

The case $s = r$ is trivial, and the case $s = r + 1$ is easy: it follows from step 3 that any $(r+1)$-dimensional subspace through a good subspace has exactly two points (hence the points of a line) in common with $\mathcal{H}$.

Suppose now $r + 1 \leq s \leq d - 1$ and assume that the assertion is true for $s - 1$ and s.

Consider an arbitrary $(s+1)$-dimensional subspace $\mathbf{W}$ through a good subspace $\mathbf{G}$. Denote by $\mathbf{W}_{s-1}$ an $(s-1)$-dimensional subspace of $\mathbf{W}$ through $\mathbf{G}$. By induction, $\mathbf{X} = \mathcal{H} \cap \mathbf{W}_{s-1}$ is a subspace of dimension $s - 1 - r$. There are exactly three s-dimensional subspaces $\mathbf{U}_0$, $\mathbf{U}_1$, $\mathbf{U}_2$ of $\mathbf{W}$ through $\mathbf{W}_{s-1}$. By induction, for any s-dimensional subspace $\mathbf{U}_i$ through $\mathbf{W}_{s-1}$ the set $\mathcal{H} \cap \mathbf{U}_i$ is an $(s-r)$-dimensional subspace $\mathbf{X}_i$ $(i = 1, 2, 3)$. If these three subspaces $\mathbf{X}_0$, $\mathbf{X}_1$, $\mathbf{X}_2$ are contained in a common subspace $\mathbf{Y}$ of dimension $s + 1 - r$, simple counting yields that any point of $\mathbf{Y}$ lies in $\mathcal{H}$. Hence we have $\mathcal{H} \cap \mathbf{W} = \mathbf{Y}$.

Assume that $\mathbf{X}_0$, $\mathbf{X}_1$, and $\mathbf{X}_2$ are not contained in a common $(s+1-r)$-dimensional subspace. Then $\langle \mathbf{X}_0, \mathbf{X}_1 \rangle$ is a subspace of dimension $s + 1 - r$ through $\mathbf{X}$. Let $\mathbf{X}_2'$ be the third $(s-r)$-dimensional subspace in $\langle \mathbf{X}_0, \mathbf{X}_1 \rangle$ through $\mathbf{X}$. This subspace $\mathbf{X}_2'$ has the property that only two of the $(s+1-r)$-dimensional subspaces through it contains points of $\mathcal{H} \cap \mathbf{W}$, namely $\langle \mathbf{X}_2', \mathbf{X}_0, \mathbf{X}_1 \rangle$ and $\langle \mathbf{X}_2', \mathbf{X}_2 \rangle$.

Finally we show that there exists an s-dimensional subspace $\mathbf{W}'$ of $\mathbf{W}$ through $\mathbf{X}_2'$ that intersects $\mathcal{H} \cap \mathbf{W}$ only in points of $\mathbf{X}_2'$. This is easy: In the quotient geometry $\mathbf{W}/\mathbf{X}_2'$ the subspaces $\langle \mathbf{X}_2', \mathbf{X}_0, \mathbf{X}_1\rangle$ and $\langle \mathbf{X}_2', \mathbf{X}_2\rangle$ are points; there is a hyperplane $\mathbf{W}'$ that contains none of the points in question of $\mathbf{W}/\mathbf{X}_2'$. $\mathbf{W}'$ is the subspace we seek. By step 1, $\mathbf{W}' \cap \mathcal{H}$ consists of the points of an $(s-r)$-dimensional subspace. Since, by construction, $\mathbf{W}' \cap \mathcal{H} = \mathbf{X}_2' \cap \mathcal{H} = \mathbf{X}$ with $\mathbf{X}$ being a subspace of dimension $s-r-1$ we get a contradiction. □

The Reed–Muller codes belong to the most studied (and most recommended to be studied) structures of coding theory. We recommend particularly the books [CaLi91], [AsKe92].

Historical remark. The Reed–Muller codes have played an important role in the application of coding theory; indeed, they have been used to encode pictures sent from satellites back to Earth.

The aim of the Mariner 9 mission in 1971 was to fly over Mars and photograph its entire surface. The pictures had to be transmitted to Earth and, obviously, during this transmission a lot of errors occurred. The data, therefore, had to be encoded by a very good code; otherwise all the details which had been detected with the extremely good optical equipment would have remained invisible to us.

The pictures had a high resolution of 700×832 pixels. Each pixel became an 8-tuple that represented a grey value.

These binary data were divided into blocks of 6 bits each; each block was encoded by a codeword of weight 32; thus one paid the price of 26 redundant bits in order to correct errors. For this, a first order Reed–Muller code of length 64 (generated by all hyperplanes of AG(6, 2)) was used, which is a 7-error correcting code. (Cf. [Hill86], pp. 9–10.)

For decoding, the so-called ‘Green Machine’ described in [MWSl83], Section 14.4, was used.

Exercises

1 Let $\mathbf{C}$ be a subset of V that satisfies either property (a) or property (b) of **5.1.2**.
Show that $\mathbf{C}$ is a t-error correcting code.

2 Show that the code described in **5.1.3** is a 1-error correcting code.
To which codeword will the vector 1100011 be decoded?

3 Show that for any quadrangle in the projective plane of order 2 there is exactly one line containing no point of the quadrangle.

4 Let **P** be a projective plane of order 2. Show that
(a) any triangle can be uniquely extended to a quadrangle,
(b) any set of five points of **P** is the union of two lines.

5 Interpret the Hamming code Ham(4) in terms of the projective space **P** = PG(3, 2).

6 Show that the matrix H whose columns consist of all the nonzero binary r-tuples has rank r.

7 Generalize **5.3.2** to t-error correcting codes.

8 We call a subset **C** of $\{0, 1\}^n$ a ***t*-error detecting code**, if the following condition is satisfied:
If e is a vector of weight at most t, then for all $c \in \mathbf{C}$ the vector $c + e$ is not a codeword.
Convince yourself that **C** is a t-error detecting code if and only if $d(\mathbf{C}) \geq t + 1$.

9 Let **C** be the subset of $\{0, 1\}^n$ defined as follows:

$$\mathbf{C} := \{(a_1, \ldots, a_n) \mid a_1 + \ldots + a_n \text{ is even}\}.$$

Show that **C** is a linear 1-error detecting code.

10 (a) Prove that by elementary row operations one can transform any generator matrix of a linear $[n, k]$-code into a matrix of the following form:

$$(E_k \mid A).$$

(b) Show that if $G = (E_k \mid A)$ is a generator matrix of a linear code **C** then

$$H = (A^T \mid E_{n-k})$$

is a parity check matrix of **C**.

11 A k-arc in a projective plane **P** is called **complete** if it is not contained in a $(k + 1)$-arc.
(a) Prove that an arc $\mathfrak{A}$ of **P** is complete if and only if through each point of **P** there is at least one line having two points of $\mathfrak{A}$.
(b) Suppose that **P** is a finite projective plane of order q. Show that a complete arc has more than $\sqrt{2q}$ points.

12 The ISBN error detecting code of book numbers is defined as follows. Any ISBN (**I**nternational **S**tandard **B**ook **N**umber) consists of nine digits for data (the first group indicates the language area, the second the publishing house, and the third is the number of the book inside the publishing house) and one check symbol. If the first nine digits are denoted by $Z_{10}, \ldots, Z_2$ then the check symbol Z_1 is computed in such a way that the number

$$10 \times Z_{10} + 9 \times Z_9 + \ldots + 2 \times Z_2 + Z_1$$

is divisible by 11. If $Z_1 = 10$ then one writes $Z_1 = X$ (Roman symbol for 10).

Questions: The ISBN code is defined over a certain alphabet. What is it? Which errors can be detected using the ISBN code?

For more information on check-digit systems the reader might consult [Gal91] and [Gal94].

13 Show that the set of all sets of points of an affine space **A** forms a vector space over GF(2), if one defines the sum of two vectors as the symmetric difference of the corresponding sets of points.

14 Let $\mathcal{C}$ be a cone with vertex V in $\mathbf{P} = \mathrm{PG}(3, q)$ where q is even. Then for any plane π of **P** that does not pass through V, the set $\mathcal{C} \cap \pi$ is an oval. Show: if N is the nucleus of such an oval, then *each* line through N is a tangent of $\mathcal{C}$.

We call any such point N a **nucleus** of the cone $\mathcal{C}$.

15 Let $\mathcal{C}$ be a cone with vertex V in $\mathbf{P} = \mathrm{PG}(3, q)$ where q is even. Then the lines of $\mathcal{C}$ form an oval in the quotient geometry $\mathbf{P}/\mathrm{V}$. Let N be the nucleus of this oval of $\mathbf{P}/\mathrm{V}$. (Then N is in **P** a line through V.) Show that any point except V on N is a nucleus of the cone $\mathcal{C}$.

16 Let $\mathcal{C}$ be a cone with vertex V in $\mathbf{P} = \mathrm{PG}(3, q)$ where q is even. Show that all nuclei of $\mathcal{C}$ lie on a common line of **P** through V.

17 Show that in a Reed–Muller code generated by the lines of $\mathrm{AG}(r, 2)$ (i.e. in a Reed–Muller code of order $r - 1$) the following is true: a point is contained in a codeword if and only if it is on an odd number of lines generating that codeword.

18 Let $\mathcal{M}$ be a set of points in $\mathbf{P} = \mathrm{PG}(d, q)$ with the property that any r-dimensional subspace of **P** contains at least one point of $\mathcal{M}$.

Show that $|\mathcal{M}| \geq q^{d-r} + \ldots + q + 1$ with equality if and only if $\mathcal{M}$ is the set of points of a $(d - r)$-dimensional subspace of **P**.

[Hint: Suppose that $|\mathcal{M}| \leq q^{d-r} + \ldots + q + 1$ and then show that

– there is an $(r - 1)$-dimensional subspace disjoint from $\mathcal{M}$,

– any r-dimensional subspace that contains an $(r - 1)$-dimensional subspace disjoint from $\mathcal{M}$ intersects $\mathcal{M}$ in exactly one point,

– $|\mathfrak{M}| = q^{d-r} + \ldots + q + 1$,
– any $(r+1)$-dimensional subspace that contains an $(r-1)$-dimensional subspace disjoint from $\mathfrak{M}$ intersects $\mathfrak{M}$ precisely in the points of a line,
– . . ..]

Remark. The case $r = 1$ is a theorem of Tallini [Tall57]; the general case was proved by Bose and Burton [BoBu66]. For a proof see also [Beu83], Section 7.3.

True or false?

☐ A code is linear if it contains the zero–vector.

☐ A code is linear if it has exactly 2^k ($k \in \mathbf{N}$) elements.

☐ A code $\mathbf{C}$ is linear, if $w(\mathbf{C}) = d(\mathbf{C})$.

☐ Any t-error correcting code has minimum weight $2t + 1$.

☐ Each codeword of a t-error correcting code has weight $2t + 1$.

☐ Each codeword of a perfect t-error correcting code has weight $2t + 1$.

Let $\mathbf{C}$ be a t-error correcting code.

☐ (a) Then, by adding a new position, one can obtain a code $\mathbf{C}^*$ with minimum weight $2(t + 1)$.

☐ (b) If $\mathbf{C}$ is linear then $\mathbf{C}^*$ is also linear.

Projects

Project 1

One can define codes not only over the field with two elements, but over any finite field $GF(q)$ with q elements. The **distance** of two n-tuples with coefficients in $GF(q)$ is defined as the number of positions in which the two vectors differ.

Try to generalize all definitions, theorems and examples of Sections **5.1**, **5.2**, **5.3** to the general situation.

Project 2

Try to compute the dimension of the Reed–Muller codes. If $\mathbf{C}$ is a Reed–Muller code in $AG(d, 2)$ of order 1, then its dimension is $\dim(\mathbf{C}) = d + 1$. For a Reed–Muller code $\mathbf{C}$ in $AG(d, 2)$ of order m one has the following formula:

$$\dim(\mathbf{C}) = \sum_{i=0}^{d} \binom{m}{i}.$$

Prove this formula.

[Hint: First try to solve the case $d = 3$, $m = 1$ and then the case $m = 1$ in general.]

You should know the following notions

Message, Hamming distance, *t*-error correcting code, minimum distance, linear code, generator matrix, weight, minimum weight, dual code, parity check matrix, syndrome, Hamming code, perfect code, extended Hamming code, MDS code, (n, s)-set, arc, cap, Reed–Muller code.

6 Applications of geometry in cryptography

Cryptography has two aims. On the one hand it provides methods that guarantee the confidentiality of information (enciphering). On the other hand it provides methods that make it possible to detect alterations of data and to verify whether the data really came from the claimed sender (authentication). Usually, such systems are based on secret keys; therefore the secure distribution and storing of secret keys is a central area of cryptography. In this chapter we shall show how geometric structures can be used for enciphering, authenticating, and for storing secret data.

Geometric cryptosystems often have essential advantages. Most importantly, the security of cryptosystems obtained from geometry is provable; their security does not rely on unproved assumptions or on unintelligible complexity – unlike most of the used algorithms today. The second advantage, no less important, is that one can obtain cryptosystems with arbitrarily high levels of security. Finally, these systems are surprisingly simple to realize. The methods presented in Sections **6.3** and **6.4** have these marvellous properties.

6.1 Basic notions of cryptography

We consider a communication model similar to that of coding theory. A **sender** wants to securely transmit **data** to a **recipient**; in such a way sender and recipient protect themselves against attacks of a third party. We distinguish two types of attacks, namely passive and active.

Performing a **passive attack**, an attacker tries (only) to *read* the transmitted message. In order to transmit data confidentially, sender and recipient must apply countermeasures to render the message unintelligible to an attacker. There are various useful countermeasures. They could be *organizational* or *physical* and consist, for instance, of requiring a stringent security check for all employees, or of transmitting the messages only in sealed envelopes. Here, we study methods of 'enciphering', which are based on mathematical structures.

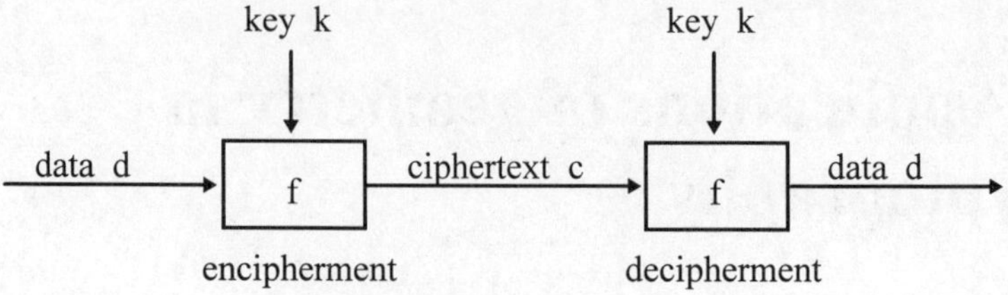

Figure 6.1 The mechanism 'encipherment'

The principle of enciphering is simple (see Figure 6.1). Using an **algorithm** f the data (**plaintext** or **cleartext**) d is **enciphered** under a secret **key** k. More precisely: for each key k there is an invertible function f_k mapping a cleartext d onto a ciphertext $c = f_k(d)$. The sender computes the **ciphertext (message)** $c = f_k(d)$ and transmits it via a possibly insecure channel to the recipient. Using the inverse function f_k^{-1} the recipient can **decipher** the ciphertext:

$$f_k^{-1}(c) = f_k^{-1}(f_k(d)) = d.$$

An algorithm f for encipherment must have the following properties:

– The recipient can easily reconstruct the original data from the received message.

– Without knowledge of the key k it is very difficult to reconstruct the cleartext corresponding to an enciphered message.

Using an **active attack** the attacker tries to *modify* the transmitted data; he may change a transmitted message, delete it, or even insert a new message. A particularly dangerous aim of the attacker is to change the sender's address. The corresponding cryptographic countermeasure is **authentication**. Although such a mechanism does not prevent an attacker from modifying or inserting his own data, it gives the recipient a means to decide whether the received message is genuine and comes from the given sender.

The principle of an authentication mechanism is as follows (cf. Figure 6.2). The sender **authenticates** the data d by applying a cryptographic algorithm f under a secret key k. Thus he gets the **authenticated message** $c = f_k(d)$, which he transmits to the recipient. The recipient **verifies** the received message c by checking whether c is a message valid under k. More precisely, he checks whether there is some cleartext that is mapped under k onto c.

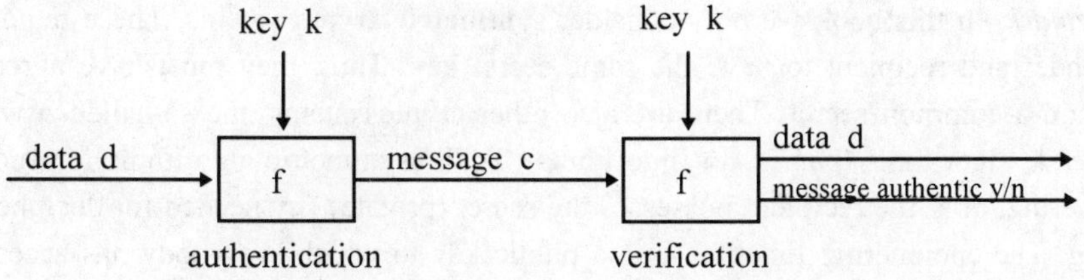

Figure 6.2 The mechanism 'authentication'

An algorithm f for authentication must have the following properties:
– The recipient can easily verify whether the received message is valid under k.
– The recipient can easily reconstruct the original data from the received message.
– Without knowledge of the key k it is very difficult for an attacker to generate a message valid under k – only a small percentage of all messages would be valid under k.

Remark. Many authentication systems have the property that the authenticated message consists of the original data with an **authentication code** attached. By looking at this special case two properties become clear:
– The transmitted message is longer than the original data; for authentication one adds redundancy.
– In this case the recipient verifies the message by checking with the key k whether the received authentication code matches with the received data. (See Section **6.3**.)

To get an idea of the security of a cryptographic system, one has to consider the possible attacks under which the algorithm remains secure. All security considerations are based on the **principle of Kerckhoffs**, which says that one has to face the possibility that the attacker knows the algorithm. The only thing the attacker must not know is the key. Only sender and recipient should know the key; for this reason the key is also called a **secret key**.

There are, in principle, three means of attack. A good algorithm must certainly be able to resist the first two; the third is only possible in extreme situations.
– The attacker knows a (often large) number of messages (**known ciphertext attack**).
– The attacker also knows a (usually small) set of data with the corresponding messages (**known plaintext attack**).
– The attacker can choose data and gets the corresponding ciphertext (**chosen plaintext attack**).

Remark. In this book we only consider **symmetric** cryptosystems. These require sender and recipient to have the same secret key. Thus, they must have agreed upon a common secret. There are also other cryptosystems, the so-called *asymmetric* algorithms (*public key* algorithms). With asymmetric algorithms for enciphering, only the recipient possesses the secret (private) key needed for deciphering. The enciphering function uses a public key to which everybody has access. An introduction to public key cryptography and an overview of asymmetric algorithms can be found in [Beu92], [DaPr89] and [Sal90].

6.2 Enciphering

In this section we present an important method for enciphering, namely **stream ciphers**. The simple idea was proposed by the American engineer G. S. Vernam (1890–1960) ([Ver26]). In order to encipher the data d they must be encoded as a binary string (i.e., a sequence of 0s and 1s):

$$d = d_1, d_2, d_3, \ldots, \qquad d_i \in \{0, 1\}.$$

As key we use a random sequence $k = k_1, k_2, k_3, \ldots$ of 0s and 1s. The sender gets the ciphertext $c = c_1, c_2, c_3, \ldots$ by adding each bit d_i to the corresponding bit of the key k_i modulo 2 (see Figure 6.3):

$$c_i = d_i + k_i \bmod 2 \qquad (i = 1, 2, \ldots).$$

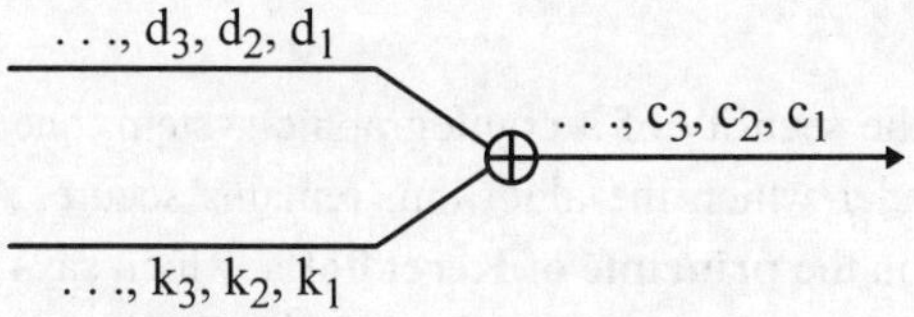

Figure 6.3 Enciphering using the one-time pad

The recipient can decipher as easily as the sender enciphers, because enciphering and deciphering are just the same. The recipient adds to each bit of the received sequence $c = c_1, c_2, c_3, \ldots$ the corresponding bit of the key and gets the original data back, since

$$c_i + k_i \bmod 2 = d_i + k_i + k_i \bmod 2 = d_i \qquad (i = 1, 2, \ldots),$$

see Figure 6.4.

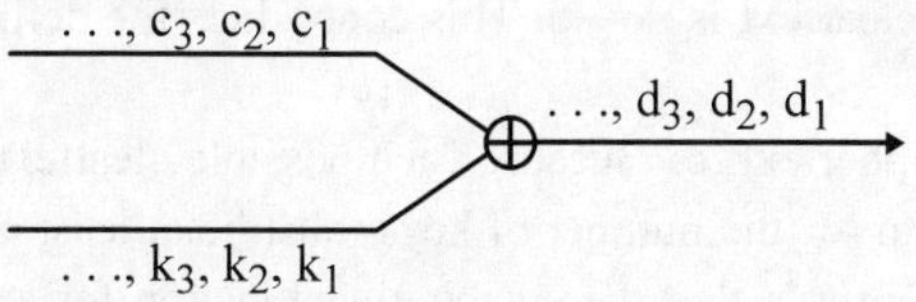

Figure 6.4 Deciphering using the one-time pad

This enciphering algorithm is called the **one-time pad**. The one-time pad has a useful property. Suppose that the sequence of key bits is truly random, if one knows arbitrarily many bits $k_1, \ldots, k_n$ then the next bit k_{n+1} can only be guessed with a probability of $\frac{1}{2}$. In this case the sequence $c_1, c_2, c_3, \ldots$ of the ciphertext also has this property.

As a consequence an attacker has a big problem. It is no use to apply complicated methods. *There is no method that is better than to guess!* Systems which have this remarkably property are therefore called **perfect**.

Despite this marvellous property, the one-time pad has a big disadvantage. The key must be transmitted to the recipient. Because the bits of the keys are independent, no bit can be calculated from the others, and so one has to transmit all bits of the key. In other words, to encipher a cleartext consisting of n bits, one has to transmit in advance – and secretly – a secret key also containing n bits. So one has reduced the problem of secretly transmitting n bits to secretly transmitting n other bits.

Although it seems as if we are back where we started, there are circumstances where this method can be used advantageously: sender and recipient can agree upon the key in advance at a predefined time, whereas there is often no choice of when the enciphered message can be sent.

But the price one has to pay for perfect secrecy is high.

Must it be so high? Is it true that in a perfect system the length of the key is necessarily equal to the length of the cleartext? (Or, equivalently, the number of keys is necessarily equal to the number of possible cleartexts?) Unfortunately, the answer to this question is 'yes'. This is the content of the famous theorem of C. Shannon (1916).

6.2.1 Theorem (Shannon, [Sha49]). *In any perfect enciphering system the number of keys is at least as large as the number of the possible cleartexts.*

Proof. Firstly, we convince ourselves that for each possible cleartext d and for each possible ciphertext c there must be at least one key mapping d onto c. Assume that there are a d and a c that cannot be mapped onto each other by any key. In this case, an attacker by observing c can learn something: he knows that the corresponding cleartext is not d. This contradicts the definition of being perfect.

Now we fix a ciphertext c'. Because each possible cleartext can be mapped by at least one key onto c', the number of keys must be at least as large as the number of cleartexts. Observe that the enciphering function for a given key is injective. □

Remark. A more formal version of the theorem and the proof can be found in [Mas86].

In practice, the enciphering method described above is often applied using sequences containing pseudo-random sequences of key bits rather than using truly random bits. These are binary sequences that look, at first glance, like random sequences, but in reality they are deterministic sequences depending on only a few parameters. These parameters provide the key, and only these values are transmitted. Of course, such systems are no longer perfectly secure.

How does one define pseudo-random? In the literature various criteria for sequences to be pseudo-random are discussed. Among these are the postulates of Golomb (see for instance [BePi82]). They apply to **periodic sequences**, i.e. sequences that are repeated for ever. The smallest positive integer n such that the sequences repeats after the nth position is called the **period** of the sequence. Any periodic sequence is **generated** by a **cycle** C, which is repeated.

Example. The sequence

$$01011\ 01011\ 01011\ldots$$

is a sequence of period 5, which has (01011) as generating cycle. Another generating cycle is (10110).

In order to formulate the postulates of Golomb, we consider a periodic sequence with generating cycle C. The first postulate is easy to state:

(**G1**) *The numbers of* 0s *and* 1s *in* C *differ by at most* 1.

Ideally, one would like the number of 0s to equal the number of 1s, but for odd n this is not possible.

To formulate the next postulate, we need the notions 'string' and 'gap'. A **string** is a sequence of 1s preceded and followed by 0s; a **gap** is a sequence of 0s preceded and followed by 1s. For *example*, the sequence

$$C = 011101100101000$$

has one gap of length 2 and two strings of length 1.

Now we can formulate the second postulate.

(G2) *For each nonnegative integer* i*, the number of strings of length* i *and the number of gaps of length* i *differ by at most* 1.

If C is a generating cycle of a periodic sequence, then the cycle $C(a)$ that is derived from C by a cyclic (left) shift of a positions is a generating cycle as well. If C denotes the above cycle, then

$$C(2) = 110110010100001.$$

For a fixed $a \neq 0$ we denote by A the number of positions in which C coincides with $C(a)$ and by D the number of positions in which C and $C(a)$ are different. Obviously, we have $A + D = n$. We say that the number

$$\frac{A-D}{n} = \frac{2A-n}{n}$$

is the **out-of-phase autocorrelation**. In the above example we get with $a = 2$ the values $A = 7$ and $D = 8$; therefore its out-of-phase autocorrelation is

$$\frac{A-D}{n} = -\frac{1}{15}.$$

Since, in general, the numbers A and D depend on a, the out-of-phase autocorrelation will depend on a as well. Golomb's third postulate reads as follows.

(G3) *For all* $a \in \{1, 2, \ldots, n-1\}$ *the out-of-phase autocorrelations are equal.*

The question arises if there are sequences fulfilling these postulates. Surprisingly enough, most of the known sequences having this property are constructed using projective spaces! For this construction we need an important tool in finite geometry, the so-called **Singer cycle**.

6.2.2 Theorem. *Let* $\mathbf{P} = PG(d, 2)$ *be a finite Desarguesian projective space of dimension* d *and order* 2. *Then* $\mathbf{P}$ *has a collineation group* Σ*, called the* ***Singer cycle****, with the following properties:*

- Σ *is a cyclic group, which means that it is generated by a single element.*
- Σ *is sharply transitive on the set of points (and on the set of hyperplanes) of* $\mathbf{P}$.

Proof. We need some algebra. By the first representation theorem (**3.4.2**) we can represent $\mathbf{P}$ as $\mathbf{P}(V)$, where V is a $(d+1)$-dimensional vector space over $K = GF(2)$. Because the field $F = GF(2^{d+1})$ is a $(d+1)$-dimensional vector space over K, w.l.o.g. we can choose $V = F = GF(2^{d+1})$. Since the field K has only one element different from zero, the points of **P** are exactly the vectors different from zero, hence the elements of $F^* = F \setminus \{0\}$.

From algebra (see e.g. [Her64]) we know that F can be constructed as follows. We take an irreducible polynomial f of degree $d+1$ over K. Then the elements of F are the polynomials in one variable x of degree at most d (including the zero-polynomial). Addition in F is the addition of polynomials. In order to calculate the product of two elements of F, one calculates the product of the corresponding polynomials g and h and then reduces the result 'modulo f'. This means that the product is the remainder of $g \cdot h$ when divided by f. Therefore, the product is also a polynomial of degree at most d.

As an *example* we construct GF(8). The polynomial $f = x^3 + x + 1$ is irreducible over GF(2) (which is easily verified in this simple example). The elements of GF(8) are the polynomials $0, 1, x, x + 1, x^2, x^2 + 1, x^2 + x, x^2 + x + 1$. In order to compute the product of the elements $x^2 + 1$ and x, we first multiply in K[x]:

$$(x^2 + 1) \cdot x = x^3 + x = x^3 + x + 1 + 1 = f + 1.$$

If we reduce modulo f we get the constant polynomial 1. (Another way of expressing this fact is $(x^2 + 1) \cdot x \equiv 1 \mod f$.)

In exercise 3 the reader is invited to construct GF(8) by this method.

The crucial point is the fact that one can choose the polynomial f in such a manner that the powers

$$x, x^2, x^3, \ldots, x^{2^{d+1}-1}$$

of x are precisely the nonzero elements of F. Polynomials with this property are called **primitive**. (See, for instance, [Her64]. The fact that the multiplicative group of the field $GF(2^{d+1})$ is cyclic corresponds to the existence of primitive polynomials of degree $d+1$ over GF(2).)

Now we return to the proof of the theorem. We are now able to define a generating element of the Singer cycle. For this, we consider the map σ from F onto itself that is defined by multiplication by x:

$$\sigma(g) := x \cdot g \qquad (g \in F).$$

We have $\sigma(0) = 0$; moreover σ is a permutation of $F \setminus \{0\}$. Therefore, if f is primitive, σ generates a cyclic group $\Sigma = \{id, \sigma, \sigma^2, \ldots\}$ of order $|F| - 1 = 2^{d+1} - 1$.

We claim that Σ is the group we are looking for. By construction, σ is a bijective map from the set of points of **P** onto itself. It remains to show that σ maps triples of collinear points onto triples of collinear points. To show this, let u, v, w be distinct elements of F* such that the corresponding points of **P** are collinear. This means that the vectors u, v, w are linearly dependent; hence we have $u + v = w$. (Observe that 1 is the only element of F different from 0). Then we have

$$\sigma(u) + \sigma(v) = x \cdot u + x \cdot v = x \cdot (u + v) = x \cdot w = \sigma(w).$$

Therefore it follows that σ is in fact a collineation of **P**. By construction, the powers of σ successively map the points of **P** onto each other.

So the theorem is proved. □

Remark. Theorem **6.2.2** also holds for finite projective space of arbitrary order q; see exercise 1.

Now we consider the projective space $\mathbf{P} = PG(d, 2)$ of dimension d and order 2. We shall continue to use the notation fixed in the proof of Theorem **6.2.2**. The points of **P** are the elements of $F^* = GF(2^{d+1})^*$. This field is obtained from $K = GF(2)$ by adjoining a root of a primitive polynomial f of degree d. The points of **P** can be identified by the polynomials in x of degree at most $d + 1$. As generating element of the Singer cycle we choose the multiplication by x mod f.

We shall label the points of **P** by the integers $1, 2, \ldots, v = 2^{d+1} - 1$, in such a way that the map $i \mapsto i + 1 \bmod v$ is our generating element.

The following theorem shows how to construct, using $PG(d, 2)$, a binary pseudo-random sequence that fulfils Golomb's postulates.

6.2.3 Theorem. *Let* $C = (a_1, a_2, \ldots, a_v)$ *be the incidence vector of a hyperplane* **H** *of* $\mathbf{P} = PG(d, 2)$ *with respect to the above labelling of the points of* **P**. *(Therefore,* $a_i = 1$ *if the point* i *lies in* **H**, *and* $a_i = 0$ *otherwise.) Then the cycle* C *fulfils Golomb's postulates.*

Proof. First, we show (**G1**). By definition, the number y of 1s in C equals the number of points in the hyperplane **H**; therefore,

$$y = 1 + 2 + 4 + \ldots + 2^{d-1} = 2^d - 1.$$

Because the number z of 0s in C equals the number of points not in **H** we get

$$z = v - y = 2^{d+1} - 1 - (2^d - 1) = 2^d.$$

Because $z - y = 1$, postulate (**G1**) holds.

In order to show **(G2)** we use the fact that the labelling of the points corresponds to the Singer cycle.

As our hyperplane we choose the hyperplane **H** spanned by the points 1, x, $x^2, \ldots, x^{d-1}$. In other words, **H** consists exactly of the polynomials of degree at most $d-1$.

In the field F each nonzero element, in particular the element x, has an inverse with respect to multiplication.

Claim 1: The polynomial inverse to x *in* F *is* $\frac{f+1}{x}$.

Since f is irreducible its constant term equals 1. Therefore, x is a factor of $f-1 = f+1$; this means that $\frac{f+1}{x}$ is a polynomial in K[x] of degree smaller than d. Moreover,

$$\frac{f+1}{x} \cdot x = f+1 \equiv 1 \bmod f,$$

which means that $\frac{f+1}{x}$ is the inverse of x in F.

Claim 2: The incidence vector C *of* **H** *has one string of length* d *and* 2^i *strings of length* $d-1-i$ $(i = 0, 1, \ldots, d-2)$.

Since x^d and $x^{-1} = \frac{f+1}{x}$ are polynomials of degree d, the points x^d and x^{-1} are not in **H**. Therefore there exists a string of length d, namely the sequence $(1, x, x^2, \ldots, x^{d-1})$.

There are exactly 2^i polynomials $h = x^{i+1} + a_i x^i + \ldots + a_1 x + 1$ of degree $i+1$ with constant term different from 0. It follows that the points

$$h, x \cdot h, x^2 \cdot h, \ldots, x^{d-2-i} \cdot h$$

are contained in **H**. In order to show that this is a string of length $d-1-i$, we have to show that this sequence of 1s cannot be extended, that is that the points $x^{-1} \cdot h$ and $x^{d-1-i} \cdot h$ are not in **H**. For this it is enough to show that the polynomials $x^{-1} \cdot h$ and $x^{d-1-i} \cdot h$ have degree d. Because h has degree $i+1$, it follows immediately that $x^{d-1-i} \cdot h$ has degree d. Moreover using

$$x^{-1} \cdot h = \frac{f-1}{x} \cdot h$$

with $h = x^{i+1} + a_i x^i + \ldots + a_1 x + 1$ we see that

$$\begin{aligned} h \cdot \frac{f-1}{x} &= \frac{f-1}{x} + (x^i + a_i x^{i-1} + \ldots + a_1)(f-1) \\ &\equiv \frac{f-1}{x} - (x^i + a_i x^{i-1} + \ldots + a_1) \bmod f. \end{aligned}$$

Because $\frac{f-1}{x}$ is a polynomial of degree d, and $i < d$, it follows that $h \cdot \frac{f-1}{x}$ has degree d.

So we have at least 2^i strings of length $d-1-i$. Because we looked at all polynomials of degree $\leq d-1$ exactly once, we have counted each possible string exactly once. Thus the claim about strings is proved.

Claim 3: The incidence vector C *of* **H** *has a gap of length* $d+1$ *and* 2^i *gaps of length* $d-1-i$ $(i=0, 1, \ldots, d-2)$.

The number of gaps of a fixed length can be calculated in a similar way to the number of strings, by observing that $x^d + h \notin$ **H**, if $h = 0$ or $h \in$ **H** (see exercise 2).

In order to show (**G3**) we use the fact that σ is a collineation of **P**. This implies that not only C is the incidence vector of a hyperplane, but also $C(a)$. More precisely, $C(a)$ is the incidence vector of the hyperplane $\mathbf{H}' = \sigma^a(\mathbf{H})$.

From this the claim follows easily: The number A of positions in which C and $C(a)$ coincide equals the number of common 1s plus the number of common 0s. In other words: this is the number of points that lie on both **H** and **H'** plus the number of points that are on neither **H** nor **H'**. Therefore

$$A = 2^{d-1} - 1 + 2^d - 2^{d-1} = 2^d - 1.$$

From this we get also

$$D = v - A = 2^{d+1} - 1 - (2^d - 1) = 2^d.$$

Therefore the out-of-phase autocorrelation is constant (more precisely, equal to $-1/(2^{d+1} - 1)$), and (**G3**) has also been shown. □

Remarks. 1. There is only one sequence known that fulfils Golomb's postulates, yet cannot be derived from a projective space in the above described way. In [PiWa84] the reader can find more information on this topic.

2. A possibly disappointing remark. The pseudo-random sequences generated using the Singer cycles fulfil the postulates of Golomb, as we have seen – but they are not suitable for serious enciphering. The reason for this is that a very small part of such a sequence determines the whole sequence. (With the aid of the polynomials one can show that these sequences can also be derived from linear shift registers, and such sequences are known to be cryptographically weak. Cf. [BePi82].)

3. Nevertheless such sequences are very useful in cryptography for at least two reasons. Firstly, they provide the elementary building blocks for more complicated and more secure (one hopes) algorithms. Secondly, they are used to 'measure' the cryptographic strength of a pseudo-random generator. The question is: how long must a linear shift register be in order to produce the output sequence of a given pseudo-random generator? If this register is short, then certainly the

pseudo-random generator is cryptographically weak; if long, the generator is (one hopes) good. For more information see [Rue86].

6.3 Authentication

In many of today's cryptography applications the principal aim is not so much the confidentiality of the plaintexts, but rather their authenticity. We call a message **authentic** if the recipient is sure that

– he receives exactly the data the sender has sent (**data integrity**) and

– the data really originate from the claimed sender (**data authenticity** in the strong sense).

Therefore one distinguishes two different types of attacks.

1. **Impersonation**: An attacker tries to insert a message claiming that it comes from the real sender.
2. **Substitution**: An attacker tries to modify a message actually sent.

An authentication protocol should provide protection against both attacks.

As protection against the attacks described **authentication systems** have been invented (cf. [Sim82]): sender and recipient share a common secret key k. Using an authentication algorithm f the sender transforms the plaintext d into the message $c = f_k(d)$ to be transmitted. This means for each key k that the mapping f_k maps the set $\mathcal{D}$ of all plaintexts onto the set of all messages (see Figure 6.5).

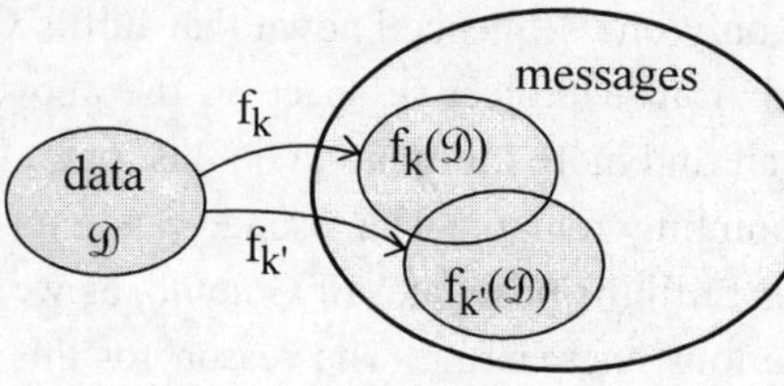

Figure 6.5 An authentication system

When the recipient receives a (possibly altered) message c' and wants to verify whether this message is authentic, he checks whether c' is a possible message under the key k. In other words: he checks whether c' is an element of $f_k(\mathcal{D}) = \{f_k(d_0) \mid d_0 \in \mathcal{D}\}$. If the answer is yes, then there is a plaintext d' with $f_k(d') = c'$. So the recipient accepts the message c' as authentic and assumes that d' was the original plaintext. Otherwise, he rejects c'.

Why is the attacker not able to calculate a message c' that would be accepted? Simple answer: because he does not know the key!

The verification method implies that an authentication system does not offer any security if the set $\mathcal{M}$ of messages and the set $\mathcal{D}$ of plaintexts have the same size. So a secure authentication system must contain more messages than plaintexts. Therefore, more information must be transmitted than without authentication.

A word on terminology. What we call 'plaintexts' are often called 'source states' or 'data' (hence the letter d); this is the information the transmitter wants to send. The data actually sent is called the 'message'.

We describe the method above again for the special case of 'authentication codes'. In this case, the message is obtained by attaching an **authentication code** (also called **message authentication code** or **MAC**, for short) $a_k(d)$ to the plaintext d. In this case the message is the pair $(d, a_k(d))$, the recipient receives a pair (d', a'). In order to verify whether (d', a') is authentic, he must verify whether (d', a') is a valid message under the key k. For this he simply calculates $a_k(d')$ and compares the result with a'. If these values coincide, he accepts the message, otherwise he rejects it.

This authentication protocol is *the* procedure to authenticate plaintexts since it can be implemented very efficiently. It is used a million times all over the world.

Definition. An authentication system is called **Cartesian** if, for all messages c, there is exactly one plaintext that is mapped onto c by a key.

In a Cartesian authentication system the plaintext can be deduced from the message without knowledge of the key. Thus, these authentication systems offer no secrecy at all. The system described above using authentication codes is Cartesian, because the transmitted message $(d, a_k(c))$ contains the plaintext d unhidden.

In the following we investigate the security of authentication systems. *For this, we only consider Cartesian authentication systems.* Most of the results can be generalized, see for instance [Sim82], [Sim92a].

The first impression might be that the security of an authentication system essentially depends on the algorithm f. This holds for most systems used in practice. But if we study how secure an authentication system can possibly be, the explicit structure of f plays no role at all, as we will see. Rather, another parameter enters the stage, namely the number κ of all possible keys. Very often, the set of keys consists of all binary *n*-tuples for some *n*; then κ equals 2^n.

For the following investigations we always assume that each key is chosen with the same probability.

From this assumption we see immediately that each attacker can modify plaintexts without being detected *with a probability of at least* $1/\kappa$: he just chooses one of the κ keys at random and calculates the message c^* belonging to his plaintext d^* by using his key.

But in reality, authentication systems offer a remarkably lower security; this fact is expressed by the famous theorem of Gilbert, MacWilliams and Sloane ([GMS74]).

In this theorem we consider the situation of an attacker being free to insert his messages at one of two occasions:

1. he can try to insert his message before any message has been sent (impersonation); or
2. he observes exactly one message and then tries to modify it (substitution).

6.3.1 Theorem (Gilbert, MacWilliams, Sloane). *Let* κ *be the number of all possible keys of an authentication system. Then an attacker can deceive with probability at least* $1/\sqrt{\kappa}$.

Proof. We suppose that there is an authentication system having κ keys and the chances for an attacker are at most $1/\sqrt{\kappa}$. We have to show that his chance of success is exactly $1/\sqrt{\kappa}$.

As described above, an attacker basically has two choices for his attack. Either he tries to insert a message before any message has been sent, or he observes one message and tries to replace this message by his own, using the information he gains from the original message. By assumption, in each case the probability for being successful is at most $1/\sqrt{\kappa}$.

First, we consider the impersonation attack. For a message c, let κ_c be the number of keys under which the message c is valid. Thus, the probability that this message will be accepted is κ_c/κ. By assumption, for each message c it holds that

$$\frac{\kappa_c}{\kappa} \leq \frac{1}{\sqrt{\kappa}}, \tag{1}$$

thus $\kappa_c \leq \sqrt{\kappa}$.

Now we analyse the substitution attack. Let c be the message observed by the attacker. We suppose that c was sent with probability $p(c)$. The attacker tries to deduce as much information as possible from c. Because the authentication system is Cartesian, he knows the corresponding plaintext d and can, at least theoretically, determine the set $\mathcal{K}(c)$ of keys mapping d onto c. He knows that the actual key used by the sender must be contained in $\mathcal{K}(c)$.

Because the keys are equally distributed and independent of the plaintexts, each key of $\mathcal{K}(c)$ could be the actual key with the same probability. Therefore, the attacker chooses an arbitrary key from $\mathcal{K}(c)$ and authenticates his plaintext; he obtains a message c'.

The probability that c' will be accepted is the number $\kappa_{c,c'}$ of keys under which c and c' are valid divided by the number κ_c of keys in $\mathcal{K}(c)$. By the choice of c' we have that $\kappa_{c,c'} \geq 1$. So, in this case, considering (1), the probability of a successful attack is at least

$$\frac{\kappa_{c,c'}}{\kappa_c} \geq \frac{1}{\kappa_c} \geq \frac{1}{\sqrt{\kappa}}.$$

For his overall probability of success p we therefore get

$$\frac{1}{\sqrt{\kappa}} \geq p \geq \sum_c p(c) \cdot \max_{\{c' \neq c\}} \frac{\kappa_{c,c'}}{\kappa_c} \geq \sum_c p(c) \cdot \frac{1}{\sqrt{\kappa}} = \frac{1}{\sqrt{\kappa}}.$$

Thus we have equality and therefore $p = 1/\sqrt{\kappa}$. From this it follows that $\kappa_{c,c'} \leq 1$ for all c, c' and $\kappa_c = \sqrt{\kappa}$ for all c. In particular $\sqrt{\kappa}$ is a positive integer. □

Definition. An authentication system with κ keys is called **perfect**, if the probability that an attacker will be able to insert a message of his own device or to substitute an authentic message is only $1/\sqrt{\kappa}$.

Our target is to precisely describe all perfect authentication systems. Statements on their structure are collected in the following corollary drawn from Gilbert, MacWilliams and Sloane.

6.3.2 Lemma. *In a perfect authentication system with κ keys the following statements are true:*
(a) *Each message is valid under exactly $\sqrt{\kappa}$ keys.*
(b) *For each plaintext there are exactly $\sqrt{\kappa}$ different messages.*
(c) *Two messages belonging to different plaintexts are valid under exactly one common key.*

Proof. Let c be a message. It follows from the proof of **6.3.1** that the number k_c of keys under which c is a valid message satisfies $\kappa_c = \sqrt{\kappa}$. This shows (a).

Two different messages belonging to the same plaintext d cannot be valid under a common key, since this implies that the map f_k maps d onto two different messages. Considering (a), it thus follows that there are $\kappa/\sqrt{\kappa} = \sqrt{\kappa}$ messages onto which d can be mapped.

Let c, c' be messages belonging to different plaintexts, and let $d' \in \mathcal{D}$ correspond to c'. The $\sqrt{\kappa}$ keys under which c is valid map d' onto different messages, because by **6.3.1** two messages are valid under at most one common key. Since there are exactly $\sqrt{\kappa}$ different messages corresponding to d', these are all messages for d'. In particular, the number $\kappa_{c, c'}$ of keys under which c and c' are valid equals 1. □

The theorem of Gilbert, MacWilliams and Sloane says that perfect authentication systems are the best possible systems for sender and recipient. From the point of view of an attacker, this is the worst situation he might face.

Now, at last we ask whether perfect authentication systems exist at all. The answer is 'yes', and, what is more: all perfect authentication systems can be constructed using geometric structures. We begin with the most important example.

6.3.3 Theorem. *Let* **P** *be a finite projective plane. We choose a line* g_0 *of* **P**, *and define an authentication system* **A** *as follows:*

- *the plaintexts of* **A** *are the points on* g_0,
- *the keys of* **A** *are the points outside* g_0,
- *the message belonging to* d *under a key* k *is the line through* k *and* d:

$$f_k(d) = kd.$$

Then **A** *is a perfect authentication system.*

Proof. Let q be the order of **P**.

First, we consider the impersonation attack: The attacker does not know any valid message and wants to authenticate some $d \in \mathcal{D}$. There are q messages corresponding to d, namely the lines through d distinct from g_0. On each of these lines there are exactly q of the q^2 keys. Thus, the probability that the attacker chooses a valid message corresponding to d is $q/q^2 = 1/q$.

We now study the substitution attack: The attacker intercepts a message c that is a line $c \neq g_0$ through a point d on g_0 (see Figure 6.6).

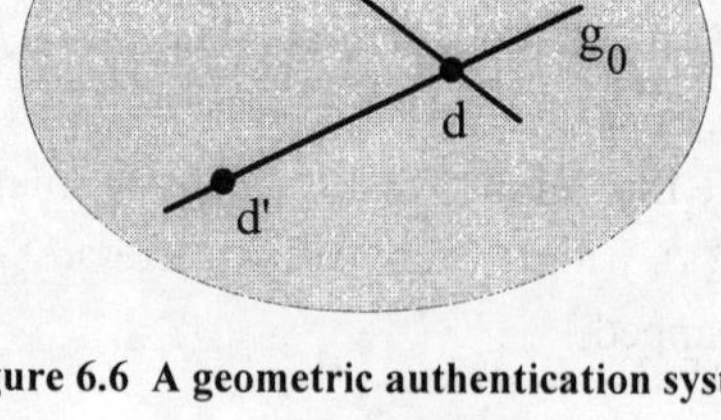

Figure 6.6 A geometric authentication system

For each possible $d' \in \mathcal{D}$, $d' \neq d$, chosen by the attacker he must find the valid message through d', in other words the line through d' and the key unknown to him.

Which information about the actual key does the attacker have? He knows the authentication system and thus that the key is a point outside g_0. Because c is a valid message, he knows furthermore that the key is one of the q points on the line c different from d. Choosing one of these points at random he can cheat with probability $1/q$. Because all potential keys are distributed with equal probability and are independent of each other, the attacker has no additional information: each one of the q points on c different from d might be the key.

Thus, in this case an attacker can also only cheat with probability $1/q$. □

Definition. A **net** is a rank 2 geometry consisting of points and lines such that through any two points there is at most one line and through each point outside a line g there is exactly one line not intersecting g.

For a **finite net** (that is a net with a finite number of points) one can prove that on each line there is a constant number q of points and through each point there is the same number r of lines. The numbers r and q are related by $r \leq q+1$. Furthermore, we can define a **parallelism** in a net by saying that two lines g and h are **parallel** if and only if $g = h$ or $g \cap h = \emptyset$. In this case we write $g \| h$. It easily follows from the axioms of a net that $\|$ is an equivalence relation; its equivalence classes are called **parallel classes**. Thus, a parallel class of a net is a set of lines such that each point of the net is on exactly one of these lines (cf. exercises 4, 5, and 6).

Examples of nets are obtained by considering only $r \leq q+1$ of the parallel classes of an affine plane of order q. But these are by far not all the nets (see [BJL85]).

Starting with a net **N** one can construct a perfect authentication system **A** = **A(N)** in a similar way to that above:

- the *plaintexts* are the parallel classes of **N**,
- the *keys* are the points of **N**,
- the *messages* are the lines of **N**. The message corresponding to a plaintext d and a key k is the line of the parallel class k through d (see exercise 7).

Conversely one can prove [DVW89] that every perfect authentication system can be obtained in this way from a net:

6.3.4 Theorem. *Every perfect authentication system is of the form* **A**(**N**) *for some net* **N**.

Proof. Let **A** be a perfect authentication system. We define a geometry **N** as follows:

- the *points* of **N** are the keys of **A**;
- the *lines* of **N** are the messages of **A**;
- a line c and a point k are *incident* if and only if the message c is valid under the key k, this means if and only if there is a d that is mapped under k onto c.

In order to prove that the geometry **N** is a net, we read the axioms of a net in terms of the underlying authentication system. The first property of a net is that through any two points there is at most one line. So we have to show that there is at most one message valid under two distinct keys. This condition follows from Lemma **6.3.2**(c).

Let (k, c) be a nonincident point–line pair. In other words, c is a message *not* valid under k. We have to show that there is exactly one message c' valid under k with the property that there is no common key for c and c'.

Let d be the plaintext belonging to c. By Lemma **6.3.2**(c) the messages that do not share a key with c are exactly the messages corresponding to d. Let d be mapped under k onto a message c'; then c' is the unique line through k that has no point in common with c.

Therefore **N** is a net. □

As corollary it follows ([GMS74], [BeRo90]) that in a perfect authentication system **A** with κ keys there are at most $\sqrt{\kappa} + 1$ plaintexts. Equality holds if and only if **A** is constructed from a projective plane. To sum up, perfect authentication systems are geometric, and the best ones are constructed from projective planes!

A corresponding theory was developed for authentication systems, in which the attacker may observe more than one message before inserting a message of his own (see [Fåk79], [Mas86]).

We consider the situation of the receiver verifying n messages with the same key. The attacker may send his message whenever he wants to, for instance before the first message or instead of the nth message. As in the above theorem one can show that the theoretical probability of success only depends on the number of keys. If κ denotes the number of keys, then the probability p that one of n messages was falsified is

$$p \geq \kappa^{-1/(n+1)}.$$

For the proof see [Fåk79], [Wal90] and [Ros93]. Authentication systems for which equality holds are called **perfect** n**-fold**.

In this scenario one has to exclude that the same message is sent twice. Otherwise, if the attacker has observed a message he can himself repeat the message later on, and the forgery would not be detected.

Examples for n-fold authentication systems can be constructed using geometry. For this we need the normal rational curves considered in Section **2.2**.

6.3.5 Theorem. *Let* $\mathbf{P}$ *be a finite projective space of dimension* $n+1$, *let* $\mathbf{H}$ *be a hyperplane, and let* P *be a point outside* $\mathbf{H}$. *Furthermore, let* $\mathfrak{D}$ *be a normal rational curve of* $\mathbf{H}$. *An authentication system* $\mathbf{A}$ *is defined as follows:*
– *the plaintexts are the points in* $\mathfrak{D}$,
– *the keys are the hyperplane not through* P,
– *the messages are the points not equal to* P. *For* $\mathrm{d} \in \mathfrak{D}$ *and a key* k *one gets the corresponding message as intersection of the line* Pd *with the hyperplane* k.
Then $\mathbf{A}$ *is a perfect* n*-fold authentication system.*

Proof. Let q denote the order of $\mathbf{P}$. First, we compute the numbers of plaintexts and keys. By definition, a normal rational curve has exactly $q+1$ points and thus $|\mathfrak{D}| = q+1$ plaintexts. The number of keys is the number of hyperplanes not through P. Altogether there are $q^{n+1} + \ldots + q + 1$ hyperplanes, of which $q^n + \ldots + q + 1$ pass through P. Therefore, there are exactly q^{n+1} keys in $\mathbf{A}$.

In order to prove that $\mathbf{A}$ is perfect n-fold we have to show that the probability of an attacker being able to forge one of the n messages equals $1/q$.

We first analyse the case in which no message has been sent. For this we consider an arbitrary plaintext, that is a point Q of the normal rational curve. Each point X on the line PQ different from P is a possible message corresponding to Q. We show that through each point X on the line PQ there is the same number q^n of keys: By exercise 10, there are exactly q^n hyperplanes through X that do not contain P, so there are exactly q^n keys through X. Thus, the probability p that X is a valid message equals the number of keys through X divided by the number of all keys. In other words,

$$p = \frac{q^n}{q^{n+1}} = \frac{1}{q}.$$

Now we consider the case that already i messages $\mathrm{P}_1, \ldots, \mathrm{P}_i$ $(1 \le i \le n)$ have been sent. We have to face the possibility that the attacker knows these points. Therefore he knows that the actual key is one of the hyperplanes through $\mathrm{P}_1, \ldots,$

P_i not through P. Let Q_j be the plaintext belonging to P_j $(1 \le j \le i)$. The points Q_j are the intersections of the lines PP_j with the hyperplane **H**.
We *claim*: *the subspace* $\mathbf{U} = \langle P_1, \ldots, P_i \rangle$ *has dimension* $i-1$, *and for each point* Q *of* $\mathfrak{D}$ *different from the points* Q_j, *the line* PQ *does not intersect* **U**.

Since the points Q_j lie on $\mathfrak{D}$, we have that $\dim\langle Q_1, \ldots, Q_i \rangle = i-1$. From this it follows that

$$\dim\langle P, P_1, \ldots, P_i \rangle = \dim\langle P, Q_1, \ldots, Q_i \rangle = i,$$

thus $\dim\langle P_1, \ldots, P_i \rangle = i-1$.

If Q is another point on $\mathfrak{D}$, then $\dim\langle Q, Q_1, \ldots, Q_i \rangle = i$. Therefore, the line PQ only intersects the subspace $\langle P, Q_1, \ldots, Q_i \rangle$ in P. In particular, PQ does not intersect the subspace $\mathbf{U} \subseteq \langle P, Q_1, \ldots, Q_i \rangle$.

Using this claim we are able to prove that the probability of an attacker being able to generate a valid message for a plaintext Q equals $1/q$: The attacker knows that the actual key is one of the hyperplanes containing the already observed messages, i.e. the hyperplanes containing **U**, but not P. There are precisely q^{n-i+1} hyperplanes fulfilling this condition. Let R be a point on the line PQ different from P. Because PQ does not intersect the subspace **U**, $\dim\langle \mathbf{U}, R \rangle = i$ holds. Furthermore, the number of hyperplanes containing this subspace, but not P, equals q^{n+i}. Therefore, each message belonging to Q corresponds to exactly q^{n-i} keys. Since

$$p = \frac{q^{n-i}}{q^{n-i+1}} = \frac{1}{q},$$

the probability p of success is also in this case just $1/q$. □

Remark. In Theorem **6.3.5** it is essential to consider only points in general position as plaintexts, which means that any $n+1$ points generate **H**. We examine the situation of an attacker having observed two messages P_1 and P_2. Let Q_1 and Q_2 be the corresponding plaintexts. The attacker does not know the actual key, but he knows that this must be one of the hyperplanes through P_1 and P_2.

If another point Q^* of the line Q_1Q_2 were a possible plaintext, the attacker could choose the point $P^* := PQ^* \cap P_1P_2$ as his message and could thus insert a new message without being detected.

This means that although the attacker does not know the key he can generate a valid message!

6.4 Secret sharing schemes

Cryptographic algorithms are often based on secret information. This is true for symmetric enciphering and authentication algorithms as well as for asymmetric algorithms.

Therefore, the management of secret data is a fundamental task in applications of cryptology. One distinguishes different aspects: generation, distribution, storage, and deletion of secret data. All these are different aspects of **key management**.

In this section we deal with the special problem of storing secret data. More precisely, we deal with the dilemma between the secrecy and availability of data. This problem is solved in an optimal way by **secret sharing schemes** ([Sha79], [Sim92b]).

What is a secret sharing scheme?

Technically speaking, secret sharing schemes are used to reconstruct secret data, when only certain parts of the secret data are available – the reconstruction is only possible for some previously defined situations, while it is impossible for all other situations.

To clarify matters we start with an example, which will serve as prototype for all secret sharing schemes dealt with in this section.

Example. We assume that the secret X is a string of m bits, e.g. a binary key of length m. We want to 'subdivide' X into 'partial secrets' ('shares') X_i such that X can be reconstructed from any two shares, while it should not be feasible to reconstruct the secret from only one share.

To construct such a secret sharing scheme (a 'threshold 2-scheme') we use a projective plane $\mathbf{P} = G(2, q)$ of order $q \geq 2^m$. We choose a line g. The secret is encoded as a point X on g. We randomly choose a line $h \neq g$ through X and on it arbitrary points $X_1, X_2, \ldots$ as shares (see Figure 6.7).

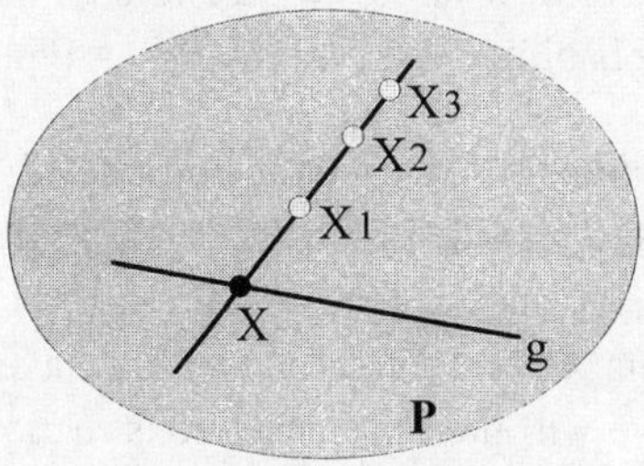

Figure 6.7 A threshold 2-scheme

To reconstruct the secret X from two shares X_i, X_j, $i \neq j$, the system computes the intersection of the line X_iX_j with g. If X_i and X_j are two valid shares then the system obtains the point X. On the other hand, knowing only one share one cannot reconstruct the secret: Given a share X_i, any of the $q+1$ points of the line g could be the secret. If an attacker only knows the line g and the point X_i, any point $\neq X_i$ on any line through X_i could be another share. Therefore, each point on g might be the secret equally likely, the probability of success for the attacker equals $1/(q+1)$.

This means, if an attacker knows only one share there is no better strategy than just guessing the secret!

We now give a precise definition of secret sharing schemes. For this we describe the fourphase life cycle of a secret sharing scheme.

1. The definition phase. In this phase the service provider formulates his requirements. He firstly has to define the 'access structure', that is to say which constellations of users shall be able to reconstruct the secret. Secondly, he must limit the probability that an illegal group of users can reconstruct the secret. Observe that no system is secure to 100%. An attacker could, for instance, guess the secret. But interestingly enough, in geometric secret sharing schemes the probability of success for an attacker can be kept as small as one likes.

The **access structure** is the set of all configurations of users allowed to reconstruct the secret. In other words: the access structure specifies the sets of participants that may legally reconstruct the secret. The access structure might be rather complex (see below): one sometimes distinguishes between different groups of users and defines how many members of each group are needed for reconstructing the secret. The **probability of deception** is the second parameter that the service provider must specify. For this he specifies an upper bound for the probability that a *illegal* set of participants can reconstruct the secret. This is necessary because there is no 100% security; any system can be defrauded with some positive probability. The user specifies how far he will tolerate an illegal reconstruction of the secret in choosing a probability p for this event. (A typical value is $p = 10^{-20}$.)

2. The mathematical phase. After having formulated the requirements it is the task of mathematicians to provide structures to realize them.

For constructing secret sharing schemes (projective) geometry with the underlying algebra has proven to be of great value. We have already described an example, subsequently we will present further constructions based on geometry for different secret sharing schemes. For detailed description of the realizability of secret sharing schemes using geometric structures see [Ker92], [Sti95].

3. Generation of the secret. Now it is the task of the service provider to choose the secret X. Then the shares are calculated by a method provided by mathematics. Finally the shares are distributed to the users.

It is crucial that the choice of the secret X – and thus of the shares – is completely under the responsibility of the service provider. It is independent of the formulated requirements and the chosen mathematical structure.

4. The application phase. In this phase the secret X will be reconstructed from a legal constellation of shares.

Remark. We distinguish two different application types of secret sharing schemes.

If the application is of type 'access', the reconstructed value is compared to the stored secret X. If the two values coincide access is provided. Thus, in this case the verifying instance (e.g. a computer) knows the secret.

There are also applications whose aim is to generate secrets. One example is the transport of a cryptographic key to a computer, where it is reconstructed. In this case the secret is not stored in the computer, but must be transported to the computer. Here we have a different problem: the computer must convince itself that the calculated value is not only an arbitrary value, but with high probability the correct secret X. For this purpose the so-called **robust** secret sharing schemes have been invented. A simple example of a robust scheme can be derived from our example. The computer requests not just two, but three shares and verifies whether all three pairs of points lie on the same line. (For details see [Sim90].)

Depending on the different types of legally constellating the participants, different types of secret sharing scheme can be distinguished. We first define the most important classes of secret sharing schemes and then describe their constructions.

(a) **Threshold schemes**. In a threshold t-scheme it is required that any t users can reconstruct the secret, but no constellation of $t-1$ or fewer users. For instance, in a threshold 2-scheme any two users can reconstruct the secret, but a single user has no chance to do this. In a threshold t-scheme, the number t is also called the **quorum**.

(b) **Compartment schemes**. The participants are partitioned into different 'compartments', which, in principle, have equal rights: In each compartment a certain quorum of users is required to let this compartment take part in the reconstruction of the secret. Moreover, a certain number of compartments must participate in the reconstruction. In other words, each compartment is a threshold scheme, and on the set of all compartments we have yet another threshold scheme.

This could be used for signing contracts in a company. There are two compartments, e.g. technical and commercial departments, and at least one signature is required from each compartment.

(c) **Multilevel schemes**. Again, the participants are partitioned into different groups, but these groups are ordered hierarchically. Each member of 'higher' order can replace a member of 'lower' order.

A special case is a **multilevel** secret sharing $(2, s)$-scheme that realizes the following access structure:

There are two groups $\mathfrak{S}$ and $\mathfrak{T}$ of participants. The secret shall only be reconstructable in the following cases:

- any s participants of $\mathfrak{S}$ can reconstruct the secret,
- any 2 participants of $\mathfrak{T}$ can reconstruct the secret,
- any $s-1$ participants of $\mathfrak{S}$ together with any participant of $\mathfrak{T}$ can reconstruct the secret.

In such a secret sharing scheme two participants of the top level $\mathfrak{T}$ have the same rights as s participants of the lower level; moreover, any user of the higher level can act as a member of the lower level.

Constructing secret sharing schemes

In the following we will present geometric constructions for the three most important classes of secret sharing schemes, namely threshold schemes, compartment schemes, and multilevel schemes.

1. Threshold schemes. To construct a threshold t-scheme one can proceed as follows: We fix a line g in $\mathbf{P} = PG(t, q)$. The points of g are the potential secrets. If the service provider chooses a point X on g as actual secret, there is a method which enables him to

- choose a hyperplane $\mathbf{H}$ (that is a $(t-1)$-dimensional subspace) through X that does not contain g, and
- choose in $\mathbf{H}$ a set $\mathfrak{T}$ of points in general position containing X. For example, one can choose $\mathfrak{T}$ as part of a normal rational curve in $\mathbf{H}$. The shares are points of $\mathfrak{T}$ different from X.

In the application phase certain partial secrets are sent to the system. The system computes the subspace through these shares (points) and intersects it with g.

If at least t legal shares are sent to the system, the constructed subspace is **H**, because the points are in general position. In this case the correct secret $\mathbf{H} \cap \mathrm{g} = \mathrm{X}$ will be obtained. This means that any t participants can reconstruct the secret.

The converse is valid as well:

6.4.1 Theorem. *If an attacker knows at most* $t-1$ *legal shares, his chance of cheating successfully is only* $1/(q+1)$.

Proof. An attacker deceives successfully if he obtains the correct secret without knowing t shares. It is clear that an attacker has an apriori probability of success of $1/(q+1)$, since he may choose one of the $q+1$ points of g at random.

The theorem claims that his chance does not increase if he knows as many as $t-1$ legal shares.

Let $\mathfrak{I}'$ be a set of at most $t-1$ shares. Then $\mathfrak{I}$ generates a subspace **U** with $\dim(\mathbf{U}) \le t-1$. Moreover, **U** does not intersect g, since $\mathfrak{I}' \cup \{\mathrm{X}\}$ is a set of independent points. An attacker knowing only g and $\mathfrak{I}'$ only knows that any further share is a point outside **U** not on g.

We show that each point X_0 on g has the same probability of being the secret: for any choice of X_0 on g, the subspace $\mathbf{W} = \langle \mathrm{X}_0, \mathbf{U} \rangle$ contains the same number of shares. Since g has exactly $q+1$ points, the attacker's probability of success is $1/(q+1)$. □

Definition. A secret sharing scheme is called **perfect** if the probability of guessing the secret has the same value for all nonlegal constellations of participants.

This means, perfect secret sharing schemes have the property that an attacker knowing only a small number of shares (not enough shares) has the same information about the secret as he would have with no share at all. In other words: perfect secret sharing schemes provide an insuperable security against insider attacks: an insider knowing at least one share only has the same extremely small probability of success as an outsider knowing nothing about the shares.

2. Compartment schemes. We restrict ourselves to the most important special case of compartment schemes: There are several user groups, namely the compartments $\mathrm{G}_1, \mathrm{G}_2, \ldots, \mathrm{G}_n$. The requirements for the access structure are as follows:

– in each compartment two participants are required to let it take part in the reconstruction of the secret;

– the participation of two compartments is sufficient to reconstruct the secret.

A corresponding secret sharing scheme can be geometrically constructed in the following way (see Figure 6.8). We fix a line g in $\mathbf{P} = \mathrm{PG}(n+2, q)$. The service provider selects a point $X \in g$ as the secret. First, he randomly chooses a line h $\neq$ g through X and n points $X_1, X_2, \ldots, X_n \neq X$ on h. Then he chooses through each point X_i a line g_i ($i = 1, 2, \ldots, n$) such that the set $\{\langle h, g_i \rangle, i = 1, 2, \ldots, n\} \cup \{\langle h, g \rangle\}$ of lines is independent (this means that it is a set of independent points of $\mathbf{P}/h$). These lines g_i correspond to the compartments G_i. Eventually each participant of compartment G_i is given a point X_i of g_i different from X_i.

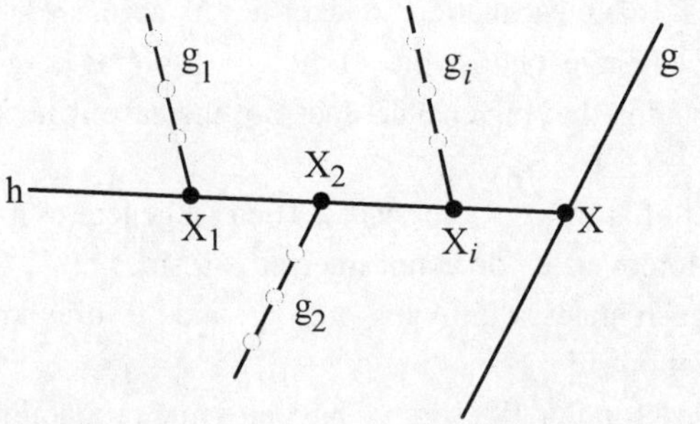

Figure 6.8 A compartment scheme

6.4.2 Theorem. *The compartment scheme described above is a perfect secret sharing scheme.*

Proof. We consider the reconstruction of the secret. The procedure is simple. One calculates the subspace $\mathbf{U}$ generated by all given points and intersects it with g.

If the constellation of participants is legal, the obtained point is the secret. If at least two points on lines g_k and g_h are given ($k \neq h$), the points X_k and X_h are contained in $\mathbf{U}$. Thus, h lies in $\mathbf{U}$ and so does $X \in \mathbf{U} \cap g$. Moreover, $g \notin \mathbf{U}$ because by assumption the planes $\langle h, g \rangle$ and $\langle h, g_i \rangle$ are independent.

Now we consider a nonlegal constellation of participants and show that each point of g can be reconstructed with the same probability.

The best situation for an attacker is to know two shares P_1, P_1' of one compartment G_1 and one share P_j of the other compartments G_j ($j = 2, \ldots, m$). Let X' be an arbitrary point of g. It is sufficient to show that

$$\mathbf{U}' := \langle P_1, P_1', P_2, \ldots, P_m, X' \rangle$$

is a subspace of dimension $m+1$ intersecting g exactly in X'.

Since the lines $g, g_1, g_2, \ldots$ are independent, the subspace $\langle \mathbf{U}', h \rangle$ has dimension $m+2$, if X' is not equal to X. So for $X' \neq X$ we have that $\dim\langle \mathbf{U}', X' \rangle$

$= m + 1$. Thus it remains to show that if $X \neq X'$ the point X is not contained in $\mathbf{U}'$.

If $m = 1$, $\mathbf{U}_1' = \langle P_1, P_1', X' \rangle$ is a plane. If X' and X were contained in $\mathbf{U}_1'$, it would contain the two skew lines g and g_1, a contradiction.

If $m = 2$ we proceed as follows:

$$\dim(\mathbf{U}_2') = \dim(\langle P_1, P_1', P_2, X' \rangle) = \dim(\langle \mathbf{U}_1', P_2 \rangle) \leq \dim(\mathbf{U}_1') + 1 = 3.$$

On the other hand, $\dim(\mathbf{U}_2') \geq \dim(\mathbf{U}_1') = 2$. Assume that $\mathbf{U}_2'$ is a plane. Then $\langle \mathbf{U}_2', h \rangle$ would be a 3-dimensional space containing the independent planes $\langle h, g \rangle$, $\langle h, g_1 \rangle$, $\langle h, g_2 \rangle$, a contradiction.

One can prove similarly the cases $m \geq 3$ (see exercise 12). □

3. Multilevel schemes. In multilevel schemes, the participants are divided into hierarchically ordered groups. Here, we only deal with twolevel schemes, the most important multilevel schemes in practice. More precisely, in a **multilevel (2, *s*)-scheme**, the participants are partitioned into two disjoint groups $\mathcal{T}$ and $\mathcal{S}$ such that the secret can only be reconstructed by the following constellations of participants:

- any set of at least two participants of $\mathcal{T}$,
- any set of at least s participants of $\mathcal{S}$,
- any participant from $\mathcal{T}$ together with at least $s - 1$ participants from $\mathcal{S}$.

We define the following system. We fix a line g in $\mathbf{P} = PG(s, q)$, whose points are the potential secrets. After selection of a secret X, the service provider chooses a line $h \neq g$ through X and a hyperplane $\mathbf{H}$ through h that only intersects g in X. The shares corresponding to the participants of $\mathcal{T}$ are points on h different from X. We denote the set of these points by $\mathcal{T}_h$. The shares belonging to the participants of $\mathcal{S}$ are points of a set $\mathcal{S}_\mathbf{H}$ of $\mathbf{H}$ with the following properties:

- $\mathcal{S}_\mathbf{H} \cup \{X\}$ is a set of points of $\mathbf{H}$ in general position. (For instance, one can choose $S_\mathbf{H} \cup \{X\}$ as a subset of a normal rational curve of $\mathbf{H}$.)
- Any subspace through $s - 1$ points of $\mathcal{S}_\mathbf{H}$ contains no point of $\mathcal{T}_h$.

As an *example* we consider the case $s = 3$. The secret sharing scheme is constructed using a 3-dimensional projective space $\mathbf{P} = PG(3, q)$. We choose a line $h \neq g$ through the point X of g, a plane π containing h, but not g, and a normal rational curve $\mathcal{K}$ of π (a conic in this case) through X with tangent h. We must choose the sets $\mathcal{T}_h \subseteq h$ and $\mathcal{S}_\pi \subseteq \mathcal{K}$ such that each line through two points of $\mathcal{S}_\pi$ intersects the line h in a point outside $\mathcal{T}_h$ (see Figure 6.9).

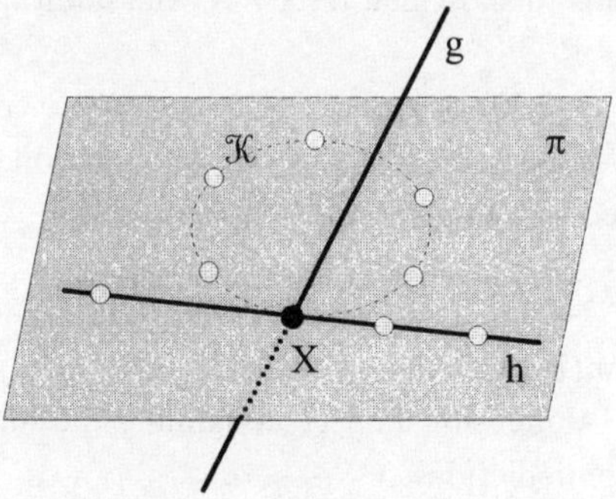

Figure 6.9 A multilevel (2, 3)-scheme

6.4.3 Theorem. *The construction above yields a perfect multilevel* (2, *s*)*-scheme.*

As *proof* we refer to exercise 13.

Remark. In [BeWe93] it is shown how to choose the points of $\mathcal{S}_h$ and $\mathcal{S}_\pi$ in an optimal way.

To end this section we summarize the essential advantages of secret sharing schemes constructed using geometry.

– In contrast to most of today's cryptographic mechanisms secret sharing schemes offer *provable security on each arbitrary level!* For each security level p there are systems for which the chance of cheating is at most p.

– Secret sharing schemes can be *implemented easily*. Because the typical deception probability demanded today is in the range between 2^{-20} and 2^{-100} there are no arithmetical problems, at least compared to the requirements for RSA (with 512 to 1024 bits) or similar algorithms (compare [BePi82], [Beu92]).

– Secret sharing schemes offer an extremely *comfortable participant management*. One can add users without changing anything in the computers used in the application phase. The removal of participants is more complicated, though. One could do this organizationally by using a blacklist. The best solution would be to withdraw all shares, and to choose a new line h and new shares. But this radical solution cannot be used too often in practice because it is very costly.

Exercises

1 Prove the theorem of Singer, **6.2.2**, in spaces of arbitrary order q.

2 Prove that the following claim holds in Theorem **6.2.3**: the incidence vector has one gap of length $d+1$ and 2^i gaps of length $d-1-i$ ($i = 0, 1, \ldots, d-2$).

3 Construct GF(8) by explicitly listing the elements and the addition and multiplication tables.

4 Let **N** be a net with line set G. If one defines the relation $\|$ on G by

$$\mathrm{g} \| \mathrm{h} \Leftrightarrow \mathrm{g} = \mathrm{h} \text{ or } \mathrm{g} \text{ and } \mathrm{h} \text{ have no point in common}$$

show that one obtains an equivalence relation with the property that there is exactly one line of each equivalence class ('parallel class') through each point of **N**.

5 Let **N** be a finite net.
(a) Show that through each point there is the same number r of lines.
(b) If $r > 2$, show that each line contains the same number of points.
(c) Is there a net with $r = 2$ in which not all lines have the same number of points?

6 Let **N** be a finite net with q points on each line and r lines through each point. Show that $r \leq q+1$ with equality if and only if **N** is an affine plane.

7 Show that from each finite net a perfect authentication system can be constructed by the method described in section **6.3**.

8 In the following way, one obtains an authentication system:
Fix a plane π_0 in $\mathbf{P} = \mathrm{PG}(3, q)$. In π_0 we choose a set G^* of lines, no three of which pass through a common point.
– The *plaintexts* are the lines of G^*,
– the *keys* are the points outside π_0,
– the *message* belonging to the key k and data d is the plane $\langle \mathrm{k}, \mathrm{d} \rangle$.
Show that the deception probability is $1/q$ if no, one or two messages are known to an attacker.

9 Generalize the previous exercise to PG(d, q).

10 Let $\mathbf{P} = \mathrm{PG}(d, q)$ be a projective space, and let **U** be a subspace of dimension i of **P**. Show that:
(a) There are exactly $q^{d-i-1} + \ldots + q + 1$ hyperplanes containing **U**.
(b) Let P be a point outside **U**. Then there are exactly q^{d-i-1} hyperplanes containing **U** not passing through P.

11 (a) Show that one obtains a perfect threshold t-scheme in the following way (cf. [Sha79]). We consider the affine plane over a field F. As secret we choose a point $(0, a_0)$ of the y-axis. Then we choose a polynomial f of degree $t-1$ with absolute term a_0. All other coefficients are chosen at random. The shares are points of the form (x, f(x)).
[Hint: The system is based on the fact that a polynomial of degree $t-1$ can easily be reconstructed from any t of its values, for instance by Lagrange interpolation.]
(b) Is the system constructed in (a) perfect?

12 Show **6.4.2** for the case $m = 3$.

13 Prove **6.4.3**.

14 Generalize the construction **6.4.2** to the situation that t groups are needed for the reconstruction, and in each group G_i a commitment of t_i users is required.

Project

In this project we study an interesting authentication system. The system is not perfect, but still provably secure. Compared to a perfect authentication system, it has a large number of possible plaintexts, making it interesting for practical applications. It is also interesting from a geometrical point of view because important structures, namely spreads, play a central role.

Let $\mathbf{P} = PG(3, q)$ with $q \equiv 3 \bmod 4$. A **spread** of **P** is a set $\mathcal{S}$ of lines with the property that each point of **P** is on exactly one line of $\mathcal{S}$. It can easily be proven that (a) each spread of **P** has q^2+1 lines and (b) each set of q^2+1 mutually skew lines form a spread.

There are lots of spreads. The set $\mathcal{S}$ of lines of the form $g_{a,b}$ and g_∞ form a spread $\mathcal{S}$:

$$g_{h,k} = \langle (b, a, 1, 0), (-a, b, 0, 1)\rangle, \qquad a, b \in GF(q),$$
$$g_\infty = \langle (1, 0, 0, 0), (0, 1, 0, 0)\rangle.$$

This spread has a special property, it contains a regulus. The set $\mathcal{R} = \{g_\infty\} \cup \{g_{a,0} \mid a \in GF(q)\}$ of lines of $\mathcal{S}$ is a regulus.

1 Show that a set of skew lines in $PG(3, q)$ is a spread if and only if it has q^2+1 lines.

2 Show that the set above defined as

$$\mathcal{S} = \{ g_{a,b} \mid a, b \in GF(q)\} \cup \{g_\infty\}$$

is a spread in PG(3, q).

3 Show that the set

$$\mathcal{R} = \{g_\infty\} \cup \{g_{a,0} \mid a \in GF(q)\}$$

is a regulus in PG(3, q).

Now we can construct the authentication system **A**. Let $\mathcal{S}$ be a spread of **P** containing a regulus $\mathcal{R}$. As *plaintexts* of **A** we choose all lines of $\mathcal{S} \setminus \mathcal{R}$, the *keys* are the points on the lines of $\mathcal{R}$; the *message* belonging to a plaintext d and a key k is the plane generated by the line d and the point k.

Convince yourself of the following hypotheses:

4 The authentication system **A** has $q^2 - q$ plaintexts and $q^2 + 1$ keys.

5 If an attacker inserts a message of his own devising at latest after the first authentic message has been sent, his probability of success is at most $2/(q+1)$.
(For details see [BeRo91].)

You should know the following notions

Algorithm, key, enciphering, one-time pad, authenticity, data integrity, data authenticity, authentication, authentication system, perfect authentication system, secret sharing scheme, threshold scheme, group scheme, hierarchical scheme, perfect secret sharing scheme, spread.

Bibliography

[AgLa85] Agrawala, A.K.; Lakshman, T.V.: Efficient decentralized consensus protocols: *IEEE Trans. Software Enging* **12** (1985), 600–607.

[AsKe92] Assmus, E.F.; Key, T.D.: *Designs and their Codes*: Cambridge University Press, Cambridge Tracts in Math. **103**, 1992.

[BaBe93] Batten, L.M.; Beutelspacher, A.: *The Theory of Linear Spaces*: Cambridge University Press, 1993.

[Baer42] Baer, R.: Homogenity of projective planes: *Amer. J. Math.* **64** (1942), 137–152.

[Baer46] Baer, R.: Projectivities with fixed points on every line of the plane: *Bull. Amer. Math. Soc.* **52** (1946), 273–286.

[Bat86] Batten, L.M.: *Combinatorics of Finite Geometries*: Cambridge University Press, 1986.

[Benz73] Benz: *Vorlesungen über Geometrie der Algebren*: Springer-Verlag, Berlin, Heidelberg, New York, 1973.

[BePi82] Beker, H.; Piper, F.C.: *Cipher Systems: the Protection of Communications*: Northwood Books, London 1982.

[Ber87] Berger, M.: *Geometry I, II*: Springer-Verlag, Berlin, Heidelberg, 1987.

[BeRo90] Beutelspacher, A.; Rosenbaum, U.: Geometric authentication systems: *Ratio Mathematica* **1** (1990), 39–50.

[BeRo91] Beutelspacher, A.; Rosenbaum, U.: Essentially l-fold secure authentication systems: *Advances in Cryptology – EUROCRYPT 90.* Lecture Notes in Computer Science **473**, Springer-Verlag, Berlin, Heidelberg, New York, London, Paris, Tokyo, Hong Kong, Barcelona, Budapest, 1991, 294–305.

[Beu82] Beutelspacher, A.: *Einführung in die endliche Geometrie I*: Bibliographisches Institut, Mannheim, Wien, Zürich, 1982.

[Beu83] Beutelspacher, A.: *Einführung in die endliche Geometrie II*: Bibliographisches Institut, Mannheim, Wien, Zürich, 1983.

[Beu86] Beutelspacher, A.: 21–6 = 15. A connection between two distinguished geometries: *Amer. Math. Monthly* **93** (1986), 29–41.

[Beu87] Beutelspacher, A.: A defense of the honour of an unjustly neglected little geometry or A combinatorial approach to the projective plane of order five: *J. Geometry* **30** (1987), 182–195.

[Beu88] Beutelspacher, A.: Enciphered geometry. Some applications of geometry to cryptography: *Ann. Discrete Math.* **37** (1988), 59–68.

[Beu90a] Beutelspacher, A.: Applications of finite geometry to cryptography*: Geometries, Codes and Cryptography*. G. Longo, M. Marchi, A. Sgarro, eds., . CISM Courses and Lectures **313**, Springer-Verlag, Wien, Berlin, Heidelberg, New York, 1990.

[Beu90b] Beutelspacher, A.: How to communicate efficiently: *J. Combinat. Theory* **54** (1990), No. 2, 312–316.

[Beu92] Beutelspacher, A.: *Cryptology:* The Mathematical Association of America, Washington, DC, 1994.

[BeWe93] Beutelspacher, A.; Wettl, F.: On 2-level secret sharing: *Designs, Codes and Cryptography* **3** (1993), 127–134.

[BJL85] Beth, T.; Jungnickel, D.; Lenz, H.: *Design Theory*: Bibliographisches Institut, Mannheim, Wien, Zürich, 1985.

[Bla83] Blahut, R.E.: *Theory and Practice of Error Control Codes*: Addison-Wesley, Reading, Mass., 1983.

[BoBu66] Bose, R.C.; Burton, R.C.: A characterization of flat spaces in a finite geometry and the uniqueness of the Hamming and MacDonald codes: *J. Combinat. Theory* **1** (1966), 96–104.

[BPBS84] Berger, M.; Pansu, P.; Berry, J.P.; Saint-Raymond, X.: *Problems in Geometry*: Springer-Verlag, New York, Berlin, Heidelberg, Tokyo, 1984.

[Brau76] Brauner, H.: *Geometrie projektiver Räume I*: Bibliographisches Institut, Mannheim, Wien, Zürich, 1976.

[BrEr48] de Bruijn, N.G.; Erdös, P.: On a combinatorial problem: *Indag. Math.* **10** (1948), 421–423.

[BrRy49] Bruck, R.H.; Ryser, H.J.: The non-existence of certain finite projective planes: *Can. J. Math.* **1** (1949), 88–93.

[BuBu88] Buekenhout, F.; Buset, D.: On the foundations of incidence geometry: *Geom. Dedicata* **25** (1988), 269–296.

[Buek69a] Buekenhout, F.: Ensembles quadratiques des espaces projectifs: *Math. Z.* **110** (1969), 306–318.

[Buek69b] Buekenhout, F.: Une caractérisation des espaces affins basée sur la notion de droite: *Math. Z.* **111** (1969), 367–371.

[Buek79] Buekenhout, F.: Diagrams for geometries and groups: *J. Combinat. Theory* (A) **27** (1979), 121–151.

[Buek81] Buekenhout, F.: The basic diagram of a geometry: *Geometries and Groups.* Lecture Notes in Mathematics **893,** Springer-Verlag, Berlin, Heidelberg, New York, 1981, 1–29.

[Buek95] Buekenhout, F.: *Handbook of Incidence Geometry. Buildings and Foundations.* North-Holland, 1995.

[BuSh74] Buekenhout, F.; Shult, E.: On the foundations of polar geometry: *Geom. Dedicata* **3** (1974), 155–170.

[CaLi80] Cameron, P.J.; Lint, van, J.H.: *Graphs, Codes and Designs*: London Math. Soc. Lecture Notes Series **43**, Cambridge University Press, 1980.

[CaLi91] Cameron, P.J.; Lint, van, J.H.: *Designs, Graphs, Codes and their Links*: London Math. Soc. Student Texts **22**, Cambridge University Press, 1991.

[Cam92] Cameron, P.J.: *Projective and Polar Spaces*: QMW Maths Notes **13**, Queen Mary and Westfield College, University of London, 1992.

[CCITT] *CCITT Recommendations X.509: the Directory – Authentication Framework*, New York, London, Sydney, Toronto, 1988.

[Cox69] Coxeter, H.S.M.: *Introduction to Geometry*: John Wiley & Sons, New York, 2nd edn, 1969.

[Cox74] Coxeter, H.S.M.: *Projective Geometry*: University of Toronto Press, 2nd edn, 1974.

[DaPr89] Davies, D.W.; Price, W.L.: *Security for Computer Networks*: John Wiley & Sons, Chichester, 1984, 2nd edn, 1989.

[Dem68] Dembowski, P.: *Finite Geometries*: Springer-Verlag, Berlin, Heidelberg, New York, 1968.

[Den83] Denning, D.: *Cryptography and Data Security*: Addison-Wesley, Reading, Mass. 1983.

[DVW89] DeSoete, M.; Vedder, K.; Walker, M.: Cartesian authentication schemes: *Advances in Cryptology – EUROCRYPT 89*. Lecture Notes in Computer Science **434**, Springer-Verlag 1990, 476–490.

[Edge55] Edge, W.L.: 31-point geometry: *Math. Gaz.* **39** (1955), 113–121.

[Fåk79] Fåk, V.: Repeated use of codes which detect deception: *IEEE Trans. Inform. Theory*, Vol. **25**, No. 2, March 1979, 233–234.

[Gal91] Gallian, J.A.: The mathematics of identification numbers: *College Math. J.* **22** (1991), 194–202.

[Gal94] Gallian, J.A.: *Contemporary Abstract Algebra*: D.C. Heath, Lexington, Mass., 3rd edn, 1994.

[GMS74] Gilbert, E.N.; MacWilliams, F.J.; Sloane, N.J.A.: Codes which detect deception: *The Bell Sys. Techn. J.* **53**, March 1974, 405–425.

[Gop88] Goppa, V.D.: *Geometry and Codes*: Kluwer Academic Publishers, Dordrecht, Netherlands, Boston, Mass., London, 1988.

[Hag71] Hagelbarger, D.W.: The application of balance symmetric incomplete block designs to switching networks: *Proceeding of the International Conference on Communications*, June 14, 15, 16, 1971, Montreal.

[HaHe76] Halder, H.-R.; Heise, W.: *Einführung in die Kombinatorik*: Carl Hanser Verlag, München, Wien, 1976.

[HeQu89] Heise, W.; Quattrocchi, P.: *Informations- und Codierungstheorie*: Springer-Verlag, Berlin, Heidelberg, 2nd edn, 1989.

[Her64] Herstein, I.N.: *Topics in Algebra*: John Wiley & Sons, New York, 2nd edn, 1964.

[Her72] Herzer, A.: Dualitäten mit zwei Geraden aus absoluten Punkten in projektiven Ebenen: *Math. Z.* **129** (1972), 235–257.

[Hes05] Hessenberg, G.: Beweis des Desarguesschen Satzes aus dem Pascalschen: *Math. Ann.* **61** (1905), 161–172.

[Hil62] Hilbert, D.: *Grundlagen der Geometrie.* B.G. Teubner, Stuttgart, 9th edn, 1962.

[Hill86] Hill, R.: *A First Course in Coding Theory*: Clarendon Press, Oxford, 1986.

[Hir79] Hirschfeld, J.W.P.: *Projective Geometries over Finite Fields*: Clarendon Press, Oxford, 1979.

[Hir85] Hirschfeld, J.W.P.: *Finite Projective Spaces of Three Dimensions*: Clarendon Press, Oxford, 1985.

[HiTh91] Hirschfeld, J.W.P.; Thas, J.A.: *General Galois Geometries*: Clarendon Press, Oxford, 1991.

[Hog94] Hogendijk, Jan P.: Mathematics in Medieval Islamic Spain. *Proceedings of the International Congress of Mathematicians,* Zürich 1994, Birkhäuser Verlag, Basel, 1995, 1568–1580.

[HuPi73] Hughes, D.R.; Piper, F.C.: *Projective Planes*: Graduate Texts in Mathematics **6**, Springer-Verlag, New York, Heidelberg, Berlin, 1973.

[KaKr88] Karzel, H.; Kroll, H.-J.: *Geschichte der Geometrie seit Hilbert*: Wissenschaftliche Buchgesellschaft, Darmstadt, 1988.

[Kall82] Kallaher, M.J.: *Affine Planes with Transitive Collineation Groups*: North-Holland, New York, 1982.

[KaPi70] Karzel, H.; Pieper, I.: Bericht über geschlitzte Inzidenzgruppen: *Jber. Deutsch. Math.-Verein.* **72** (1970), 70–114.

[Ker92] Kersten, A.: Shared Secret Schemes aus geometrischer Sicht. *Mitt. Math. Sem. Univ. Giessen* **208** (1992).

[KSW73] Karzel, H.; Sörensen, D.; Windelberg, D.: *Einführung in die Geometrie*: Vandenhock & Ruprecht, Göttingen, 1973.

[Lam91] Lam, C.W.H.: The search for a finite projective plane of order 10: *Amer. Math. Monthly* **98** (1991), 305–318.

[Lang65] Lang, S.: *Algebra*: Addison-Wesley, Reading, Mass., 1965.

[Lenz54] Lenz, H.: Zur Begründung der analytischen Geometrie: *Bayerische Akad. Wiss.* **2** (1954), 17–72.

[Lenz65] Lenz, H.: *Vorlesungen über projektive Geometrie*: Akad. Verl. Ges. Geest & Portig, Leipzig, 1965.

[Lin69] Lingenberg, R.: *Grundlage der Geometrie I*: Bibliographisches Institut, Mannheim, 1969.

[Lint82] Lint, van, J.H.: *Introduction to Coding Theory*: Springer-Verlag, New York, Heidelberg, Berlin, 1982.

[Mas86] Massey, J.L.: Cryptography – a selective survey: *Alta Frequenza* **LV**, **1** (1986), 4–11.

[Mey76] Meyberg, K.: *Algebra*, Teil 2: Carl Hanser Verlag, München - Wien, 1976.

[MiPi87a] Mitchell, C.J.; Piper, F.C.: The cost of reducing key-storage requirements in secure networks: *Computers & Security* **6** (1987), 339–341.

[MiPi87b] Mitchell, C.J.; Piper, F.C.: Key storage in secure networks: *J. Discrete Applied Math.* **21** (1988), 215–228.

[Mou02] Moulton, F.R.: A simple non-desarguesian plane geometry: *Trans. Amer. Math. Soc.* **3** (1902), 192–195.

[Mul54] Muller, D.E.: Application of Boolean algebra to switching circuit design and to error detection: *IEEE Trans. Computers* **3** (1954), 6–12.

[MWSl83] MacWilliams, F.J.; Sloane N.J.A.: *The Theory of Error-Correcting Codes*: North-Holland, Amsterdam, New York, Oxford, 1983.

[Pas82] Pasch, M.: *Vorlesungen über neuere Geometrie*: Teubner, Leipzig, Berlin, 1882.

[Pas94] Pasini, A.: *Diagram Geometries*: Oxford University Press, 1994.

[PaTh85] Payne, S.E.; Thas, J.A.: *Finite Generalized Quadrangles*: Pitman, New York 1985.

[Ped63] Pedoe, D.: *An Introduction to Projective Geometry*: Macmillan, New York, 1963.

[Pick55] Pickert, G.: *Projektive Ebenen*: Springer-Verlag, Berlin, Göttingen, Heidelberg, 1955.

[PiWa84] Piper, F.C.; Walker, M.: Binary sequences and Hadamard designs: *Geometrical Combinatorics*, F.C. Holroyd, R.J. Wilson (eds.). Research Notes in Mathematics **114**, Pitman, Boston, Mass., 1984.

[Qvi52] Qvist, B.: Some remarks concerning curves of the second degree in a finite plane: *Ann. Acad. Sci. Fenn.* **134** (1952), 1–27.

[Reed54] Reed, I.S.: A class of multiple-error-correcting codes and the decoding scheme: *IEEE Trans. Inform. Theory* **4** (1954), 38–49.

[Ros93] Rosenbaum, U.: A lower bound on authentication after having observed a sequence of messages: *J. Cryptology* **6** (1993), 135–156.

[Rue86] Rueppel, R.: *Analysis and Design of Stream Ciphers*: Springer-Verlag, Berlin, Heidelberg, New York, London, Paris, Tokyo, 1986.

[Sal90] Salooma, A.: *Public-Key Cryptography*: Springer-Verlag, Berlin, Heidelberg, New York, London, Paris, Tokyo, Hong Kong, Barcelona, 1990.

[Schn96] Schneier, B.: *Applied Cryptography: Protocols, Algorithms and Source Code in C*: John Wiley & Sons, New York, 2nd edn, 1996.

[Schu91] Schulz, R.H.: *Codierungstheorie – eine Einführung*: Verlag Vieweg, Braunschweig, Wiesbaden, 1991.

[Seg54] Segre, B.: Sulle ovali nei piani lineari finiti: *Atti Accad. Naz. Lincei Rendic.* **17** (1954), 141–142.

[Seg61] Segre, B.: *Lectures on Modern Geometry*: Cremonese, Roma, 1961.

[Sha49] Shannon, C.E.: Communication theory of secrecy systems: *Bell. Sys. Tech. J.* **10** (1949), 657–715.

[Sha79] Shamir, A.: How to share a secret: *Commun. ACM* **22** (1979), 612–613.

[Sim82] Simmons, G.J.: A game theoretical model of digital message authentication: *Congressus Numerantium* **34** (1982), 413–424.

[Sim84] Simmons, G.J.: Authentication theory / Coding theory: *Advances in Cryptology: Proceedings of CRYPTO* 1984. Lecture Notes in Computer Science **196**, Springer-Verlag, New York, Berlin, Heidelberg, London, Paris, Tokyo, Hong Kong, 1984, 411–432.

[Sim90] Simmons, G.J.: How to (really) share a secret: *Advances in Cryptology – CRYPTO 88.* Lecture Notes in Computer Science **403**, Springer-Verlag, New York, Berlin, Heidelberg, London, Paris, Tokyo, Hong Kong, 1990, 390–448.

[Sim92a] Simmons, G.J.: *Contemporary Cryptology*: IEEE Press, New York, 1992.

[Sim92b] Simmons, G.J.: An introduction to shared secret/shared control schemes: *Contemporary Cryptology,* G.J. Simmons, ed., IEEE Press, New York, 1992, 441–497.

[Sing38] Singer, J.: A theorem in finite projective geometry and some applications to number theory: *Trans. Amer. Math. Soc.* **43** (1938), 377–385.

[SiSm89] Simmons, G.J.; Smeets, B.: A paradoxical result in unconditionally secure authentication codes – and an explanation: *IMA Conference on Cryptography and Coding,* Dec. 18–20, 1989, Cirencester, England, Clarendon Press, Oxford.

[Stei13] Steinitz, E.H.: Bedingt konvergente Reihen und konvexe Systeme: *J. Reine Angew. Math.* **143** (1913), 128–175.

[Sti95] Stinson, D.R.: *Cryptography. Theory and Practice*: CRC Press, Boca Raton, Fl., 1995.

[Tall56] Tallini, G.: Sulle k-calotte di uno spazio lineare finito: *Ann. Mat. Pura Appl.* **42** (1956), 119–164.

[Tall57] Tallini, G.: Caratterizazione grafica delle quadriche ellittiche negli spazi finiti: *Rend. Mat. Roma* **16** (1957), 328–351.

[Tam72] Tamaschke, O.: *Projektive Geometrie II.* Bibliographisches Institut, Mannheim, Wien, Zürich, 1972.

[Tec87] Tecklenburg, H.: A proof of the theorem of Pappus in finite Desarguesian affine planes: *J. Geometry* **30** (1987), 172–181.

[Tits56] Tits, J.: Les groupes de Lie exceptionnels et leur interprétation géométrique: *Bull. Soc. Math. Belg.* **8** (1956), 48–81.

[Tits74] Tits, J.: *Buildings of Spherical Type and Finite BN-Pairs.* Springer-Verlag, Berlin Heidelberg, New York, 1974.

[Ver26] Vernam, G.S.: Cipher printing telegraph systems for secret wire and radio telegraphic communications: *J. AIEE* **45** (1926), 109–115.

[VeYo16] Veblen, O.; Young, J.W.: *Projective Geometry*: 2 vols. Ginn & Co., Boston, Mass., 1916.

[Wal90] Walker, M.: Information-theoretic bounds for authentication schemes: *J. Cryptology* **2** (1990), 131–143.

Index of notation

$(a_0: \ldots : a_d)$ homogeneous coordinates 66
$(a_1, \ldots, a_d)$ inhomogeneous coordinates 68
$\|$ parallelism 28
$\mathbf{A} = \mathbf{P}\setminus\mathbf{H}_\infty$ affine space 27
AG(d, F) d-dimensional affine space over the division ring F 69
AG(d, q) d-dimensional affine space over the finite field of order q 69
Aut(F) group of automorphisms of F 132
$\langle\mathcal{X}\rangle$ span of the set $\mathcal{X}$ 10
$\chi(\mathcal{M})$ characteristic vector of the set $\mathcal{M}$ of points of **A** 203
$\mathbf{C}^\perp$ dual code 187
$d(\mathbf{C})$ minimum distance of the code **C** 184
d(v, w) distance of the vectors v, w 182
dim(**P**) dimension of a projective space **P** 18
$\dim_V$ dimension of a vector space 56
D_O group of dilatations (cental collineations with axis **H** and centre O $\notin$ **H**) 110
$\mathcal{F}$ flag 3
Γ group of collineations of **A** 118
G = (Ω, I) geometry 1
Γ_O group of collineations of **A** that fix the point O 118
$\mathbf{H}_\infty$ hyperplane at infinity 27
Ham(r) Hamming code 191
Ham(r)* extended Hamming code 195
I incidence relation 2
$\max_{d-1}(r, q)$ maximum n such that an $(n, d-1)$-set in PG($r-1$, q) exists 199
P projective space 7
P/Q quotient geometry 21
P(V) projective space over V 56
$\mathcal{P}^*$ points of the affine space **P****H** 107
$\mathbf{P}^\Delta$ dual of the projective plane **P** 9
PG(d, F) d-dimensional projective space over the division ring F 59
PG(d, q) d-dimensional projective space over the finite field of order q 59
$\mathcal{Q}$ quadratic set 137
$\mathcal{Q}$ quadric 162
$\mathcal{Q}_P$ tangent space of $\mathcal{Q}$ at P 137
$\theta_t(q)$ $q^t + \ldots + q + 1$ 25
rad($\mathcal{Q}$) radical of the quadratic set $\mathcal{Q}$ 139
Res($\mathcal{F}$) residue of the flag $\mathcal{F}$ 35
S_r(v) Hamming sphere 183
T(**H**) translations (group of central collineations with axis **H** and centre on **H**) 105
T(**H**) group of translations of **A** = **P****H** 118
T(P, **H**) group of central collineations with axis **H** and centre P $\in$ **H** 109
U = **P**($\mathcal{U}$) subspace 10
$\mathcal{U}$(**P**) set of subspaces of **P** 18
$\mathcal{U}^*$(**P**) set of nontrivial subspaces of **P** 18
$w(\mathbf{C})$ minimum weight of the code **C** 186

General index